D0554860

ROBOTICS:
A BIBLIOGRAPHY WITH INDEXES

ROBOTICS:
A BIBLIOGRAPHY WITH INDEXES

PETER J. BENNE (EDITOR)

Nova Science Publishers, Inc.
New York

Senior Editors: Susan Boriotti and Donna Dennis
Coordinating Editor: Tatiana Shohov
Office Manager: Annette Hellinger
Graphics: Wanda Serrano
Editorial Production: Jennifer Vogt, Matthew Kozlowski and Maya Columbus
Circulation: Ave Maria Gonzalez, Indah Becker and Vladimir Klestov
Communications and Acquisitions: Serge P. Shohov
Marketing: Cathy DeGregory

Library of Congress Cataloging-in-Publication Data
Available Upon Request

ISBN: 1-59033-296-2.

Copyright © 2002 by Nova Science Publishers, Inc.
           400 Oser Ave, Suite 1600
           Hauppauge, New York 11788-3619
           Tele. 631-231-7269   Fax 631-231-8175
           e-mail: Novascience@earthlink.net
           Web Site: http://www.novapubishers.com

Printed in the United States of America

CONTENTS

PREFACE

Robotics has worked its way from science fiction to science without the fiction to wide–spread applications not only in manufacturing but in medical research and other fields as well. Progress had been accelerated by the faster-than-light advances in miniaturization. This book provides important citations, primarily to the book literature, with access through author, subject and title indexes.

GENERAL BIBLIOGRAPHY

13th International Symposium on Industrial Robots and Robots 7: conference proceedings, April 17-21, 1983, Chicago, Illinois / sponsored by Robotics International of SME in cooperation with Robot Institute of America.
Edition Information: 1st ed.
Published/Created: Dearborn, Mich.: Robotics International of SME, c1983.
Related Authors: Robotics International of SME. Robot Institute of America. Robots Conference (7th: 1983: Chicago, Ill.)
Description: 2 v.: ill.; 28 cm.
ISBN: 0872631141 (set: pbk.)
Contents: v. 1. Applications worldwide -- v. 2. Future directions.
Notes: Includes bibliographical references and index.
Subjects: Robots, Industrial-- Congresses.
LC Classification: TS191.8 .I57 1983
Dewey Class No.: 670.42/7 19

1981 report on the industrial process controls & robotics market in the United Kingdom.
Published/Created: New York, NY: American European Consulting Co., c1981.
Related Authors: American European Consulting Company.
Description: 167 leaves: ill.; 28 cm.
Notes: Cover title.
Subjects: Automatic control equipment industry--Great Britain. Robot industry--Great Britain. Market surveys--Great Britain.
LC Classification: HD9696.A963 G718 1981
Dewey Class No.: 380.1/456298/0941 19

1983 Rochester FORTH Applications Conference, June 7-11, 1983 / hosted by the Laboratory for Laser Energetics, the University of Rochester, and the Institute for Applied Forth Research, Inc.; [conference chairman, Lawrence P. Forsley; proceedings editor, Diane D. Ranocchia].
Published/Created: Rochester, N.Y.: The Institute, c1983.
Related Authors: Forsley, Lawrence P. Ranocchia, Diane D. University of Rochester. Laboratory for Laser Energetics. Institute for Applied FORTH Research.
Description: 301 p.: ill.; 28 cm.
ISBN: 0914593005 (pbk.)
Notes: Includes bibliographies.
Subjects: FORTH (Computer program language)--Congresses. Computer programming--Congresses. Robotics--Congresses.
LC Classification: QA76.73.F24 R59 1983
Dewey Class No.: 001.64/24 19

1985 SME World Congress on the Human Aspects of Automation: conference,

September 8-11, 1985, Cambridge,
Massachusetts / sponsored and
published by the Association for
Finishing Processes of SME, Computer
and Automated Systems Association of
SME, Robotics International of SME.
Edition Information: [1st ed.].
Published/Created: Dearborn, Mich.:
Society of Manufacturing Engineers,
[c1985]
Related Authors: Society of
Manufacturing Engineers. Association
for Finishing Processes of SME.
Computer and Automated Systems
Association of SME. Robotics
International of SME.
Description: 1 v. (various pagings): ill.;
28 cm.
ISBN: 0872631966 (pbk.)
Notes: "SME technical papers."
Subjects: Technological innovations--
Economic aspects--Congresses.
Business--Data processing--Congresses.
LC Classification: HC79.T4 S64 1985
Dewey Class No.: 658/.05 19

1988 IEEE Workshop on Languages for
Automation: symbiotic and intelligent
robotics: University of Maryland,
College Park, Maryland, USA, August
29-31, 1988 / sponsor the Computer
Society in cooperation with IEEE SMC
Society ... [et al.].
Published/Created: Washington, D.C.:
IEEE Computer Society Press; Los
Angeles, CA: Order from Computer
Society, c1988.
Description: ix, 233 p.: ill.; 28 cm.
ISBN: 0818608900 (pbk.) 0818688904
(hard) 0818648902 (microfiche)
LC Classification: IN PROCESS

1989 World Conference on Robotics
Research: the next five years and
beyond: conference proceedings, May
7-11, 1989, Gaithersburg, Maryland /
sponsored by Society of Manufacturing
Engineers, Robotics International of
SME.
Published/Created: Dearborn, Mich.:

SME: RI, c1989.
Related Authors: Society of
Manufacturing Engineers. Robotics
International of SME.
Description: 2 v.: ill.; 28 cm.
ISBN: 0872633616
Notes: "Creative manufacturing
engineering program"--V. 2. Cover title
(v. 2). Includes bibliographical
references.
Subjects: Robotics--Congresses.
Robotics--Research--Congresses.
LC Classification: TJ210.3 .W67 1989
Dewey Class No.: 670.42/72 20

1990 Japan-U.S.A. Symposium on Flexible
Automation: proceedings: a Pacific Rim
Comference, Kyoto, Japan, July 9-13,
1990 / sponsored by Institute of
Systems, Control, and Information
Engineers, American Society of
Mechanical Engineers; participating
societies, Japan Society of Mechanical
Engineers ... [et al.].
Published/Created: Kyoto: ISCIE, 1990.
Related Authors: Shisutemu Seigyo
Jˉohˉo Gakkai. American Society of
Mechanical Engineers.
Description: 3 v. (xxiii, 1407 p.): ill.; 28
cm.
Notes: Includes bibliographical
references and index.
Subjects: Flexible manufacturing
systems--Congresses. Robotics--
Congresses.
LC Classification: TS155.6 .J36 1990
Dewey Class No.: 670.42/7 20

1991 RI/SME robotics research &
development laboratory directory /
Barbara P. Knape, project director.
Edition Information: 1st ed.
Published/Created: Detroit, MI:
Robotics International of the Society of
Manufacturing Engineers, c1991.
Related Authors: Knape, Barbara P.
Robotics International of SME.
Description: 126 p.; 28 cm.
Subjects: Robotics laboratories--
Directories.

LC Classification: TJ210.5 .A15 1991
Dewey Class No.: 629.8/92/072 20

1994 5th International Symposium on Micro
Machine and Human Science
proceedings: Nagoya Congress Center,
October 2-4, 1994 / hosted by Nagoya
University; cosponsored by IEEE
Industrial Electronics Society, City of
Nagoya, Chubu Industrial Advancement
Center; technically cosponsored by
IEEE Robotics & Automation Society ...
[et al.]; in cooperation with Chubu
Bureau of International Trade &
Industry of MITI ... [et al.].
Published/Created: [New York]:
Institute of Electrical and Electronics
Engineers; Piscataway, N.J.: IEEE
Service Center, c1994.
Related Authors: Nagoya Daigaku.
IEEE Industrial Electronics Society.
Nagoya-shi (Japan) Ch‾ubu Seisansei
Honbu. IEEE Robotics and Automation
Society. Japan. Ch‾ubu Ts‾ush‾o
Sangy‾okyoku.
Description: ix, 204 p.: ill.; 28 cm.
ISBN: 0780320956 (softbound)
0780320964 (microfiche)
Notes: "IEEE catalog number:
94TH0707-0." Includes bibliographical
references and index.
Subjects: Mechatronics--Congresses.
Micromechanics--Congresses.
LC Classification: TJ163.12 .I57 1994
Dewey Class No.: 621 21

1995 IEEE/RSJ International Conference on
Intelligent Robots and Systems: human
robot interaction and cooperative robots:
proceedings, August 5-9, 1995,
Pittsburgh, Pennsylvania, USA / co-
sponsored by IEEE Industrial
Electronics Society ... [et al.].
Published/Created: Los Alamitos,
Calif.: IEEE Computer Society Press,
c1995.
Related Authors: IEEE Industrial
Electronics Society.
Description: 3 v.: ill.; 28 cm.
ISBN: 0818671084 (softbound)

0780330064 (casebound) 0780330072
(microfiche)
Notes: "IEEE catalog number
95CB35836"--T.p. verso. Includes
bibliographical references and indexes.
Subjects: Robotics--Congresses.
Artificial intelligence--Congresses.
Intelligent control systems--Congresses.
LC Classification: TJ210.3 .I447 1995
Dewey Class No.: 629.8/92 20

1995 INRIA/IEEE Symposium on Emerging
Technologies and Factory Automation:
proceedings, ETFA '95, Paris, France,
October 10-13, 1995 / sponsored by
IEEE Industrial Electronics Society,
INRIA institut national de recherche en
informatique et en automatique.
Published/Created: Los Alamitos, CA:
IEEE Computer Society Press, c1995.
Related Authors: IEEE Industrial
Electronics Society. Institut national de
recherche en informatique et en
automatique (France)
Description: 3 v.: ill.; 28 cm.
ISBN: 0780325354 (paper) 0780325362
(microfiche)
Notes: "IEEE catalog number
95TH8056"--T.p. verso. Includes
bibliographical references.
Subjects: Automation--Congresses.
Artificial intelligence--Congresses.
Robotics--Congresses.
LC Classification: T59.5 .I49 1995
Dewey Class No.: 670.42/7 21

1996 IEEE International Workshop on
Variable Structure Systems, VSS '96:
proceedings, December 5-6, 1996,
Institute of Industrial Science,
University of Tokyo, Roppongi, Tokyo,
Japan / co-sponsors, IEEE Industrial
Electronics Society, Institute of
Industrial Science, University of Tokyo;
technical co-sponsors, International
Federation of Automatic Control ... [et
al.].
Published/Created: [New York]:
Institute of Electrical and Electronics
Engineers, c1996.

Related Authors: IEEE Industrial
Electronics Society. T⁻oky⁻o Daigaku.
Seisan Gijutsu Kenky⁻ujo.
Description: viii, 243 p.: ill.; 30 cm.
ISBN: 0780337182 (softbound)
0780337190 (microfiche)
Notes: "IEEE catalog number:
96TH8245"--T.p. verso. Includes
bibliographical references and index.
Subjects: Automatic control--
Congresses. Industrial electronics--
Congresses. Robotics--Congresses.
LC Classification: TJ212.2 .I3253 1996
Dewey Class No.: 629.8 21

1997 8th International Conference on
     Advanced Robotics: proceedings, ICAR
     '97, Hyatt Regency, Monterey,
     California, U.S.A., July 7-9, 1997.
     Published/Created: [New York]:
     Institute of Electrical and Electronics
     Engineers; Piscataway, NJ: IEEE
     Service Center, c1997.
     Related Authors: Institute of Electrical
     and Electronics Engineers.
     Description: xx, 1032 p.: ill.; 28 cm.
     ISBN: 0780341600 (softbound)
     Notes: "IEEE catalog number
     97TH8308"--T.p. verso. Includes
     bibliographical references.
     Subjects: Robotics--Congresses.
     LC Classification: TJ210.3 .I5765 1997
     Dewey Class No.: 629.8/92 21

1997 IEEE 6th International Conference on
     Emerging Technologies and Factory
     Automation proceedings, EFTA '97,
     UCLA Conference Center, Los Angeles,
     California, September 9-12, 1997 /
     sponsored by the IEEE Industrial
     Electronics Society in technical
     cosponsorship with Society of
     Instrument and Control Engineers,
     Japan in cooperation with
     Micromechanical Systems Panel of the
     ASME Dynamic Systems and Control
     Division.
     Published/Created: [New York]:
     Institute of Electrical and Electronics
     Engineers, c1997.

Related Authors: IEEE Industrial
Electronics Society. Keisoku Jid⁻o
Seigyo Gakkai (Japan) American
Society of Mechanical Engineers.
Dynamic Systems and Control Division.
Micromechanical Systems Panel.
Description: xxi, 573 p.: ill.; 28 cm.
ISBN: 0780341929 (softbound)
0780341937 (microfiche)
Notes: "IEEE catalog number
97TH8314"--T.p. verso. Includes
bibliographical references and index.
Subjects: Automation--Congresses.
Artificial intelligence--Congresses.
Robotics--Congresses. Computer
integrated manufacturing systems--
Congresses.
LC Classification: T59.5 .I277 1997
Dewey Class No.: 670.42/7 21

1997 IEEE International Symposium on
     Assembly and Task Planning
     (ISATP'97): towards flexible and agile
     assembly and manufacturing, August 7-
     9, 1997, Marina del Rey Hotel, Marina
     del Rey, California / sponsored by IEEE
     Robotics and Automation Society; co-
     sponsored by CIRP, International
     Institute for Production Engineering
     Research; supported by LG Industrial
     Systems Co., Ltd., Samsung Electronics
     Co., Ltd.
     Published/Created: [New York, N.Y.]:
     Institute of Electrical and Electronics
     Engineers; Piscataway, N.J.: IEEE
     Customer Service, c1997.
     Related Authors: IEEE Robotics and
     Automation Society. International
     Institution for Production Engineering
     Research.
     Description: x, 293 p.: ill.; 28 cm.
     ISBN: 0780338200 (softbound)
     0780338219 (microfiche)
     Notes: "IEEE catalog number:
     97TH8264"--T.p. verso. Includes
     bibliographical references and index.
     Subjects: Assembly-line methods--
     Congresses. Production planning--
     Congresses. Robots, Industrial--
     Congresses. Flexible manufacturing

systems--Congresses.
LC Classification: TS178.4 .I34 1997
Dewey Class No.: 670.42 21

1997 IEEE International Symposium on
Computational Intelligence in Robotics
and Automation, CIRA '97: "towards
new computational principles for
robotics and automation": proceedings,
July 10-11, 1997, Monterey, California,
USA / sponsored by IEEE Robotics and
Automation Society, IEEE Neural
Network Council.
Published/Created: Los Alamitos,
Calif.: IEEE Computer Society Press,
c1997.
Related Authors: IEEE Robotics and
Automation Society. IEEE Neural
Networks Council.
Description: xiii, 420 p.: ill.; 28 cm.
ISBN: 0818681381 0818681403
(microfiche)
Notes: "IEEE Computer Society Press
order number PR08138"--T.p. verso.
"IEEE Order Plan catalog number
97TB100176"--T.p. verso. Includes
bibliographical references and index.
Subjects: Robotics--Congresses.
Automatic control--Congresses.
Computational intelligence--
Congresses.
LC Classification: TJ210.3 .I442 1997
Dewey Class No.: 629.8/92 21

1998 IEEE/RSJ International Conference on
Intelligent Robots and Systems:
proceedings: innovatiions in theory,
practice, and applications: October 13-
17, 1998, Victoria Conference Centre,
Victoria, B.C., Canada / sponsoring
organizations, IEEE Industrial
Electronics Society ... [et al.].
Published/Created: Piscataway, New
Jersey: IEEE, c1998.
Related Authors: IEEE Industrial
Electronics Society.
Description: 3 v.: ill.; 28 cm.
ISBN: 0780344650 (softbound)
0780344669 (casebound) 0780344677
(microfiche)

Notes: "IROS '98"--Cover p. [1]. "IEEE
Catalog Number: 98CH36190"--verso
of T.p. Includes bibliographical
references and author indexes.
Subjects: Robotics--Congresses.
Artificial intelligence--Congresses.
Intelligent control systems--Congresses.
LC Classification: TJ210.3 .I447 1998
Dewey Class No.: 629.8/92 21

1999 7th IEEE International Conference on
Emerging Technologies and Factory
Automation proceedings, ETFA'99,
October 18-21, 1999, UPC, Barcelona,
Catalonia, Spain / sponsored by IEEE
Industrial Electronics Society, IEEE
Robotics and Automation Society,
Universitat Politècnica de Catalunya;
edited by J.M. Fuertes.
Published/Created: Piscataway, N.J.:
Institute of Electrical and Electronics
Engineers, [1999]
Related Authors: Fuertes, J. M. (Josep
M.) IEEE Industrial Electronics Society.
IEEE Robotics and Automation Society.
Universidad Politécnica de Catalunya.
Description: 2 v. (xxiv, 1573 p.): ill.; 30
cm.
ISBN: 0780356705 (softbound)
Notes: "IEEE catalog number
99TH8467"--T.p. verso. Includes
bibliographical references and index.
Subjects: Computer integrated
manufacturing systems--Congresses.
Robotics--Congresses. Artificial
intelligence--Congresses. Production
control--Automation--Congresses.
Scheduling--Congresses.

1999 IEEE/ASME International Conference
on Advanced Intelligent Mechatronics
proceedings: AIM '99, September 19-
23, 1999, Renaissance Atlanta Hotel,
Atlanta, Georgia, USA / co-sponsored
by IEEE Industrial Electronics Society
(IES), IEEE Robotics and Automation
Society (RAS), American Society of
Mechanical Engineers (ASME).
Published/Created: [Piscataway, N.J.]:
IEEE, [c1999]

Related Authors: IEEE Industrial
Electronics Society. American Society
of Mechanical Engineers. IEEE
Robotics and Automation Society.
Description: xix, 1049 p.: ill.; 28 cm.
ISBN: 0780350383 (softbound ed.)
0780350391 (microfiche ed.)
0780350405 (CD-ROM)
Notes: Cover title. Includes
bibliographical references and index.
Subjects: Mechatronics--Congresses.
LC Classification: TJ163.12 .I33 1999
Dewey Class No.: 621 21

2000 IEEE/RSJ International Conference on
Intelligent Robots and Systems (IROS
2000): proceedings: October 31-
November 5, 2000, Kagawa University,
Takamatsu, Japan / sponsored by, IEEE
Industrial Electronics Society, IEEE
Robotics and Automation Society,
Robotics Society of Japan, Society of
Instrument and Control Engineers, New
Technology Foundation and Kagawa
University.
Published/Created: Piscataway, NJ:
IEEE, c2000.
Description: 3 v.; 30 cm.
ISBN: 0780363485 (softbound edition)
0780363493 (casebound edition)
0780363507 (microfiche edition)

2001 IEEE international symposium on
computational intelligence on robotics
and automation.
Published/Created: New York City, NY:
IEEE, 2001.
Description: p.; cm.
ISBN: 0780372034 (softbound edition)

6th Mediterranean Electrote[c]hnical
Conference: 22-24 May 1991,
"Cankarjev dom"--Cultural and
Congress Center, Ljubljana, Slovenia,
Yugoslavia: proceedings / edited by
Baldomir Zajc, Franc Solina.
Published/Created: Piscataway, NJ,
USA: IEEE Service Center, c1991.
Related Authors: Zajc, Baldomir.
Solina, Franc.

Description: 2 v. (xxxii, 1584 p.): ill.;
30 cm.
ISBN: 087942656X (set) 0879426551
(set: pbk.) 0879426578 (set: microfiche)
Notes: Cover 1991 Mediterranean
Electrotechnical Conference. "IEEE
catalog number: 91CH2964-5." Includes
bibliographical references and index.
Subjects: Electronics--Congresses.
Telecommunication--Congresses.
Artificial intelligence--Congresses.
Robotics--Congresses. Power
electronics--Congresses.
LC Classification: TK7801 .M43 1991
Dewey Class No.: 621.381 20

7th International Workshop on Robotics in
Alpe-Adria-Danube region RAAD
1998, Smolenice Castle, Slovakia, June
26-28, 1998: proceedings / edited by
Karol Dobrovodský.
Published/Created: Bratislava: ASCO
Art & Science, 1998.
Description: p.; cm.
ISBN: 8096796275
Notes: 1555 GN-C332

A Competitive assessment of the U.S.
robotics industry / prepared by Capital
Goods and International Construction
Sector Group.
Published/Created: Washington, D.C.:
U.S. Dept. of Commerce, International
Trade Administration: For sale by the
Supt. of Docs., U.S. G.P.O., [1987]
Related Authors: United States.
International Trade Administration.
Capital Goods and International
Construction Sector Group.
Description: xiii, 85 p.; 28 cm.
Notes: Shipping list no.: 87-197-P.
"March 1987." S/N 003-009-00499-3
Item 231-B-1 Includes bibliographical
references.
Subjects: Robot industry--United States.
Robotics--United States. Competition,
International.
LC Classification: HD9696.R623 U615
1987
Dewey Class No.: 338.4/7629892 19

Govt. Doc. No.: C 61.2:R 57

Actuators for control / edited by Hiroyasu
Funakubo.
Published/Created: New York: Gordon
and Breach Science Publishers, c1991.
Related Authors: Funakubo, Hiroyasu,
1927-
Description: xvii, 429 p.: ill.; 24 cm.
ISBN: 288124694X :
Notes: Translation of: Seigyoy⁻o
akuchu⁻eta. Includes bibliographical
references and index.
Subjects: Actuators.
Series: Precision machinery and
robotics, 0889-860X; v. 2
LC Classification: TJ223.A25 S4513
1991
Dewey Class No.: 629.8/95 19

Adams, Martin David.
Sensor modelling, design and data
processing for autonomous navigation /
Martin David Adams.
Published/Created: Singapure: River
Edge, NJ: World Scientific, 1998.
Description: xvii, 232 p.: ill.; cm.
ISBN: 9810234961 (alk. paper)
Notes: Includes bibliographical
references (p. 209-221) and index.
Subjects: Mobile robots. Robots--
Control systems.
Series: World Scientific series in
robotics and intelligent systems; vol. 13
LC Classification: TJ211.415 .A33 1998
Dewey Class No.: 629.8/92 21

Advanced guided vehicles: aspects of the
Oxford AGV Project / editors, Stephen
Cameron, Penelope Probert.
Published/Created: Singapore; River
Edge, N.J.: World Scientific Pub.,
c1994.
Related Authors: Cameron, Stephen.
Probert, Penelope.
Description: xi, 267 p.: ill.; 23 cm.
ISBN: 981021393X
Notes: Includes bibliographical
references (p. 249-263) and index.
Subjects: Robots--Control systems.

Tactile sensors--Industrial applications.
Series: World Scientific series in
robotics and automated systems; vol. 9
LC Classification: TJ211.35 .A24 1994
Dewey Class No.: 629.8/92 20

Advanced robot control: proceedings of the
International Workshop on Nonlinear
and Adaptive Control, Issues in
Robotics, Grenoble, France, Nov. 21-
23, 1990 / C. Canudas de Wit (ed.).
Published/Created: Berlin; New York:
Springer-Verlag, c1991.
Related Authors: Canudas de Wit,
Carlos A.
Description: ix, 314 p.: ill.; 24 cm.
ISBN: 3540541691 (Berlin)
0387541691 (New York)
Notes: Includes bibliographical
references.
Subjects: Robots--Control systems--
Congresses.
Series: Lecture notes in control and
information sciences; 162
LC Classification: TJ211.35 .I58 1990
Dewey Class No.: 629.8/92 20

Advanced robotics & intelligent machines /
edited by J.O. Gray & D.G. Caldwell.
Published/Created: London: Institution
of Electrical Engineers, c1996.
Related Authors: Gray, J. O., 1937-
Caldwell, D. G. (Darwin G.) Institution
of Electrical Engineers.
Description: xxiii, 374 p.: ill.; 24 cm.
ISBN: 0852968531
Notes: Includes bibliographical
references and index.
Subjects: Robotics. Artificial
intelligence.
Series: IEE control engineering series;
v. 51.
LC Classification: TJ211 .A37 1996
Dewey Class No.: 629.8/92 20

Advanced robotics, 1989: proceedings of the
4th International Conference on
Advanced Robotics, Columbus, Ohio,
June 13-15, 1989 / Kenneth J. Waldron,
editor.

Published/Created: Berlin; New York: Springer-Verlag, c1989.
Related Authors: Waldron, Kenneth J. Ohio State University. American Society of Mechanical Engineers.
Description: ix, 687 p.: ill.; 25 cm.
ISBN: 0387517510 (U.S.)
Notes: Sponsored by the Ohio State University and the American Society of Mechanical Engineers. Includes bibliographical references.
Subjects: Robotics--Congresses.
LC Classification: TJ210.3 .I5765 1989
Dewey Class No.: 629.8/92 20

Advanced software in robotics: proceedings of an international meeting held in Liège, Belgium, May 4-6, 1983 / edited by André Danthine and Michel Géradin.
Published/Created: Amsterdam; New York: North-Holland; New York, N.Y.: Sole distributors for the U.S.A. and Canada, Elsevier Science Pub. Co., 1984.
Related Authors: Danthine, A. Géradin, Michel, 1945- Université de Liège. Association des ingénieurs électriciens sortis de l'Institut Montefiore.
Description: ix, 369 p.: ill.; 23 cm.
ISBN: 0444868143
Notes: "University of Liège; A.I.M., Association des ingénieurs électriciens sortis de l'Institut Montefiore." Includes bibliographical references and index.
Subjects: Robotics--Computer programs--Congresses.
LC Classification: TJ211 .A38 1984
Dewey Class No.: 629.8/92 19

Advanced tactile sensing for robotics / edited by Howard R. Nicholls.
Published/Created: Singapore; River Edge, N.J.: World Scientific Pub., c1992.
Related Authors: Nicholls, Howard R.
Description: xvii, 294 p.: ill.; 23 cm.
ISBN: 9810208707
Notes: Includes bibliographical references and index.
Subjects: Robots--Control systems.

Tactile sensors--Industrial applications.
Series: World Scientific series in robotics and automated systems; vol. 5
LC Classification: TJ211.35 .A25 1992
Dewey Class No.: 629.8/92 20

Advanced technologies: architecture, planning, civil engineering / edited by M.R. Beheshti, K. Zreik.
Published/Created: Amsterdam; New York: Elsevier, 1993.
Related Authors: Beheshti, M. R. (Reza R.) Zreik, K. (Khaldoun)
Description: xii, 466 p.: ill.; 25 cm.
ISBN: 044481566X
Notes: Page facing t.p.: Fourth EuropIA International Conference on the Application of Artificial Intelligence, Robotics, and Image Processing to Architecture, Building Engineering, Civil Engineering, Urban Design, and Urban Planning, Delft, the Netherlands, 21-24 June 1993. Includes bibliographical references and index.
Subjects: Computer-aided engineering--Congresses. Civil engineering--Data processing--Congresses. Structural engineering--Data processing--Congresses. Architectural design--Data processing--Congresses. Artificial intelligence--Congresses.
LC Classification: TA345 .E94 1993
Dewey Class No.: 624/.0285 20

Advances in artificial intelligence, II: seventh European Conference on Artificial Intelligence, ECAI-86, Brighton, U.K., July 20-25, 1986 / edited by Ben du Boulay, David Hogg, Luc Steels.
Published/Created: Amsterdam; New York: North-Holland; New York, N.Y., U.S.A.: Sole distributors for the U.S.A. and Canada, Elsevier Science Pub. Co., 1987.
Related Authors: Du Boulay, Ben. Hogg, David. Steels, Luc.
Description: xiv, 682 p.; 24 cm.
Notes: Includes bibliographies and index.

Subjects: Artificial intelligence--
Congresses. Expert systems (Computer
science)--Congresses. Robotics--
Congresses. Learning--Congresses.
Knowledge, Theory of--Congresses.
LC Classification: Q334 .E97 1986
Dewey Class No.: 006.3 19

Advances in artificial intelligence:
proceedings of the Sixth European
Conference on Artificial Intelligence,
ECAI-84, Pisa, Italy, September 5-7,
1984 / edited by Tim O'Shea.
Published/Created: Amsterdam; New
York: North Holland; New York, N.Y.,
U.S.A.: Sole distributors for the U.S.A.
and Canada, Elsevier Science Pub. Co.,
1985.
Related Authors: O'Shea, Tim, 1949-
European Coordinating Committee for
Artificial Intelligence.
Description: xi, 423 p.: ill.; 23 cm.
ISBN: 0444876111
Notes: "Organised under the auspices of
the European Coordinating Committee
for Artificial Intelligence. Includes
bibliographies and index.
Subjects: Artificial intelligence--
Congresses. Expert systems (Computer
science)--Congresses. Robotics--
Congresses. Cognition--Congresses.
Knowledge, Theory of--Congresses.
LC Classification: Q334 .E97 1984
Dewey Class No.: 006.3 19

Advances in artificial life: 6th European
Conference, ECAL 2001, Prague, Czech
Republic, September 10-14, 2001:
proceedings / Jozef Kelemen, Petr
Sosík, eds.
Published/Created: Heildelberg; New
York: Springer, c2001.
Related Authors: Kelemen, Jozef, 1951-
Sosík, Petr, 1967-
Description: xix, 724 p.: ill.; 24 cm.
ISBN: 3540425675 (pbk.: alk. paper)
Notes: Includes bibliographical
references and index.
Subjects: Biological systems--Computer
simulation--Congresses. Biological

systems--Simulation methods--
Congresses. Robotics--Congresses.
Artificial intelligence--Congresses.
Series: Lecture notes in computer
science; 2159. Lecture notes in
computer science. Lecture notes in
artificial intelligence.
LC Classification: QH324.2 .E87 2001
Dewey Class No.: 570/.1/13 21

Advances in artificial life: Third European
Conference on Artificial Life, Granada,
Spain, June 4-6, 1995: proceedings / F.
Morán ... [et al.], eds.
Published/Created: Berlin; New York:
Springer, 1995.
Related Authors: Morán, F. (Frederico)
Description: xiii, 960 p.: ill.; 24 cm.
ISBN: 3540594965 (Berlin: acid-free
paper)
Notes: Includes bibliographical
references and index.
Subjects: Biological systems--Computer
simulation--Congresses. Biological
systems--Simulation methods--
Congresses. Artificial intelligence--
Congresses. Robotics--Congresses.
Series: Lecture notes in computer
science; 929 Lecture notes in computer
science. Lecture notes in artificial
intelligence
LC Classification: QH324.2 .E87 1995
Dewey Class No.: 574/.01/13 20

Advances in artificial life; 5th European
Conference, ECAL'99, Lausanne,
Switzerland, September 13-17, 1999:
proceedings / Dario Floreano, Jean-
Daniel Nicoud, Francesco Mondada,
(eds.).
Published/Created: Berlin; New York:
Springer, c1999.
Related Authors: Floreano, Dario.
Nicoud, Jean-Daniel. Mondada,
Francesco.
Description: xvi, 737 p.: ill.; 24 cm.
ISBN: 3540664521 (softcover: alk.
paper)
Notes: Errata slip inserted. Includes
bibliographical references.

Subjects: Biological systems--Computer simulation--Congresses. Biological systems--Simulation methods--Congresses. Robotics--Congresses. Artificial intelligence--Congresses. Series: Lecture notes in computer science; 1674. Lecture notes in computer science. Lecture notes in artificial intelligence.
LC Classification: QH324.2 .E87 1999
Dewey Class No.: 570/.1/13 21

Advances in automation and robotics.
Published/Created: Greenwich, Conn.: JAI Press, c1985-
Description: 2 v.: ill.; 24 cm. Vol. 1 (1985)- Ceased with v. 2 (1990). Cf. Letter from publisher.
Notes: "Theory and applications." SERBIB/SERLOC merged record
Subjects: Automation--Periodicals. Robotics--Periodicals.
LC Classification: TJ212 .A364
Dewey Class No.: 629.8/06 19

Advances in factories of the future, CIM, and robotics / edited by Michel Cotsaftis, François Vernadat.
Published/Created: Amsterdam; New York: Elsevier, 1993.
Related Authors: Cotsaftis, Michel. Vernadat, F. International Conference on CAD/CAM, Robotics, and Factories of the Future (8th: 1992: Metz, France)
Description: xii, 537 p.: ill.; 25 cm.
ISBN: 0444898565 (acid-free paper)
Notes: Papers presented at the eighth International Conference on CAD/CAM Robotics and Factories of the Future, held in Metz, France, on Aug. 17-19, 1992. Includes bibliographical references and indexes.
Subjects: Computer integrated manufacturing systems. Flexible manufacturing systems. Robots, Industrial.
Series: Manufacturing research and technology; 16
LC Classification: TS155.6 .A385 1993

Dewey Class No.: 670.42 20

Advances in laboratory automation robotics.
Published/Created: Hopkinton, Mass.: Zymark Corp., 1984-
Related Authors: Hawk, Gerald L. Strimaitis, Janet R. International Symposium on Laboratory Robotics.
Description: v.: ill.; 24 cm. 1984-
ISSN: 1073-2195 CODEN: ALOREY
Summary: Vols. for 1984- contain selected papers presented at the International Symposium on Laboratory Robotics.
Notes: Edited by Gerald L. Hawk and Janet R. Strimaitis. SERBIB/SERLOC merged record Indx'd selectively by: Chemical abstracts 0009-2258
Subjects: Laboratories--Data processing. Laboratories--Automation. Robotics. Laboratories--Automation--Congresses. Robots, Industrial--Congresses.
LC Classification: Q183.A1 A38
Dewey Class No.: 629.8/92 19

Advances in manufacturing systems: design, modeling, and analysis / edited by R.S. Sodhi; assistant editors, M. Zhou and S. Das.
Published/Created: Amsterdam; New York: Elsevier, 1994.
Related Authors: Sodhi, R. S. (Raj S.) International Conference on CAD/CAM, Robotics, and Factories of the Future (9th: 1993: Newark, N.J.)
Description: xiv, 462 p.: ill.; 25 cm.
ISBN: 0444819711 (acid-free paper)
Notes: "Ninth International Conference on CAD/CAM, Robotics, and Factories of the Future (CARS & FOFS 93) held in Newark, New Jersey, USA on August 17-20, 1993"--Foreword. Includes bibliographical references.
Subjects: Production engineering. Production management. Manufacturing processes. Production control.
Series: Manufacturing research and technology; 22
LC Classification: TS176 .A343 1994

Dewey Class No.: 670.42 20

Advances in manufacturing: decision, control, and information technology / [edited by] S.G. Tzafestas.
Published/Created: New York: Springer, 1999.
Related Authors: Tzafestas, S. G., 1939- European Robotics, Intelligent Systems, and Control Conference (3rd: 1998: Athens, Greece)
Description: xxii, 424 p.: ill.; 24 cm.
ISBN: 1852331267 (alk. paper)
Notes: Selected papers from the Third European Robotics, Intelligent Systems, and Control Conference, held June 22-25, 1998 in Athens, Greece. Includes bibliographical references and index.
Subjects: Manufacturing processes-- Planning. Process control. Decision support systems.
Series: Advanced manufacturing
LC Classification: TS183.3 .A39 1999
Dewey Class No.: 658.5 21

Advances in robotics, 1992: presented at the Winter Annual Meeting of the American Society of Mechanical Engineers, Anaheim, California, November 8-13, 1992 / sponsored by the Dynamic Systems and Control Division, ASME; edited by H. Kazerooni.
Published/Created: New York, N.Y.: American Society of Mechanical Engineers, c1992.
Related Authors: Kazerooni, H. (Homayoon) American Society of Mechanical Engineers. Dynamic Systems and Control.
Description: v, 249 p.: ill.; 28 cm.
ISBN: 0791811077 (pbk.)
Notes: Includes bibliographical references and index.
Subjects: Robotics--Congresses.
Series: DSC (Series); vol. 42.
LC Classification: TJ210.3 .A54 1992
Dewey Class No.: 629.8/92 20

Advances in robotics, mechatronics and haptic interfaces, 1993: presented at the 1993 ASME Winter Annual Meeting, New Orleans, Louisiana, November 28-December 3, 1993 / sponsored by the Dynamic Systems and Control Division, ASME; edited by H. Kazerooni, J. Edward Colgate, Bernard D. Adelstein.
Published/Created: New York, N.Y.: American Society of Mechanical Engineers, c1993.
Related Authors: Kazerooni, H. (Homayoon) Colgate, J. Edward (James Edward) Adelstein, Bernard D. American Society of Mechanical Engineers. Dynamic Systems and Control Division.
Description: vi, 367 p.: ill.; 29 cm.
ISBN: 0791810194
Notes: Includes bibliographical references and index.
Subjects: Robotics--Congresses. Mechatronics--Congresses. Virtual reality--Congresses.
Series: DSC (Series); vol. 49.
LC Classification: TJ210.3 .A54 1993
Dewey Class No.: 629.8/92 20

Advances in robotics.
Published/Created: New York,: Wiley, c1985-
Description: Series unnumbered. Began in 1985.
ISSN: 0749-1603
Notes: Description based on: Recent advances in robotics, published 1985. Indx'd selectively by: Mathematical reviews 0025-5629 1987-
LC Classification: UNC
Dewey Class No.: 629 11

Advances in robotics: the ERNET perspective: proceedings of the Research Workshop of ERNET-- European Robotics Network, Darmstadt, Germany, September 9-10, 1996 / edited by Claudio Bonivento, Claudio Melchiorri, Henning Tolle.
Published/Created: Singapore; River Edge, NJ: World Scientific, c1996.

Related Authors: Bonivento, Claudio.
Melchiorri, Claudio. Tolle, H.
Description: x, 293 p.: ill.; 22 cm.
ISBN: 9810227639 (alk. paper)
Notes: Includes bibliographical
references.
Subjects: Robotics--Congresses.
Robotics--Research--Europe--
Congresses.
LC Classification: TJ210.3 .E86 1996
Dewey Class No.: 629.8/92 21

Aerospace automation and fastening:
October 11-13, 1990, Arlington, Texas:
creative manufacturing engineering
program.
Published/Created: Dearborn, Mich. (1
SME Dr., Dearborn 48121): Society of
Manufacturing Engineers, c1990.
Related Authors: Society of
Manufacturing Engineers.
Description: 1 v. (unpaged): ill.; 28 cm.
Notes: Cover title. "For presentation at a
creative manufacturing engineering
program." Includes bibliographical
references.
Subjects: Aerospace industries--
Automation--Congresses. Fasteners--
Congresses. Robotics, Industrial--
Congresses.
LC Classification: TL671.28 .A34 1990
Dewey Class No.: 629.134/2 20

Ager, Richard T.
Introduction to robotics: a flexible 8- to
12-day technology module.
Published/Created: Pasco, WA (1620 E.
Hillsboro St., Pasco 99301): Marcraft
International Corp., c1992.
Description: 1 v. (various foliations):
ill.; 28 cm.
Notes: Cover Marcraft technology
education module.
Subjects: Robotics--Programmed
instruction.
Series: Robotic technology series
Technology learning activity; TE-2200
LC Classification: TJ211.26 .A33 1992
Dewey Class No.: 629.8/92/077 20

Agrotique 89: proceedings of the second
international conference, Bordeaux /
edited by J.P. Sagaspe & A. Villeger.
Published/Created: Marseille: Teknea,
c1989.
Related Authors: Sagaspe, J. P.
Villeger, A.
Description: 402 p.: ill.; 23 cm.
ISBN: 2877170128
Notes: Articles in English and French.
Includes bibliographical references and
index.
Subjects: Robotics--Congresses.
Agriculture--Automation--Congresses.
LC Classification: S678.65 .A47 1989
Dewey Class No.: 631.3 20

Akman, Varol.
Unobstructed shortest paths in
polyhedral environments / Varol
Akman.
Published/Created: Berlin; New York:
Springer-Verlag, c1987.
Description: 103 p.: ill.; 25 cm.
ISBN: 0387176292 (pbk.: U.S.)
Notes: Based on the author's thesis (Ph.
D.), Rensselaer Polytechnic, 1985.
Bibliography: p. 93-103.
Subjects: Robotics. Polyhedral
functions. Algorithms.
Series: Lecture notes in computer
science; 251
LC Classification: TJ211 .A42 1987
Dewey Class No.: 629.8/92 19

Albus, James Sacra.
Brains, behavior, and robotics / by
James S. Albus.
Published/Created: Peterborough, N.H.:
BYTE Books, c1981.
Description: x, 352 p.: ill.; 25 cm.
ISBN: 0070009759
Notes: Includes index. Bibliography: p.
341-347.
Subjects: Artificial intelligence.
LC Classification: Q335 .A44 1981
Dewey Class No.: 001.53/5 19

Albus, James Sacra.
Engineering of mind: an introduction to

the science of intelligent systems /
James S. Albus, Alexander M. Meystel.
Published/Created: New York: Wiley,
c2001.
Related Authors: Meystel, A. (Alex)
Description: xv, 411 p.: ill., map; 25 cm.
ISBN: 0471438545 (acid-free paper)
Notes: Includes bibliographical
references (p. 379-400) and index.
Subjects: Intelligent control systems.
Robotics. Artificial intelligence.
Series: Wiley series on intelligent
systems
LC Classification: TJ217.5 .A428 2001
Dewey Class No.: 629.8/92 21

Albus, James Sacra.
Peoples' capitalism: the economics of
the robot revolution / James Sacra
Albus.
Published/Created: College Park, Md.:
New World Books, 1976.
Description: vi, 157 p.: graphs; 23 cm.
Notes: Includes bibliographical
references.
Subjects: Robotics--Economic aspects--
United States. Capitalism--United
States. United States--Economic
conditions--1971-1981.
LC Classification: HC110.A9 A7
Dewey Class No.: 330.12/2

Aleksander, Igor.
Decision and intelligence / Igor
Aleksander, Henri Farreny, and Malik
Ghallab.
Published/Created: London: Kogan
Page; Englewood Cliffs, N.J.: Prentice-
Hall, c1987.
Related Authors: Farreny, Henri.
Ghallab, Malik.
Description: 203 p.: ill.; 24 cm.
ISBN: 0137820798 (Prentice-Hall)
Notes: Translation of: Fonction,
décision et intelligence. Includes index.
Bibliography: p. 193-200.
Subjects: Robotics. Robots, Industrial.
Series: Robots. English; v. 6.
LC Classification: TJ211 .R57313 1983
vol. 6 1987

Dewey Class No.: 629.8/92 s 629.8/92
19

Aleksander, Igor.
Decision and intelligence / Igor
Aleksander, Henri Farreny and Malik
Ghallab; [translated by Meg Tombs].
Published/Created: London: Kogan
Page, 1986.
Related Authors: Farreny, Henri.
Ghallab, Malik.
Description: 203 p.: ill.; 24 cm.
ISBN: 0850386519
Notes: Translation of: Fonction,
décision et intelligence. Includes index.
Bibliography: p. 193-199.
Subjects: Robotics. Artificial
intelligence. Expert systems (Computer
science)
Series: Robots. English; v. 6.
LC Classification: TJ211 .R57313 1983
vol. 6
Dewey Class No.: 006.3/3 19

Aleksander, Igor.
Designing intelligent systems: an
introduction / Igor Aleksander.
Published/Created: New York:
UNIPUB, c1984.
Description: 166 p.: ill.; 23 cm.
ISBN: 0890590435 :
Notes: Includes index. Bibliography: p.
159-161.
Subjects: Artificial intelligence.
Mathematical models. Robotics. System
theory.
LC Classification: Q335 .A442 1984
Dewey Class No.: 001.53/5 19

Aleksander, Igor.
Reinventing man: the robot becomes
reality / Igor Aleksander and Piers
Burnett.
Edition Information: 1st Amer. ed.
Published/Created: New York: Holt,
Rinehart, and Winston, 1984, c1983.
Related Authors: Burnett, Piers.
Description: 301 p.: ill.; 24 cm.
ISBN: 0030638577 :
Notes: Includes index. Bibliography: p.

291-296.
Subjects: Robotics. Artificial
intelligence.
LC Classification: TJ211 .A43 1984
Dewey Class No.: 629.8/92 19

Alford, J. Michael (James Michael)
Employing robotics in small
manufacturing firms: strategic
implications / by J. Michael Alford;
with a foreword by Ernest F. Hollings.
Edition Information: Original ed.
Published/Created: Malabar, Fla.: R.E.
Krieger Pub. Co., 1988.
Description: xv, 186 p.: ill.; 24 cm.
ISBN: 0894642227
Notes: Includes indexes. Bibliography:
p. 175-184.
Subjects: Robots, Industrial.
LC Classification: TS191.8 .A44 1988
Dewey Class No.: 629.8/92 19

Algorithmic and computational robotics:
new directions: the fourth Workshop on
the Algorithmic Foundations of
Robotics / edited by Bruce Randall
Donald, Kevin M. Lynch, Daniela Rus.
Published/Created: Natick, Mass.: A K
Peters, c2001.
Related Authors: Donald, Bruce R.
Lynch, Kevin (Kevin M.) Rus, Daniela.
Description: vi, 390 p.: ill.; 29 cm.
ISBN: 156881125X (alk. paper)
Notes: Includes bibliographical
references.
Subjects: Robotics--Congresses.
Algorithms--Congresses.
LC Classification: TJ210.3 .W664 2000
Dewey Class No.: 629.8/92 21

Algorithmic and geometric aspects of
robotics / edited by Jacob T. Schwartz,
Chee-Keng Yap.
Published/Created: Hillsdale, N.J.: L.
Erlbaum Associates, 1987.
Related Authors: Schwartz, Jacob T.
Yap, Chee-Keng.
Description: xii, 305 p.: ill.; 24 cm.
ISBN: 0898595541
Notes: Includes bibliographies and

indexes.
Subjects: Robotics. Robot vision.
Series: Advances in robotics (Hillsdale,
N.J.); vol. 1.
LC Classification: TJ211 .A44 1987
Dewey Class No.: 629.8/92 19

Algorithmic foundations of robotics /
WAFR 94, the Workshop on the
Algorithmic Foundations of Robotics;
edited by Ken Goldberg ... [et al.].
Published/Created: Wellesley, Mass.:
A.K. Peters, c1995.
Related Authors: Goldberg, Ken.
Description: xi, 555 p.: ill.; 29 cm.
ISBN: 1568810458
Notes: Workshop held 17-19 Feb. 1994
in San Francisco, Calif. Includes
bibliographical references and index.
Subjects: Robotics--Congresses.
Algorithms--Congresses.
LC Classification: TJ211 .W665 1994
Dewey Class No.: 629.8/92 20

Algorithms for robotic motion and
manipulation: 1996 Workshop on the
Algorithmic Foundations of Robotics /
edited by Jean-Paul Laumond, Mark
Overmars.
Published/Created: Wellesley, Mass.: A
K Peters, c1997.
Related Authors: Laumond, J.-P. (Jean-
Paul) Overmars, Mark H., 1958-
Description: xii, 468 p.: ill.; 29 cm.
ISBN: 1568810679
Notes: Workshop was held in July 1996
in Toulouse, France. Includes
bibliographical references and index.
Subjects: Robots--Motion. Algorithms.
LC Classification: TJ211.4 .W64 1996
Dewey Class No.: 629.8/92 21

Allen, Peter K.
Robotic object recognition using vision
and touch / by Peter K. Allen.
Published/Created: Boston: Kluwer
Academic Publishers, c1987.
Description: viii, 172 p.: ill.; 24 cm.
ISBN: 0898382459
Notes: Includes index. Bibliography: p.

[151]-156.
Subjects: Robotics. Robot vision.
Pattern recognition systems.
Series: Kluwer international series in
engineering and computer science;
SECS 34. Kluwer international series in
engineering and computer science.
Robotics.
LC Classification: TJ211 .A45 1987
Dewey Class No.: 629.8/92 19

An Investigation of micromechanical
structures, actuators, and sensors:
proceedings, IEEE Micro Robots and
Teleoperators Workshop 1987, Hyannis,
Massachusetts, November 9-11 /
sponsored by the IEEE Robotics and
Automation Council.
Published/Created: New York, NY:
Institute of Electrical and Electronics
Engineers, 1987.
Related Authors: IEEE Robotics and
Automation Council.
Description: ca. 160 p.: ill.; 28 cm.
Notes: "IEEE catalog number
87TH0204-8." Includes bibliographies.
Subjects: Robotics--Congresses.
Remote control--Congresses. Miniature
objects--Congresses.
LC Classification: TJ210.3 .I445 1987
Dewey Class No.: 629.8/92 19

An, Chae H.
Model-based control of a robot
manipulator / Chae H. An, Christopher
G. Atkeson, John M. Hollerbach.
Published/Created: Cambridge, Mass.:
MIT Press, c1988.
Related Authors: Atkeson, Christopher
G. Hollerbach, John M.
Description: 233 p.: ill.; 24 cm.
ISBN: 0262011026
Notes: Includes index. Bibliography: p.
[211]-226.
Subjects: Robotics. Manipulators
(Mechanism)--Automatic control.
Series: The MIT Press series in artificial
intelligence
LC Classification: TJ211 .A49 1988

Dewey Class No.: 629.8/92 19

Android epistemology / edited by Kenneth
M. Ford, Clark Glymour & Patrick J.
Hayes.
Published/Created: Menlo Park: AAAI
Press; Cambridge, Mass.: MIT Press,
c1995.
Related Authors: Ford, Kenneth M.
Glymour, Clark N. Hayes, Patrick J.
Description: xvii, 316 p.: ill.; 24 cm.
ISBN: 0262061848
Notes: Includes bibliographical
references and index.
Subjects: Androids. Robotics. Artificial
intelligence.
LC Classification: TJ211 .A54 1995
Dewey Class No.: 629.8/92 20

Angeles, Jorge, 1943-
Fundamentals of robotic mechanical
systems: theory, methods, and
algorithms / Jorge Angeles.
Edition Information: 2nd ed.
Published/Created: New York: Springer,
2002.
Description: p. cm.
ISBN: 038795368X (alk. paper)
Notes: Includes bibliographical
references and index.
Subjects: Robotics.
Series: Mechanical engineering series
(Berlin, Germany)
LC Classification: TJ211 .A545 2002
Dewey Class No.: 629.8/92 21

Angeles, Jorge, 1943-
Fundamentals of robotic mechanical
systems: theory, methods, and
algorithms / Jorge Angeles.
Published/Created: New York: Springer,
c1997.
Description: xix, 510 p.: ill.; 25 cm.
ISBN: 0387945407 (alk. paper)
Notes: Includes bibliographical
references (p. [493]-504) and index.
Subjects: Robotics.
Series: Mechanical engineering series
(Berlin, Germany)
LC Classification: TJ211 .A545 1997

Dewey Class No.: 629.8/92 20

Annual Workshop on Space Operations
    Applications and Research (SOAR ...)
    Published/Created: Washington, D.C.:
    National Aeronautics and Space
    Administration, Office of Management,
    Scientific and Technical Information
    Division; Springfield, Va.: For sale by
    the National Technical Information
    Service, 1991-
    Related Authors: United States.
    National Aeronautics and Space
    Administration. Scientific and Technical
    Information Division. United States.
    National Aeronautics and Space
    Administration. Scientific and Technical
    Information Program. United States.
    National Aeronautics and Space
    Administration. United States. Air
    Force.
    Description: v.: ill.; 28 cm. Issued in 2
    or more v. 4th ('90)-
    Notes: "Proceedings of a workshop ..."
    Sponsored by: National Aeronautics and
    Space Administration and the U.S. Air
    Force. Vol. for 1990 issued by: National
    Aeronautics and Space Administration,
    Office of Management, Scientific and
    Technical Information Division; 1991-
    by: National Aeronautics and Space
    Administration, Office of Management,
    Scientific and Technical Information
    Program. SERBIB/SERLOC merged
    record
    Subjects: Space stations--Congresses.
    Astronautics--Systems engineering--
    Congresses. Human-machine systems--
    Congresses. Artificial intelligence--
    Congresses. Space robotics--
    Congresses.
    Series: NASA conference publication
    LC Classification: TL797 .W67a
    Dewey Class No.: 629.47/4 20
    Govt. Doc. No.: NAS 1.55:

Annual Workshop on Space Operations
    Automation and Robotics (SOAR ...)
    Published/Created: Washington, D.C.:
    National Aeronautics and Space

Administration, Scientific and Technical
    Information Division; Springfield, VA:
    For sale by the National Technical
    Information Service, 1987-1990.
    Related Authors: United States.
    National Aeronautics and Space
    Administration. Scientific and Technical
    Information Division. United States.
    National Aeronautics and Space
    Administration. Scientific and Technical
    Information Branch. United States.
    National Aeronautics and Space
    Administration. United States. Air
    Force.
    Description: 3 v.: ill.; 28 cm. 1st ('87)-
    3rd ('89).
    Notes: Vol. for 1987 issued by: National
    Aeronautics and Space Administration,
    Scientific and Technical Information
    Branch; 1988-1989 by: National
    Aeronautics and Space Administration,
    Scientific and Technical Information
    Division. Sponsored by NASA and the
    U.S. Air Force. SERBIB/SERLOC
    merged record
    Subjects: Space stations--Automation--
    Congresses. Space robotics--
    Congresses.
    Series: NASA conference publication
    LC Classification: TL797 .W67a
    Dewey Class No.: 629.44/2 20
    Govt. Doc. No.: NAS 1.55:

Applications of AI, machine vision and
    robotics / edited by Kim L. Boyer,
    Louise Stark, Horst Bunke.
    Published/Created: Singapore; New
    Jersey: World Scientific, [1993]
    Related Authors: Boyer, Kim L. Stark,
    Louise. Bunke, Horst.
    Description: 259 p.: ill.; 26 cm.
    ISBN: 9810221509
    Notes: Includes bibliographical
    references and index.
    Subjects: Robot vision. Artificial
    intelligence.
    Series: Series in machine perception and
    artificial intelligence; vol. 17
    LC Classification: TJ211.3 .A67 1993

Dewey Class No.: 629.8/92 20

Applications of artificial intelligence in engineering X / editor, R.A. Adey, G. Rzevski, C. Tasso.
Published/Created: Southampton; Boston: Computational Mechanics Publications, c1995.
Related Authors: Adey, R. A. Rzevski, G. (George) Tasso, Carlo.
Description: 591 p.: ill.; 24 cm.
ISBN: 1853123161 (Southampton) 156252240X (Boston)
Notes: "Tenth International Conference on Applications of Artificial Intelligence in Engineering, AIENG/95." Includes bibliographical references and index.
Subjects: Computer-aided engineering--Congresses. Engineering design--Data processing--Congresses. Genetic algorithms--Congresses. Expert systems (Computer science)--Congresses. Neural networks--Congresses. Robotics--Congresses.
LC Classification: TA345 .I555 1995

Applications of artificial intelligence.
Published/Created: Bellingham, Wash., USA: SPIE, International Society for Optical Engineering, c1984-
Related Authors: Society of Photo-optical Instrumentation Engineers. Applications of Artificial Intelligence Conference.
Description: v.: ill.; 28 cm. Vol. for May 3-4, 1984 carries no numbering but constitutes 1. May 3-4, 1984- Ceased in 1991?
ISSN: 0893-9810
Notes: SERBIB/SERLOC merged record Split into: Applications of artificial intelligence. Machine vision and robotics; and: Applications of artificial intelligence. Knowledge-based systems.
Subjects: Artificial intelligence--Congresses. Expert systems (Computer science)--Congresses. Image processing--Congresses.

Series: Proceedings of SPIE--the International Society for Optical Engineering
LC Classification: Q334 .A65
Dewey Class No.: 006.3 19

Applications of artificial intelligence. Machine vision and robotics.
Published/Created: Bellingham, Wash.: SPIE--the International Society for Optical Engineering, c1992-
Related Authors: Society of Photo-optical Instrumentation Engineers.
Description: v.: ill.; 28 cm. 10 (22-24 Apr. 1992)-
ISSN: 1019-0716
Notes: SERBIB/SERLOC merged record
Subjects: Artificial intelligence--Congresses. Artificial intelligence--Industrial applications Congresses.
Series: Proceedings of SPIE--the International Society for Optical Engineering.
LC Classification: Q334 .A66
Dewey Class No.: 006.3/7 20

Applications of control and robotics: proceedings of the IASTED international conference, Orlando, Florida, January 8-10, 1996 / editor, M.H. Hamza.
Published/Created: Anaheim, Calif.: IASTED: ACTA Press, c1995.
Related Authors: Hamza, M. H. International Association of Science and Technology for Development.
Description: ii, 219 p.: ill.; 28 cm.
ISBN: 0889861897
Notes: "Proceedings of the IASTED International Conference on Applications of Control and Robotics. Sponsor, International Association of Science and Technology for Development--IASTED." Includes bibliographical references and index.
Subjects: Automatic control--Congresses. Robotics--Congresses.
LC Classification: TJ212.2 .I27 1996

Dewey Class No.: 629.8 20

Applications of learning & planning
methods / editor, Nikolaos G.
Bourbakis.
Published/Created: Singapore; New
Jersey: World Scientific, c1991.
Related Authors: Bourbakis, Nikolaos
G.
Description: xx, 365 p.: ill.; 23 cm.
ISBN: 9810205465
Notes: Includes bibliographical
references.
Subjects: Machine learning. Artificial
intelligence. Robotics.
Series: Series in computer science; vol.
26.
LC Classification: Q325.5 .A66 1991
Dewey Class No.: 006.3/1 20

Applications of mini-computers to control
and robotic systems; [proceedings.
Published/Created: New York, Institute
of Electrical and Electronics Engineers,
1973]
Related Authors: University of
Wisconsin--Milwaukee. Institute of
Electrical and Electronics Engineers.
Description: x, 115 l. illus. 28 cm.
Notes: Includes bibliographical
references.
Subjects: Minicomputers--Congresses.
Automatic control--Congresses.
Robotics--Congresses.
LC Classification: TK7888.3 .M55 1973
Dewey Class No.: 629.8/95

Applications of neural adaptive control
technology / editors, Jens Kalkkuhl ...
[et al.].
Published/Created: Singapore; River
Edge, NJ: World Scientific, c1997.
Related Authors: Kalkkuhl, Jens.
Description: vii, 307 p.: ill.; 23 cm.
ISBN: 9810231512 (alk. paper)
Notes: Includes bibliographical
references and index.
Subjects: Adaptive control systems.
Neural networks (Computer science)
Series: World scientific series in

robotics and intelligent systems; vol. 17
LC Classification: TJ217 .A66 1997
Dewey Class No.: 629.8/36 21

Applying robotics in the aerospace industry:
conference, March 27-29, 1984, St.
Louis, Missouri / sponsored and
published by Robotics International of
SME.
Edition Information: 1st ed.
Published/Created: Dearborn, Mich.:
Robotics International of SME, c1984.
Related Authors: Robotics International
of SME.
Description: 205 p. in various pagings:
ill.; 28 cm.
ISBN: 0872631451 (pbk.)
Notes: Includes bibliographies.
Subjects: Aerospace industries--
Automation--Congresses. Robots,
Industrial--Congresses.
LC Classification: TL671.28 .A67 1984
Dewey Class No.: 629.134/2 19

Archives of control sciences / Polish
Academy of Sciences, Committee of
Automatic Control and Robotics.
Published/Created: Warszawa: Polish
Scientific Publishers PWN, 1992-
Related Authors: Polska Akademia
Nauk. Komitet Automatyki i Robotyki.
Description: v.: ill.; 24 cm. Issues also
have whole numbering. Vol. 1, no. 1/2-
ISSN: 1230-2384
Notes: SERBIB/SERLOC merged
record Indx'd selectively by: Computer
& control abstracts 0036-8113 1992-
Electrical & electronics abstracts 0036-
8105 1992- Physics abstracts 0036-8091
1992-
Subjects: Automatic control--
Periodicals. Robotics--Periodicals.
LC Classification: TJ212 .A7
Dewey Class No.: 629.8/05 20

Ardayfio, David D., 1942-
Fundamentals of robotics / David D.
Ardayfio.
Published/Created: New York: M.
Dekker, c1987.

Description: x, 430 p.: ill.; 24 cm.
ISBN: 082477440X
Notes: Includes bibliographies and
index.
Subjects: Robotics.
Series: Mechanical engineering (Marcel
Dekker, Inc.); 57.
LC Classification: TJ211 .A73 1987
Dewey Class No.: 629.8/92 19

Arkin, Ronald C., 1949-
Behavior-based robotics / Ronald C.
Arkin.
Published/Created: Cambridge, Mass.:
MIT Press, c1998.
Description: xiv, 490 p.: ill.; 24 cm.
ISBN: 0262011654 (alk. paper)
Notes: Includes bibliographical
references (p. [445]-476) and indexes.
Subjects: Autonomous robots.
Intelligent control systems.
Series: Intelligent robots and
autonomous agents
LC Classification: TJ211 .A75 1998
Dewey Class No.: 629.8/92 21

Armstrong, Brian Stewart Randall.
Dynamics for robot control: friction
modeling and ensuring excitation during
parameter identification / by Brian
Stewart Randall Armstrong.
Published/Created: Stanford, CA: Dept.
of Computer Science, Stanford
University, [1988]
Related Authors: Stanford University.
Computer Science Dept.
Description: xii, 184 p.: ill.; 28 cm.
Notes: "May 1988." Thesis (Ph.D.)--
Stanford University, 1988.
Bibliography: p. 181-184.
Subjects: Robotics. Friction. Parameter
estimation. Robots--Dynamics. Robots--
Motion.
Series: Report (Stanford University.
Computer Science Dept.); no. STAN-
CS-88-1205.
LC Classification: MLCM 93/02213 (Q)

Armstrong-Hélouvry, Brian, 1958-
Control of machines with friction / by

Brian Armstrong-Hélouvry.
Published/Created: Boston: Kluwer
Academic Publishers, c1991.
Description: xi, 173 p.: ill.; 25 cm.
ISBN: 0792391330
Notes: Includes bibliographical
references and index.
Subjects: Tribology. Machine design.
Series: Kluwer international series in
engineering and computer science;
SECS 128. Kluwer international series
in engineering and computer science.
Robotics.
LC Classification: TJ1075 .A67 1991
Dewey Class No.: 621.8/9 20

Artificial ethology / [edited by] Owen
Holland and David McFarland.
Published/Created: Oxford; New York:
Oxford University Press, 2001.
Related Authors: Holland, Owen.
McFarland, David.
Description: 261 p.: ill.; 24 cm.
ISBN: 0198510578 (pbk.)
Notes: Based on proceedings of the
Artificial Ethology Workshop, held at
Puerto Chico, Playa Blanca de Yaiza,
Lanzerote in July 1998. Includes
bibliographical references (p. 236-255).
Subjects: Animal behavior--Simulation
methods--Congresses. Robotics--
Congresses.
LC Classification: QL751.65.S55 A78
2001
Dewey Class No.: 591.5/01/13 21

Artificial intelligence & robotics in military
and paramilitary markets.
Published/Created: Norwalk, Conn.,
U.S.A. (6 Prowitt St., Norwalk 06855):
International Resource Development,
c1985.
Related Authors: International Resource
Development, inc.
Description: ix, 273 leaves; 29 cm.
Notes: "February 1985."
Subjects: Artificial intelligence--
Military applications. Robotics--
Military applications. Defense
industries--United States. Market

surveys--United States. United States--
Armed Forces--Procurement.
Series: Report (International Resource
Development, inc.); #637.
LC Classification: UG479 .A77 1985
Dewey Class No.: 623/.028/563 19

Artificial intelligence and information-
control systems of robots: proceedings
of the Third International Conference on
Artificial Intelligence and Information-
Control Systems of Robots, Smolenice,
Czechoslovakia, June 11-15, 1984 /
edited by Ivan Plander.
Published/Created: Amsterdam; New
York: North-Holland; New York, N.Y.,
U.S.A.: Sole distributors for the U.S.A.
and Canada, Elsevier Science Pub. Co.,
1984.
Related Authors: Plander, Ivan.
Nauchnyi sovet po probleme
"Iskusstvennyi intellekt" (Akademiia
nauk SSSR) U´stav technickej
kibernetiky SAV.
Description: xvi, 413 p.: ill.; 23 cm.
ISBN: 0444875336 (U.S.)
Notes: Sponsored by the Sciencetific
Board for Artificial Intelligence of the
Systems Analysis Committee of the
Presidium of the Academy of Sciences
of the USSR and the Institute of
Technical Cybernetics of the Slovak
Academy of Sciences. Includes
bibliographies.
Subjects: Artificial intelligence--
Congresses. Robotics--Congresses.
Electronic data processing--Congresses.
LC Classification: Q334 .I55 1984
Dewey Class No.: 001.53/5 19

Artificial intelligence and information-
control systems of robots -87:
proceedings of the Fourth International
Conference on Artificial Intelligence
and Information-Control Systems of
Robots, Smolenice, Czechoslovakia, 19-
23 October, 1987 / edited by Ivan
Plander.
Published/Created: Amsterdam; New
York: North-Holland; New York, N.Y.,

U.S.A.: Sole distributors for the U.S.A.
and Canada, Elsevier Science Pub. Co.,
1987.
Related Authors: Plander, Ivan.
Nauchnyi sovet po probleme
"Iskusstvennyi intellekt" (Akademiia
nauk SSSR) Ústav technickej
kibernetiky SAV.
Description: xvii, 499 p.: ill.; 23 cm.
ISBN: 0444703039
Notes: "Organised by the Scientific
Board for Artificial Intelligence of the
Academy of Sciences of the USSR and
the Institute of Technical Cybernetics of
the Slovak Academy of Sciences."
Includes bibliographies.
Subjects: Artificial intelligence--
Congresses. Robotics--Congresses.
Electronic data processing--Congresses.
Expert systems (Computer science)--
Congresses. Perceptrons--Congresses.
LC Classification: Q334 .I55 1987
Dewey Class No.: 006.3 19

Artificial intelligence and information-
control systems of robots '89:
proceedings of the Fifth International
Conference on Artificial Intelligence
and Information-Control Systems of
Robots, Strbské Pleso, Czechoslovakia,
6-10 November 1989 / edited by Ivan
Plander.
Published/Created: Amsterdam; New
York: North-Holland; New York, NY,
U.S.A.: Distributors for the U.S. and
Canada, Elsevier Science Pub. Co.,
1989.
Related Authors: Plander, Ivan.
Nauchnyi sovet po probleme
"Iskusstvennyi intellekt" (Akademiia
nauk SSSR) Ústav technickej
kibernetiky SAV.
Description: xv, 452 p.: ill.; 23 cm.
ISBN: 0444883177
Notes: "Organised by the Scientific
Board for Artificial Intelligence of the
Academy of Sciences of the USSR and
the Institute of Technical Cybernetics of
the Slovak Academy of Sciences."
Includes bibliographical references and

index.
Subjects: Artificial intelligence--
Congresses. Robotics--Congresses.
Expert systems (Computer science)--
Congresses.
LC Classification: Q334 .I55 1989
Dewey Class No.: 006.3 20

Artificial intelligence and information-
control systems of robots '94:
proceedings of the Sixth International
Conference on Artificial Intelligence
and Information-Control Systems of
Robots, Smolenice Castle, Slovakia,
Sep 12-16, 1994 / editor, Ivan Plander.
Published/Created: Singapore; River
Edge, NJ: World Scientific, c1994.
Related Authors: Plander, Ivan.
Description: xvi, 413 p.: ill.; 23 cm.
ISBN: 981021877X
Notes: Includes bibliographical
references and index.
Subjects: Artificial intelligence--
Congresses. Robotics--Congresses.
Expert systems (Computer science)--
Congresses.
LC Classification: Q334 .I55 1994
Dewey Class No.: 629.8/92 20

Artificial intelligence in engineering:
robotics and processes / edited by J.S.
Gero.
Published/Created: Amsterdam; New
York: Elsevier; Southampton; Boston:
Computational Mechanics Publications,
1988.
Related Authors: Gero, John S.
International Conference on the
Applications of Artificial Intelligence in
Engineering (3rd: 1988: Palo Alto,
Calif.)
Description: 403 p.: ill.; 24 cm.
ISBN: 0444704701 (Elsevier)
1853120111 (Computational Mechanics
Publication UK) 0931215986
(Computational Mechanics Publications
USA)
Notes: Papers presented at the 3rd
International Conference on the
Applications of Artificial Intelligence in

Engineering held in Palo Alto, Calif.,
Aug. 1988. Includes bibliographical
references and index.
Subjects: Process control--Data
processing--Congresses. Expert systems
(Computer science)--Congresses.
Robotics--Congresses. Computer-aided
design--Congresses.
LC Classification: TS156.8 .A74 1988
Dewey Class No.: 629.8/95 20

Artificial intelligence in manufacturing,
assembly, and robotics / edited by Horst
O. Bunke.
Published/Created: München: R.
Oldenbourg, 1988.
Related Authors: Bunke, Horst.
Description: 182 p.: ill.; 25 cm.
ISBN: 3486209205
Notes: Includes bibliographical
references and indexes.
Subjects: Assembly-line methods--
Automation. Robots, Industrial.
Artificial intelligence--Industrial
applications.
LC Classification: TS178.4 .A77 1988
Dewey Class No.: 670.42 20

Artificial intelligence research reports.
Carnegie-Mellon: catalogue /
introduction by Allen Newell.
Published/Created: New York, N.Y.:
Scientific DataLink, [c1984]
Related Authors: Carnegie-Mellon
University. Computer Science Dept.
Carnegie-Mellon University. Robotics
Institute.
Description: 2 v.; 30 cm.
ISBN: 0913251062 (set)
Contents: pt. 1. Complex information
processing, 1954-1983 -- pt. 2.
Computer Science Department,
Robotics Institute, 1964-1983.
Subjects: Artificial intelligence--
Bibliography--Catalogs.
LC Classification: Z7405.A7 A782
1984 Q335
Dewey Class No.: 016.0063 20

Artificial vision for robots / edited by I.
Aleksander.
Edition Information: Collected ed.
Published/Created: New York:
Chapman & Hall, 1984.
Related Authors: Aleksander, Igor.
Description: 233 p.: ill.; 23 cm.
ISBN: 0412004518 :
Notes: Includes bibliographical
references.
Subjects: Robotics. Artificial vision.
Robots, Industrial.
LC Classification: TJ211 .A77 1984
Dewey Class No.: 629.8/92 19

Asada, H. (Haruhiko)
Direct-drive robots: theory and practice
/ Haruhiko Asada, Kamal Youcef-
Toumi.
Published/Created: Cambridge, Mass.:
MIT Press, c1987.
Related Authors: Youcef-Toumi,
Kamal.
Description: 262 p.: ill.; 24 cm.
ISBN: 0262010887
Notes: Includes index. Bibliography: p.
[249]-257.
Subjects: Robotics. Manipulators
(Mechanism)
LC Classification: TJ211 .A78 1987
Dewey Class No.: 629.8/92 19

Asada, H. (Haruhiko)
Robot analysis and control / H. Asada
and J.-J. E. Slotine.
Published/Created: New York, N.Y.: J.
Wiley, c1986.
Related Authors: Slotine, J.-J. E. (Jean-
Jacques E.)
Description: xi, 266 p.: ill.; 25 cm.
ISBN: 0471830291
Notes: "A Wiley Interscience
publication." Includes index.
Bibliography: p. 235-261.
Subjects: Robotics.
LC Classification: TJ211 .A79 1986
Dewey Class No.: 629.8/92 19

Asimov, Isaac, 1920-
How did we find out about robots? /

Isaac Asimov; illustrated by David
Wool.
Published/Created: New York: Walker,
1984.
Related Authors: Wool, David, ill.
Description: 62 p.: ill.; 22 cm.
ISBN: 0802765637
Summary: Traces the development of
robots from the automatic clock to the
microchip and discusses scientific and
industrial uses of robots today.
Notes: Includes index.
Subjects: Robotics--Juvenile literature.
Robots, Industrial--Juvenile literature.
Robots.
Series: Asimov, Isaac, 1920- How did
we find out--series.
LC Classification: TJ211.2 .A76 1984
Dewey Class No.: 629.8/92 19

Asimov, Isaac, 1920-
Robots, machines in man's image / Isaac
Asimov and Karen A. Frenkel.
Edition Information: 1st ed.
Published/Created: New York:
Harmony Books, c1985.
Related Authors: Frenkel, Karen A.
Description: 246 p.: ill.; 24 cm.
ISBN: 0517551101
Notes: Includes index. Bibliography: p.
237-240.
Subjects: Robotics--History.
LC Classification: TJ211 .A83 1985
Dewey Class No.: 629.8/92 19

Assistive technology and artificial
intelligence: applications in robotics,
user interfaces, and natural language
processing / Vibhu O. Mittal ... [et al.],
eds.
Published/Created: Berlin; New York:
Springer, c1998.
Related Authors: Mittal, Vibhu O.
Description: x, 273 p.: ill.; 24 cm.
ISBN: 3540647902 (softcover: acid-free
paper)
Notes: Includes bibliographical
references and index.
Subjects: Robotics. User interfaces
(Computer systems) Natural language

processing (Computer science)
Artificial intelligence.
Series: Lecture notes in computer
science; 1458 Lecture notes in computer
science. Lecture notes in artificial
intelligence.
LC Classification: TJ211 .A85 1998
Dewey Class No.: 617/.03 21

Autofact Europe: conference proceedings,
13-15 September 1983, Geneva,
Switzerland / sponsored by Computer
and Automated Systems Association of
SME, in cooperation with Robotics
International of SME [and] Society of
Manufacturing Engineers.
Edition Information: 1st ed.
Published/Created: Dearborn, Mich.:
CASA/SME, c1983.
Related Authors: Computer and
Automated Systems Association of
SME. Robotics International of SME.
Society of Manufacturing Engineers.
Description: 380 p. in various pagings:
ill.; 28 cm.
ISBN: 0872631230 (pbk.)
Notes: Includes bibliographical
references and index.
Subjects: Manufacturing processes--
Automation--Congresses. CAD/CAM
systems--Congresses. Robots,
Industrial--Congresses.
LC Classification: TS183 .A93 1983
Dewey Class No.: 670.42/7 19

Autofact west: proceedings / CAD/CAM
VIII, November 17-20, 1980, Anaheim,
California.
Edition Information: 1st ed.
Published/Created: Dearborn, Mich.:
Society of Manufacturing Engineers,
c1980.
Related Authors: Society of
Manufacturing Engineers. Computer
and Automated Systems Association of
SME. Robotics International of SME.
Description: 2 v.: ill.; 28 cm.
ISBN: 087263065X (pbk.: v. 1)
0872630668 (pbk.: v. 2)
Notes: "Sponsored by Society of

Manufacturing Engineers, Computer
and Automated Systems Association of
SME, [and] Robotics International of
SME." Includes bibliographical
references and index.
Subjects: CAD/CAM systems--
Congresses.
LC Classification: TS176 .C3 1980
Dewey Class No.: 670.42/7 19

Automach Australia '85: conference
proceedings, July 2-5, 1985, Melbourne,
Australia / sponsored by Computer and
Automated Systems Association of
SME, Robotics International of SME,
Society of Manufacturing Engineers and
its Australasian chapters, chapter 169
Adelaide ... [et al.].
Edition Information: 1st ed.
Published/Created: Dearborn, Mich.:
SME, c1985.
Related Authors: Computer and
Automated Systems Association of
SME. Robotics International of SME.
Society of Manufacturing Engineers.
Description: 570 p. in various pagings:
ill.; 28 cm.
ISBN: 0872631915 (pbk.)
Notes: Includes bibliographies.
Subjects: Production management--
Congresses. Automation--Congresses.
LC Classification: TS155.A1 A873
1985
Dewey Class No.: 658.5 19

Automation and robotics in construction X:
proceedings of the 10th International
Symposium on Automation and
Robotics in Construction (ISARC),
Houston, Texas, U.S.A., 24-26 May,
1993 / edited by George H. Watson,
Richard L. Tucker, Jewell K. Walters.
Published/Created: Amsterdam; New
York: Elsevier, 1993.
Related Authors: Watson, George H.
Tucker, Richard L. Walters, Jewell K.
Description: xiii, 558 p.: ill.; 25 cm.
ISBN: 0444815236 (acid.free paper)
Notes: "Organized by the Construction
Industry Institute." Includes

bibliographical references.
Subjects: Building--Automation--
Congresses. Robotics--Congresses.
LC Classification: TH437 .I57 1993
Dewey Class No.: 690 20

Automation and robotics in construction XI:
proceedings of the 11th International
Symposium on Automation and
Robotics in Construction (ISARC),
Brighton, UK, 24-26 May, 1994 / edited
by Denis A. Chamberlain.
Published/Created: Amsterdam; New
York: Elsevier, c1994.
Related Authors: Chamberlain, Denis A.
Description: xvi, 716 p.: ill.; 25 cm.
ISBN: 0444820442 (alk. paper)
Notes: Includes bibliographical
references.
Subjects: Building--Automation--
Congresses. Robotics--Congresses.
LC Classification: TH437 .I57 1994
Dewey Class No.: 624 20

Automation and robotics in the textile and
apparel industries / edited by Gordon A.
Berkstresser III, David R. Buchanan.
Published/Created: Park Ridge, N.J.,
U.S.A.: Noyes Publications, c1986.
Related Authors: Berkstresser, Gordon
A. Buchanan, David R.
Description: xii, 328 p.: ill.; 25 cm.
ISBN: 0815510772 :
Notes: Includes bibliographies.
Subjects: Textile factories--Automation.
Clothing factories--Automation. Robots,
Industrial.
Series: Textile series
LC Classification: TS1449 .A9 1986
Dewey Class No.: 677 19

Automation, miniature robotics, and sensors
for nondestructive testing and
evaluation / Yoseph Bar-Cohen,
technical editor.
Published/Created: Columbus, Ohio:
American Society for Nondestructive
Testing, c2000.
Description: 481 p.; 29 cm.
ISBN: 157117043X

Series: Topics on nondestructive
evaluation series; v. 4
LC Classification: *

Autonomous robots.
Published/Created: [Dordrecht]: Kluwer
Academic Publishers, c1994-
Related Authors: Bekey, George A.
Description: v.: ill.; 26 cm. Vol. 1, no. 1
(1994)-
ISSN: 0929-5593
Notes: Title from cover. Editor: George
A Bekey. SERBIB/SERLOC merged
record
Subjects: Autonomous robots--
Periodicals. Robotics--Periodicals.
LC Classification: TJ211.495 .A98
Dewey Class No.: 629.8/92 21

AutoWeld Conference.
Edition Information: 1st ed.
Published/Created: Dearborn, Mich.:
Robotics International of SME, c1983.
Related Authors: Robotics International
of SME.
Description: 1 v. (various pagings): ill.;
28 cm.
ISBN: 0872631370 (pbk.)
Notes: Includes bibliographies.
Subjects: Welding--Automation.
LC Classification: TS227.2 .A98 1983
Dewey Class No.: 671.5/2 19

Ayres, Robert U.
Robotics, applications and social
implications / Robert U. Ayres, Steven
M. Miller; with contributions from Vary
Coates ... [et al.].
Published/Created: Cambridge, Mass.:
Ballinger Pub. Co., c1983.
Related Authors: Miller, Steven M.
Description: xx, 339 p.: ill.; 24 cm.
ISBN: 0884108910
Notes: Includes bibliographies and
index.
Subjects: Robotics.
LC Classification: HD9696.R622 A97
1983
Dewey Class No.: 338.4/7629892 19

Baker, Christopher W.
    Robots among us: the challenges and
    promises of robotics / Christopher W.
    Baker.
    Published/Created: Brookfield, Conn.:
    Millbrook Press, 2002.
    Description: 48 p.: ill.; 21 cm.
    ISBN: 0761319697 (lib. bdg.)
    Summary: Describes the development
    of various types of robots, the
    challenges inherent in making them
    more and more complex, and uses of
    robots now and in the future.
    Notes: Includes index.
    Subjects: Robots--Juvenile literature.
    Robots. Automata.
    Series: New century technology
    LC Classification: TJ211.2 .B35 2002
    Dewey Class No.: 629.8/92 21

Balafoutis, C. A. (Constantinos A.)
    Dynamic analysis of robot
    manipulators: a Cartesian tensor
    approach / C.A. Balafoutis and R.V.
    Patel.
    Published/Created: Boston: Kluwer
    Academic Publishers, c1991.
    Related Authors: Patel, Rajnikant V.
    Description: xii, 292 p.: ill.; 25 cm.
    ISBN: 0792391454 (acid-free paper)
    Notes: Includes bibliographical
    references and index.
    Subjects: Manipulator (Mechanism)
    Robotics.
    Series: The Kluwer international series
    in engineering and computer science;
    SECS 131
    LC Classification: TJ211 .B335 1991
    Dewey Class No.: 629.8/92 20

Baldwin, Margaret.
    Robots and robotics / Margaret Baldwin
    & Gary Pack.
    Published/Created: New York: Watts,
    1984.
    Related Authors: Pack, Gary.
    Description: 61 p., [1] p. of plates: ill.;
    23 cm.
    ISBN: 0531047059
    Summary: Describes robotics--the

science of building robots--and gives
    advice on careers in that field. Also
    defines robots and describes their
    function in industry, outer space, and
    the home.
    Notes: Includes index. Bibliography: p.
    57-58.
    Subjects: Robotics--Juvenile literature.
    Robots--Juvenile literature. Robotics.
    Robots.
    Series: A Computer-awareness first
    book
    LC Classification: TJ211 .B34 1984
    Dewey Class No.: 629.8/92 19

Barrett, Norman S.
    Robots / N.S. Barrett; [illustrated by
    Mike Saunders, Stuart Willard].
    Published/Created: London; New York:
    F. Watts, 1985.
    Related Authors: Saunders, Mike, ill.
    Willard, Stuart, ill.
    Description: 32 p.: col. ill.; 26 cm.
    ISBN: 0531049477 (U.S.)
    Summary: Photos and simple text
    describe how robots perform tasks such
    as making cars and flying planes.
    Notes: Includes index.
    Subjects: Robots--Juvenile literature.
    Robots. Robotics.
    Series: Picture library
    LC Classification: TJ211.2 .B37 1985
    Dewey Class No.: 629.8/92 19

Berger, Fredericka.
    Robots: what they are, what they do / by
    Fredericka Berger; illustrated by Tom
    Huffman.
    Edition Information: 1st ed.
    Published/Created: New York:
    Greenwillow Books, c1992.
    Related Authors: Huffman, Tom, ill.
    Description: 47 p.: ill. (some col.); 22
    cm.
    ISBN: 0688098630 (TR): 0688098649
    (LE)
    Summary: Examines different kinds of
    robots, what they do, and how they
    work.
    Notes: "17D3"--P. [4] of cover. "24D5

(LIB)"--P. [4] of cover.
Subjects: Robots--Juvenile literature.
Robotics--Juvenile literature. Robots.
Robotics.
Series: Greenwillow read-alone books.
LC Classification: TJ211.2 .B47 1992
Dewey Class No.: 629.8/92 20

Berger, Phil.
The state-of-the-art robot catalog / Phil
Berger.
Edition Information: 1st ed.
Published/Created: New York: Dodd,
Mead, c1984.
Description: xi, 148 p.: ill.; 24 cm.
ISBN: 0396083617 (pbk.) :
Notes: Includes indexes.
Subjects: Robots--Catalogs. Robotics--
History.
LC Classification: TJ211 .B48 1984
Dewey Class No.: 629.8/92 19

Berk, A. A.
Practical robotics and interfacing for the
Spectrum / A.A. Berk.
Published/Created: London; New York:
Granada; [Dobbs Ferry, N.Y.]:
Distributed in the U.S.A. by Sheridan
House, 1984.
Description: x, 147 p.: ill.; 24 cm.
ISBN: 0246125764 (pbk.) :
Notes: Includes index.
Subjects: Robotics.
LC Classification: TJ211 .B49 1984
Dewey Class No.: 629.8/92 19

Beyer, Mark (Mark T.)
Robotics / by Mark Beyer.
Published/Created: New York:
Children's Press, c2002.
Description: p. cm.
ISBN: 051623918X (library binding)
0516240072 (pbk.)
Summary: Takes a look at the current
state and future of the field of robotics,
emphasizing that robots are already
playing an important role in making our
lives easier.
Contents: Robotic nation -- Our Bodies,
Our Bots -- RoboToys -- Robots among

us.
Notes: Includes bibliographical
references (p.) and index.
Subjects: Robotics--Juvenile literature.
Robotics.
Series: Life in the future
LC Classification: TJ211.2 .B49 2002
Dewey Class No.: 629.8/92 21

Bhanu, Bir.
Genetic learning for adaptive image
segmentation / Bir Bhanu, Sungkee Lee.
Published/Created: Boston: Kluwer
Academic Publishers, c1994.
Related Authors: Lee, Sungkee, 1956-
Description: xix, 271 p.: ill.; 24 cm.
ISBN: 0792394917 (acid-free paper)
Notes: Includes bibliographical
references (p. [261]-267) and index.
Subjects: Computer vision. Image
processing. Machine learning.
Series: The Kluwer international series
in engineering and computer science;
287. Robotics
LC Classification: TA1634 .B47 1994
Dewey Class No.: 006.3/7 20

Billard, Mary.
All about robots / by Mary Billard;
illustrated by Walter Wright.
Published/Created: New York: Platt &
Munk, c1982.
Related Authors: Wright, Walter, 1937-
ill.
Description: 45 p.: ill. (some col.); 31
cm.
ISBN: 044847493X :
Summary: Discusses the history of
robots, their current use in industry,
research, and space, and their future.
Subjects: Robots--Juvenile literature.
Robots, Industrial--Juvenile literature.
Robots. Robotics.
LC Classification: TJ211 .B55 1982
Dewey Class No.: 629.8/92 19

Biological neural networks in invertebrate
neuroethology and robotics / edited by
Randall D. Beer, Roy E. Ritzmann,
Thomas McKenna.

Published/Created: Boston: Academic Press, c1993.
Related Authors: Beer, Randall D. Ritzmann, Roy E. McKenna, Thomas M.
Description: xi, 417 p.: ill.; 24 cm.
ISBN: 0120847280 (acid-free paper)
Notes: Includes bibliographical references and index.
Subjects: Neural networks (Neurobiology) Invertebrates--Physiology.
Series: Neural networks, foundations to applications
LC Classification: QP363.3 .B56 1993
Dewey Class No.: 591.1/88 20

Bio-robotics, information technology, and intelligent control for bioproduction systems 2000: a proceedings volume from the 2nd IFAC/CIGR international workshop, Sakai, Osaka, Japan / edited by S. Shibusawa, M. Monta, and H. Murase.
Edition Information: 1st ed.
Published/Created: New York: Pergamon, 2001.
Related Authors: Shibusawa, S. (Sakae) Monta, M. (Mitsuji) Murase, H. (Haruhiko), 1948-
Description: p. cm.
ISBN: 0080435556
Subjects: Robotics--Congresses. Agricultural instruments--Congresses. Agriculture--Automation--Congresses. Artificial intelligence--Technological innovations Congresses.
LC Classification: S678.65 .I32 2000
Dewey Class No.: 631.3 21

Biorobotics: methods and applications / edited by Barbara Webb and Thomas R. Consi.
Published/Created: Menlo Park, CA: AAAI Press/MIT Press, 2001.
Related Authors: Webb, Barbara (Barbara H.) Consi, Thomas R., 1956-
Description: xiv, 208 p.; 23 cm.
ISBN: 026273141X (pbk.: alk. paper)
Notes: Includes bibliographical references (p. [175]-196).
Subjects: Animal behavior--Simulation methods. Senses and sensation--Simulation methods. Robotics.
LC Classification: QL751.65.S55 B56 2001
Dewey Class No.: 591.5/01/13 21

Biped locomotion: dynamics, stability, control, and application / M. Vukobratovi´c ... [et al.].
Published/Created: Berlin; New York: Springer-Verlag, c1990.
Related Authors: Vukobratovi´c, Miomir.
Description: xiv, 349 p.; 25 cm.
ISBN: 0387174567 (U.S.: alk. paper)
Notes: Translated from the Serbo-Croatian (Cyrillic). Includes bibliographical references and index.
Subjects: Robots--Dynamics. Human locomotion.
Series: Scientific fundamental of robotics; 7 Communications and control engineering series
LC Classification: TJ211.4 .B57 1990
Dewey Class No.: 629.8/92 20

Blankenship, John, 1948-
Apple II/IIe robotic arm projects / John Blankenship.
Published/Created: Englewood Cliffs, N.J.: Prentice-Hall, c1985.
Description: viii, 149 p.: ill.; 25 cm.
ISBN: 0130383244: 0130383163 (pbk.) :
Subjects: Robotics. Apple II (Computer)--Programming. Apple IIe (Computer)--Programming.
LC Classification: TJ211 .B56 1985
Dewey Class No.: 629.8/92 19

Bone, Jan.
Opportunities in robotics careers / Jan Bone; foreword by Donald A. Vincent.
Published/Created: Lincolnwood, Ill.: VGM Career Horizons, c1993.
Description: xiii, 145 p.; 21 cm.
ISBN: 0844240575 (hardcover): 0844240583 (soft) :

Notes: Includes bibliographical references (p. 143-145).
Subjects: Robotics--Vocational guidance.
Series: VGM opportunities series
LC Classification: TJ211.25 .B66 1993
Dewey Class No.: 629.8/92/02373 20

Bone, Jan.
Opportunities in robotics careers / Jan Bone; foreword by Donald A. Vincent.
Published/Created: Lincolnwood, Ill., U.S.A.: VGM Career Horizons, c1987.
Description: xi, 147 p.: ill.; 20 cm.
ISBN: 0844260207
Subjects: Robotics--Vocational guidance.
LC Classification: TJ211.25 .B66 1987
Dewey Class No.: 629.8/92/02373 19

Bowling, Charles M. (Charles McKaughan)
Principles and elements of thought construction, artificial intelligence, and cognitive robotics / by Charles M. Bowling.
Edition Information: 1st ed.
Published/Created: Houston, Tex.: CSY Pub., c1987.
Description: ii, 151 p.: ill.; 26 cm.
ISBN: 0945541007
Notes: Includes index.
Subjects: Artificial intelligence. Robotics. Cognition. Thought and thinking.
LC Classification: Q335 .B69 1987
Dewey Class No.: 006.3 19

Boyet, Howard.
Heath's robot "HERO": 68 experiments: fundamentals and applications / by Howard Boyet.
Published/Created: N.Y., N.Y. (14 E. 8th St., N.Y. 10003): Microprocessor Training, c1983.
Related Authors: Microprocessor Training Inc.
Description: x, 159 p.: ill.; 28 cm.
Subjects: Robotics. Microprocessors--Programming.
LC Classification: TJ211 .B68 1983

Dewey Class No.: 629.8/92 19

Boyet, Howard.
HERO 1: advanced programming experiments / written by Howard Boyet; revised and edited by Ron Johnson.
Published/Created: Benton Harbon, Mich.: Health Co., c1984.
Related Authors: Johnson, Ron, 1941-
Description: xiv, 296 p. in various pagings; 28 cm.
ISBN: 0871190362 (pbk.) :
Subjects: Robotics. Robots. Microprocessors--Programming.
LC Classification: TJ211 .B683 1984
Dewey Class No.: 629.8/92 19

Bringsjord, Selmer.
What robots can and can't be / by Selmer Bringsjord.
Published/Created: Dordrecht; Boston: Kluwer Academic, c1992.
Description: xiv, 380 p.: ill.; 23 cm.
ISBN: 0792316622 (alk. paper)
Notes: Includes bibliographical references (p. 354-371) and indexes.
Subjects: Robotics.
Series: Studies in cognitive systems; v. 12
LC Classification: TJ211 .B69 1992
Dewey Class No.: 629.8/92 20

Brooks, Rodney Allen.
Cambrian intelligence: the early history of the new AI / Rodney A. Brooks.
Published/Created: Cambridge, Mass.: MIT Press, c1999.
Description: xii, 199 p.: ill.; 23 cm.
ISBN: 0262522632 (pbk.: alk. paper)
Notes: "A Bradford book." Includes bibliographical references (p. [187]-199]).
Subjects: Robotics. Artificial intelligence.
LC Classification: TJ211 .B695 1999
Dewey Class No.: 629.8/9263 21

Brooks, Rodney Allen.
Flesh and machines: how robots will change us / Rodney A. Brooks.

Edition Information: 1st ed.
Published/Created: New York:
Pantheon Books c2002.
Description: x, 260 p.: ill.; 25 cm.
ISBN: 0375420797
Notes: Includes bibliographical
references and index.
Subjects: Robotics. Human-machine
systems. Artificial intelligence.
LC Classification: TJ211 .B697 2002
Dewey Class No.: 629.8/92 21

Buchsbaum, Frank, 1923-
Design and application of small
standardized components data book 757
/ [Frank Buchsbaum, Ferdinand
Freudenstein, Peter J. Thornton].
Published/Created: New York: Stock
Drive Products; Mineola, NY:
Distributed by Educational Products,
c1983.
Related Authors: Freudenstein,
Ferdinand, 1926- Thornton, Peter J.
Stock Drive Products (Firm)
Description: 784 p.: ill.; 20 cm.
ISBN: 0960987819 :
Notes: "Volume 2." Includes
bibliographies and index.
Subjects: Machine design. Power
transmission. Robotics.
LC Classification: TJ233 .B79 1983
Dewey Class No.: 621.8/15 19

Bundy, Alan.
Catalogue of artificial intelligence tools
/ edited by Alan Bundy.
Edition Information: 2nd rev. ed.
Published/Created: Berlin; New York:
Springer-Verlag, c1986.
Description: 168 p.; 25 cm.
ISBN: 0387168931 (U.S.)
Notes: Includes bibliographical
references and index.
Subjects: Artificial intelligence--Data
processing. Robotics. Computer
programming. Computer software--
Catalogs.
Series: Symbolic computation. Artificial
intelligence
LC Classification: Q336 .B86 1986

Dewey Class No.: 006.3 19

Bundy, Alan.
Catalogue of artificial intelligence tools
/ edited by Alan Bundy.
Published/Created: Berlin; New York:
Springer-Verlag, 1984.
Description: xxv, 150 p.; 25 cm.
ISBN: 0387139389 (U.S.: pbk.)
Notes: Includes bibliographical
references and index.
Subjects: Artificial intelligence--Data
processing. Robotics. Computer
programming. Computer programs--
Catalogs.
Series: Symbolic computation. Artificial
intelligence
LC Classification: Q336 .B86 1984
Dewey Class No.: 001.53/5 19

Burdick, Joel Wakeman.
Kinematic analysis and design of
redundant robot manipulators / by Joel
Wakeman Burdick IV.
Published/Created: Stanford, Calif.:
Dept. of Computer Science, Stanford
University, c1988.
Description: ix, 258 p.: ill.; 28 cm.
Notes: "March 1988." Thesis (Ph.D.)--
Stanford University. Bibliography: p.
252-258.
Subjects: Manipulators (Mechanism)
Robotics.
Series: Report (Stanford University.
Computer Science Dept.); no. STAN-
CS-88-1207.
LC Classification: TJ211 .B84 1988
Dewey Class No.: 670.42/72 20

Bürger, Erich.
Dictionary of robot technology in four
languages: English, German, French,
Russian / compiled by Erich Bürger in
collaboration with Günter Korzak.
Published/Created: Amsterdam; New
York: Elsevier; Berlin: VEB Verlag
Technik; New York, NY: Distributors
for the U.S. and Canada, Elsevier
Science, 1986.
Related Authors: Korzak, Günter.

Description: 275 p.; 25 cm.
ISBN: 0444995196 :
Notes: Includes indexes.
Subjects: Robotics--Dictionaries--
Polyglot. Dictionaries, Polyglot.
LC Classification: TJ210.4 .B87 1986
Dewey Class No.: 629.8/92/03 19

Bürger, Erich.
Robotika: anglicko-nemecko-
francúzsko-rusko-slovenský slovník /
spracoval Erich Bürger s autorským
kolektívom.
Published/Created: Bratislava: ALFA;
Berlin: VEB Verlag Technik, 1988.
Description: 345 p.; 25 cm.
Subjects: Robotics--Dictionaries--
Polyglot. Dictionaries, Polyglot.
Series: Edícia prekladových slovníkov
LC Classification: TJ210.4 .B84x 1988

Burger, Wilhelm.
Qualitative motion understanding /
Wilhelm Burger, Bir Bhanu.
Published/Created: Boston: Kluwer
Academic Publishers, c1992.
Related Authors: Bhanu, Bir.
Description: xiii, 210 p.: ill.; 25 cm.
ISBN: 0792392515 (acid-free paper)
Notes: Includes bibliographical
references (p. [199]-206) and index.
Subjects: Robots--Motion. Computer
vision. Artificial intelligence.
Series: Kluwer international series in
engineering and computer science;
SECS 184. Kluwer international series
in engineering and computer science.
Robotics.
LC Classification: TJ211.4 .B87 1992
Dewey Class No.: 629.8/92 20

Burks, Arthur W. (Arthur Walter), 1915-
Robots and free minds / Arthur W.
Burks.
Published/Created: Ann Arbor: College
of Literature, Science, and the Arts,
University of Michigan, 1986.
Description: vi, 97 p.: ill.; 23 cm.
Notes: Bibliography: p. 97.
Subjects: Robotics. Computers and

civilization.
LC Classification: TJ211 .B86 1986
Dewey Class No.: 629.8/92 19

Burnett, Betty, 1940-
Cool careers without college for math
and science wizards / Betty Burnett.
Published/Created: New York: Rosen
Pub. Group, 2002.
Description: 144 p.: col. ill.; 25 cm.
ISBN: 0823935027 (library binding)
Contents: Agriculture technician --
Earth science technician -- Wastewater
technician -- Hazardous materials
technician -- Aerospace (Avionics)
technician -- Medical lab technician --
Robotics technician -- Cost estimator --
Inspector and tester -- Surveying
technician -- Fiber optics technician --
Electronics technician -- Chemical
technician -- Civil engineering
technician.
Notes: Includes bibliographical
references and index.
Subjects: Mathematics--Vocational
guidance--Juvenile literature. Science--
Vocational guidance--Juvenile
literature. Mathematics--Vocational
guidance. Science--Vocational
guidance. Vocational guidance.
Series: Cool careers without college
LC Classification: QA10.5 .B85 2002
Dewey Class No.: 602.3 21

Burns, William C.
Practical robotics: systems, interfacing,
and applications / William C. Burns, Jr.,
Janet Evans Worthington.
Published/Created: Englewood Cliffs,
N.J.: Prentice Hall, c1986.
Related Authors: Worthington, Janet
Evans, 1942-
Description: vii, 296 p.: ill.; 25 cm.
ISBN: 0835957799 :
Notes: "A Reston book." Includes
index.
Subjects: Robotics.
LC Classification: TJ211 .B87 1986
Dewey Class No.: 629.8/92 19

CAD and robotics in architecture and
construction: proceedings of the joint
international conference at Marseilles,
25-27 June 1986 / o[r]ganized by
IIRIAM, GAMSAU, CSTB.
Published/Created: London: Kogan
Page; New York: Nichols Pub. Co.,
[c1986]
Related Authors: Institute of Robotics
and Artificial Intelligence of Marseille
(France) Groupe d'application des
méthodes scientifiques à l'architecture et
à l'urbanisme (Marseille, France) Centre
scientifique et technique du bâtiment
(France)
Description: 287 p.: ill.; 24 cm.
ISBN: 0893972584 (Nichols: pbk.) :
Notes: Includes bibliographies.
Subjects: Computer-aided design--
Congresses. Architectural design--Data
processing--Congresses. Engineering
design--Data processing--Congresses.
Building--Automation--Congresses.
LC Classification: TA174 .C25 1986
Dewey Class No.: 720/.28/5 19

CAD/CAM robotics and factories of the
future: 3rd International Conference on
CAD/CAM Robptocs amd Factories of
the Future (CARS & FOF '88):
proceedings / Biren Prasad, editor.
Published/Created: Berlin; New York:
Springer-Verlag, c1989.
Related Authors: Prasad, Birendra.
Description: 3 v.: ill. 24 cm.
ISBN: 0387511326 (U.S.: v. 1)
0387511334 (U.S.: v. 2) 0387511342
(U.S.: v. 3) 0387511350 (U.S.: set)
Contents: v. 1. Integration of design,
analysis, and manufacturing -- v. 2.
Automation of design, analysis, and
manufacturing -- v. 3. Robotics and
plant automation.
Notes: Conference held Aug. 14-17,
1988, Southfield, Mich.
Subjects: CAD/CAM systems--
Congresses. Robots, Industrial--
Congresses. Automation--Congresses.
LC Classification: TS155.6 .I5817 1988

Dewey Class No.: 670/.285 20

CAD/CAM, robotics, and autonomous
factories: proceedings of INCARF-93,
Indian Institute of Technology, New
Delhi, India, December 16-19, 1993 /
editors, B.L. Juneja, K.K. Pujara, R.
Sagar.
Published/Created: New Delhi; New
York: Tata McGraw Hill, 1993-1994.
Related Authors: Juneja, B. L. Pujara,
K. K. (Kewal Krishan) Sagar, R.
(Rakesh) Indian Institute of
Technology, Delhi.
Description: 4 v.: ill.; 23 cm.
ISBN: 0074621742 (v. 1) :
Notes: Includes bibliographical
references.
Subjects: CAD/CAM systems--
Congresses. Manufacturing processes--
Automation--Congresses. Robotics--
Congresses. Computer-aided
engineering--Congresses.
LC Classification: TS155.6 .I5818 1993
Dewey Class No.: 670 20

CAD/CAM, robotics, and factories of the
future '90: 5th International Conference
on CAD/CAM, Robotics, and Factories
of the Future (CARS and FOF'90)
proceedings / Suren N. Dwivedi, Alok
K. Verma, John E. Sneckenberger.
Published/Created: Berlin; New York:
Springer-Verlag, c1991-
Related Authors: Dwivedi, Suren N.
Verma, Alok K. Sneckenberger, John E.
International Society for Productivity
Enhancement.
Description: v. <1-2: ill.; 25 cm.
ISBN: 3540533982 (set) 0387533982
(set) 3540533990 (Berlin Heidelberg
New York: v. 1) 0387533990 (New
York Berlin Heidelberg: v. 1)
3540534008 (Berlin Heidelberg New
York: v. 2) 0387534008 (New York
Berlin Heidelberg: v. 2)
Notes: "International Society for
Productivity Enhancement." Includes
bibliographical references.
Subjects: CAD/CAM systems--

Congresses. Flexible manufacturing systems--Congresses. Robotics--Congresses. Manufacturing processes--Automation--Congresses.
LC Classification: TS155.6 .I5818 1990
Dewey Class No.: 670.42/7 20

CAD/CAM, robotics, and factories of the future: proceedings of the 14th International Conference, November 30, December 1-3, 1998 / editors, P. Radhakrishnan ... [et al.].
Published/Created: New Delhi: Narosa Pub. House, c1999.
Related Authors: Radhakrishnan, P. (Pezhingattil), 1944- PSG College of Technology. Dept. of Mechanical and Production Engineering.
Description: xxv, 966 p.: ill.; 25 cm.
ISBN: 8173192847
Notes: "Organised by Department of Mechanical and Production Engineering, PSG College of Technology, Coimbatore, India, sponsored by International Society for Productivity Enchancement, USA." Includes bibliographical references and index.
Subjects: CAD/CAM systems--Congresses.. Robotics--Congresses.
LC Classification: TS155.6 .I5818 1998
Dewey Class No.: 670/.285 21

CAD/CAM, robotics, and factories of the future: proceedings of the Fourth International Conference on CAD/CAM, Robotics, and Factories of the Future, Indian Institute of Technology, New Delhi, India, December 19-22, 1989 / chief editor, B.L. Juneja; editors, K.K. Pujara, R. Sagar.
Published/Created: New Delhi; New York: Tata McGraw Hill Pub. Co., c1989.
Related Authors: Juneja, B. L. Pujara, K. K. (Kewal Krishan) Sagar, R. (Rakesh) Indian Institute of Technology. Dept. of Mechanical Engineering.

Description: 3 v.: ill.; 25 cm.
ISBN: 0074601776: 0074601784: 0074601687 :
Notes: "Mechanical Engineering Department, Indian Institute of Technology, New Delhi." Includes bibliographical references.
Subjects: CAD/CAM systems--Congresses. Manufacturing processes--Automation--Congresses. Robotics--Congresses. Computer-aided engineering--Congresses.
LC Classification: TS155.6 .I5818 1989
Dewey Class No.: 670.42/7 20

Cams, gears, robot, and mechanism design: presented at the 1990 ASME Design Technical Conferences--21st Biennial Mechanisms Conference, Chicago, Illinois, September 16-19, 1990 / sponsored by the Mechanisms Committee of the Design Engineering Division, ASME; edited by A. Pisano, M. McCarthy, S. Derby.
Published/Created: New York, N.Y.: American Society of Mechanical Engineers, c1990.
Related Authors: Pisano, A. (Al) McCarthy, M. (Mike) Derby, Stephen J. American Society of Mechanical Engineers. Mechanisms Committee.
Description: viii, 441 p.: ill.; 28 cm.
ISBN: 0791805247
Notes: Includes bibliographical references.
Subjects: Machinery--Design and construction--Congresses. Cams--Design and construction--Congresses. Gearing--Congresses. Robotics--Congresses.
Series: DE (Series) (American Society of Mechanical Engineers. Design Engineering Division); vol. 26.
LC Classification: TJ230 .M44 1990
Dewey Class No.: 621.8/15 20

Castle of Dr. Brain [computer file].
Edition Information: Ver#1.1.
Published/Created: Coarsegold, CA: Sierra, c1991.

Related Authors: Cole, Corey. Sierra On-Line, Inc.
Description: 2 computer disks: sd., col.; 3 1/2 in. + 1 puzzle book (128 p.) + 2 manuals.
Computer File Info.: Computer program.
Summary: An adventure game involving logic puzzles and experimentation with time, astronomy, robotics, codes and ciphers, and math at three levels of difficulty.
Notes: Title from disk label. Game designer, Corey Cole. Issued in container with previously published book: Fantastic book of logic puzzles / by Muriel Mandell; illustrated by Elise Chanowitz. New York: Sterling Pub., c1986. Issued also for EGA (16 color) graphics displays. "INT#1.16.92"--Disk label. System requirements: IBM-compatible PC with 80286 or better microprocessor; 640K RAM; MS-DOS; VGA display; hard disk; mouse recommended. Supports Thunderboard, ProAudio Spectrum, AdLib, Roland MT-32/LAPC-1/CM-32L, Sound Blaster, and Sound Source.
Subjects: Computer adventure games--Juvenile software. Scientific recreations. Mathematical recreations. Computer games--Software.
LC Classification: GV1469.22
Dewey Class No.: 793.3 12

Caudill, Maureen.
In our own image: building an artificial person / Maureen Caudill.
Published/Created: New York: Oxford University Press, 1992.
Description: viii, 242 p.: ill.; 25 cm.
ISBN: 019507338X (acid-free paper) :
Notes: Includes bibliographical references (p. 225-230) and index.
Subjects: Robotics. Artificial intelligence.
LC Classification: TJ211 .C28 1992
Dewey Class No.: 629.8/92 20

Cetron, Marvin J.
Encounters with the future [sound recording] / Marvin Cetron with Thomas O'Toole.
Published/Created: Chicago, Ill.: Nightingale-Conant, p1983.
Related Authors: O'Toole, Thomas, 1941-
Description: 6 sound cassettes + 1 descriptive sheet.
Summary: A scientific survey of what will happen to man and to his world from now to the year 2000 and beyond.
Contents: Scientific forecasting and tomorrow's medical miracles. A new age of cybernetics and communication -- The robotics revolution and you. Major shifts in employment -- A tough look at education, reeducation and training. Social changes and consumer legislation -- Changing roles and attitudes in family life. Energy and economics -- The linkage and transfer of technology. Potential international conflicts -- Changes and opportunities around the globe. Future missions in outer space.
Notes: Title on container: Encounters with the future into the 21st century and beyond. In container (30 cm.)
Subjects: Twentieth century--Forecasts. Twenty-first century--Forecasts.
LC Classification: CB161 RZA 0462

Chacko, George Kuttickal, 1930-
Robotics/artificial intelligence/productivity: U.S.--Japan concomitant coalitions / George K. Chacko.
Published/Created: Princeton, N.J.: Petrocelli Books, c1986.
Description: xvii, 335 p.: ill.; 25 cm.
ISBN: 0894332287
Notes: Includes bibliographical references and index.
Subjects: Flexible manufacturing systems--United States. Flexible manufacturing systems--Japan. Robotics--United States. Robotics--Japan. Computer industry--United

States--Automation. Computer industry-
-Japan--Automation. Automobile
industry and trade--United States--
Automation. Automobile industry and
trade--Japan--Automation.
LC Classification: HD9725 .C48 1986
Dewey Class No.: 658.5/14 19

Chester, Michael.
Robots: facts behind the fiction /
Michael Chester.
Published/Created: New York:
Macmillan; London: Collier Macmillan,
c1983.
Description: 90 p.: ill.; 25 cm.
ISBN: 0027182207
Summary: Describes the development
of robots, which paralleled that of
computers, how robots work, and the
many functions they fulfill, with
emphasis on robots built by amateur
hobbyists.
Notes: Includes index. Bibliography: p.
85-86.
Subjects: Robotics--Juvenile literature.
Robots.
LC Classification: TJ211 .C526 1983
Dewey Class No.: 629.8/92 19

Chin, Felix, 1937-
Automation and robots: a selected
bibliography of books / Felix Chin.
Published/Created: Monticello, Ill.:
Vance Bibliographies, [1982]
Description: 19 p.; 29 cm.
Notes: Cover title. "May 1982."
Subjects: Automation--Bibliography.
Robotics--Bibliography.
Series: Public administration series--
bibliography, 0193-970X; P-969
LC Classification: Z5167 .C47 1982
T59.5
Dewey Class No.: 016.6298 19

Chorafas, Dimitris N.
Microprocessors for management,
CAD, CAM, and robotics / Dimitris N.
Chorafas.
Published/Created: New York:
Petrocelli Books, c1982.

Description: xi, 307 p.: ill.; 24 cm.
ISBN: 0894331833
Notes: Includes index.
Subjects: Microprocessors. CAD/CAM
systems.
LC Classification: QA76.5 .C48 1982
Dewey Class No.: 001.64 19

Choudhary, Alok N. (Alok Nidhi), 1961-
Parallel architectures and parallel
algorithms for integrated vision systems
/ by Alok N. Choudhary and Janak H.
Patel.
Published/Created: Boston: Kluwer
Academic Publishers, c1990.
Related Authors: Patel, Janak H., 1948-
Description: xvii, 157 p.: ill.; 25 cm.
ISBN: 0792390784
Notes: Includes bibliographical
references (p. [147]-151) and index.
Subjects: Computer vision. Parallel
processing (Electronic computers)
Computer architecture.
Series: Kluwer international series in
engineering and computer science;
SECS 108. Kluwer international series
in engineering and computer science.
Robotics.
LC Classification: TA1632 .C47 1990
Dewey Class No.: 006.3/7 20

Coiffet, Philippe.
An introduction to robot technology /
Philippe Coiffet and Michel Chirouze;
[translated by Meg Tombs].
Published/Created: New York:
McGraw-Hill, 1983.
Related Authors: Chirouze, Michel.
Description: 197 p.: ill.; 25 cm.
ISBN: 0070106894
Notes: Translation of: Eléments de
robotique. Includes index. Bibliography:
p. 191.
Subjects: Robotics. Robots, Industrial.
LC Classification: TJ211 .C61713 1983
Dewey Class No.: 629.8/92 19

Collective robotics: First International
Workshop, CRW '98, Paris, France, July
4-5, 1998: proceedings / Alexis

Drogoul, Milind Tambe, Toshio Fukuda (eds.).
Published/Created: Berlin; New York: Springer, c1998.
Related Authors: Drogoul, Alexis, 1968- Tambe, Milind, 1965- Fukuda, T. (Toshio), 1948-
Description: 161 p.: ill.; 24 cm.
ISBN: 3540647686 (acid-free paper)
Notes: Includes bibliographical references.
Subjects: Robotics--Congresses.
Series: Lecture notes in computer science; 1456. Lecture notes in computer science. Lecture notes in artificial intelligence.
LC Classification: TJ210.3 .C79 1998
Dewey Class No.: 629.8/92 21

Competitive automation--new frontiers, new opportunities: symposium proceedings / 26th International Symposium on Industrial Robots, 4-6 October 1995, Singapore; organised by International Federation of Robotics, Singapore Industrial Automation Association; sponsored by Economic Development Board Singapore, National Science & Technology Board.
Published/Created: [Singapore?]: MEP, [1996]
Related Authors: International Federation of Robotics. Singapore Industrial Automation Association.
Description: 623 p.: ill.; 31 cm.
ISBN: 1860580009
Notes: Includes bibliographical references and index.
Subjects: Robots, Industrial--Congresses.
LC Classification: TS191.8 .I57 1995
Dewey Class No.: 670.42/72 21

Complex robotic systems / Pasquale Chiacchio and Stefano Cjiaverini, eds.
Published/Created: London; New York: Springer, c1998.
Related Authors: Chiacchio, Pasquale, 1963- Chiaverini, Stefano, 1961-
Description: xi, 177 p.: ill.; 24 cm.

ISBN: 3540762655 (pbk.: alk. paper)
Notes: Includes bibliographical references.
Subjects: Robotics. Robots--Control systems.
Series: Lecture notes in control and information sciences; 233
LC Classification: TJ211 .C6218 1998
Dewey Class No.: 629.8/92 21

Computational intelligence for modelling, control & automation: neural networks & advanced control strategies / edited by Masoud Mohammadian.
Published/Created: Amsterdam; Washington, DC: IOS Press; Tokyo: Ohmsha, c1999.
Related Authors: Mohammadian, Masoud. International Conference on Computational Intelligence for Modelling, Control and Automation (1999: Vienna, Austria)
Description: xi, 392 p.: ill.; 24 cm.
ISBN: 9051994737 (IOS Press) 905199477X (IOS Press, complete set) 4274902765 (Ohmsha)
Notes: The book consists of 64 accepted research papers of the 1999 International Conference on Computational Intelligence for Modelling, Control and Automation in the areas of neural networks and advanced control strategies for manufacturing, robotics and automation. "The 1999 International Conference on Computational Intelligence for Modelling, Control and Automation - CIMCA '99 was held from 17th to 19th of February 1999 in Vienna, Austria"-- P. v. Includes bibliographical references and author index.
Subjects: Neural networks (Computer science)--Congresses. Automatic control--Congresses. Computational intelligence--Congresses.
Series: Concurrent systems engineering series; v. 54
LC Classification: QA76.87 .C6653 1999

Computational processes in human vision: an interdisciplinary perspective / edited by Zenon W. Pylyshyn.
Published/Created: Norwood, N.J.: Ablex Pub. Corp., c1988.
Related Authors: Pylyshyn, Zenon W., 1937-
Description: xi, 489 p.: ill.; 24 cm.
ISBN: 0893914606
Notes: Includes bibliographies and indexes.
Subjects: Vision.
Series: The Canadian Institute for Advanced Research series in artificial intelligence and robotics
LC Classification: QP475 .C63 1988
Dewey Class No.: 152.1/4 19

Computer aided design, engineering, and manufacturing: systems techniques and applications / editor, Cornelius Leondes.
Published/Created: Boca Raton, FL: CRC Press, 2001-
Related Authors: Leondes, Cornelius T.
Description: 7 v.: ill.; 26 cm.
ISBN: 084930993X (v. 1) 0849309956 (v. 3)
Notes: Includes bibliographical references and index.
Subjects: CAD/CAM systems.
LC Classification: TS155.6 .C6455 2001
Dewey Class No.: 670/.285 21

Computer architectures for robotics and automation / edited by James H. Graham.
Published/Created: New York: Gordon and Breach Science Publishers, c1987.
Related Authors: Graham, James H.
Description: viii, 238 p.: ill.; 24 cm.
ISBN: 2881241549
Notes: Based on presentations made at 2 workshops held at the 1985 and 1986 IEEE International Conference on Robotics and Automation. Includes bibliographies and index.
Subjects: Robotics--Congresses.
Automatic control--Congresses. Robot vision--Congresses. Computer

architecture--Congresses.
LC Classification: TJ210.3 .C65 1987
Dewey Class No.: 629.8/92 19

Computer vision, virtual reality, and robotics in medicine: first international conference, CVRMed '95, Nice, France, April 3-6, 1995: proceedings / Nicholas Ayache, ed.
Published/Created: Sophia-Antipolis, France: INRIA; Paris: INSERM; Berlin; New York: Springer-Verlag, c1995.
Related Authors: Ayache, Nicholas.
International Conference on Computer Vision, Virtual Reality, and Robotics in Medicine (1st: 1995: Nice, France)
Description: xiv, 567 p.: ill.; 24 cm.
ISBN: 0387591206 (New York: acid-free paper) 3540591206 (Berlin: acid-free paper)
Notes: Contributions to the program of the First International Conference on Computer Vision, Virtual Reality, and Robotics in Medicine (CVRMed '95)"--Preface. Includes bibliographical references and index.
Subjects: Computer vision in medicine--Congresses. Virtual reality in medicine--Congresses. Robotics in medicine--Congresses.
Series: Lecture notes in computer sciences; 905
LC Classification: R859.7.C67 I58 1995
Dewey Class No.: 610/.285/6 20

Computer vision: an overview.
Published/Created: Sacramento, CA: Business/Technology Books, c1984.
Related Authors: Business/Technology Books (Firm)
Description: xii, 150 p.: ill.; 28 cm.
ISBN: 0899341802 (pbk.)
Notes: Includes bibliographies.
Subjects: Computer vision.
Series: Robotics and artificial intelligence applications series; v. 3
LC Classification: TA1632 .C659 1984
Dewey Class No.: 001.53/5 19

Computer-based natural language
processing: an overview.
Published/Created: Sacramento, CA:
Business/Technology Books, c1984.
Description: ix, 44 p.: ill.; 28 cm.
ISBN: 0899341829 (pbk.)
Notes: Bibliography: p. 41-42.
Subjects: Interactive computer systems.
Programming languages (Electronic
computers)--Semantics. Parsing
(Computer grammar) Natural language
processing (Computer science)
Series: Robotics and artificial
intelligence applications series; v. 4
LC Classification: QA76.9.I58 C66
1984
Dewey Class No.: 001.53/5 19

Computer-integrated manufacturing and
robotics / presented at the Winter
Annual Meeting of the American
Society of Mechanical Engineers, New
Orleans, Louisiana, December 9-14,
1984; sponsored by the Production
Engineering Division, ASME; edited by
Ming C. Leu, Miguel R. Martinez.
Published/Created: New York, N.Y.
(345 E. 47th St., New York 10017):
ASME, c1984.
Related Authors: Leu, M. C. Martinez,
Miguel R. American Society of
Mechanical Engineers. Production
Engineering Division.
Description: vi, 370 p.: ill.; 26 cm.
Notes: Includes bibliographies.
Subjects: Computer integrated
manufacturing systems--Congresses.
Robots, Industrial--Congresses.
Series: PED; v. 13.
LC Classification: TS176 .A4 1984
Dewey Class No.: 670.42/7 19

Computer-integrated manufacturing
systems.
Published/Created: London, UK:
Butterworth Scientific Ltd., c1988-
1998.
Description: 11 v.: ill.; 30 cm. Vol. 1,
no. 1 (Feb. 1988)-v. 11, no. 4 (Oct.
1998).

ISSN: 0951-5240 CODEN: CMASEU
Notes: Title from cover.
SERBIB/SERLOC merged record
Absorbed in 1999 by: Robotics and
computer-integrated manufacturing.
Indx'd selectively by: Computer &
control abstracts 0036-8113 Feb. 1988-
Electrical & electronic abstracts 0036-
8105 Feb. 1988- Physics abstracts 0036-
8091 Feb. 1988-
Subjects: Computer integrated
manufacturing systems--Periodicals.
LC Classification: TS155.6 .C6548
Dewey Class No.: 670.42/7 19

Computer-integrated surgery: technology
and clinical applications / edited by
Russell H. Taylor ... [et al.].
Published/Created: Cambridge, Mass.:
MIT Press, c1996.
Related Authors: Taylor, Russell H.
Description: xix, 736 p., 44 col. plates:
ill. (some col.); 29 cm.
ISBN: 026220097X
Notes: Includes bibliographical
references and index.
Subjects: Surgery--Data processing.
Surgery--Computer simulation. Robot
hands. Surgery, Operative--methods.
Computer Systems. Human-Machine
Systems. Image Processing, Computer-
Assisted. Robotics.
LC Classification: RD29.7 .C65 1996
Dewey Class No.: 617.9/0285 20

Computers in engineering, 1988:
proceedings of the 1988 ASME
International Computers in Engineering
Conference and Exhibition, July 31-
August 4, 1988, San Francisco,
California / sponsored by Computers in
Engineering Division, ASME; editors,
V.A. Tipnis, E.M. Patton; associate
editors, D.W. Bennett ... [et al.].
Published/Created: New York, N.Y.
(345 E. 47th St., New York 10017):
American Society of Mechanical
Engineers, c1988.
Related Authors: Tipnis, Vijay A.
Patton, E. M. American Society of

Mechanical Engineers. Computers in
Engineering Division.
Description: 3 v.: ill.; 28 cm.
Contents: v. 1. Expert systems, artificial
intelligence, knowledge-based systems -
- v. 2. Computer-aided design,
computer-aided manufacturing,
computer-aided engineering, robotics,
computers in ME education -- v. 3.
Finite element technology, energy
systems, computational
thermohydraulics.
Notes: Includes bibliographies.
Subjects: Engineering--Data processing-
-Congresses. Computers--Congresses.
Robotics--Congresses.
LC Classification: TA345 .A86 1988
Dewey Class No.: 620/.0028/5 19

Computers in stereotactic neurosurgery /
edited by Patrick J. Kelly and Bruce A.
Kall.
Published/Created: Boston: Blackwell
Scientific Publications, 1992.
Related Authors: Kelly, Patrick J.,
1941- Kall, Bruce A.
Description: xii, 365 p.: ill.; 26 cm.
ISBN: 0865421455
Notes: Includes bibliographical
references and index.
Subjects: Stereoencephalotomy--Data
processing. Image processing. Robotics.
Computer Systems. Diagnosis, Surgical-
-methods. Image Interpretation,
Computer-Assisted. Image Processing,
Computer--Assisted. Neurosurgery--
instrumentation. Neurosurgery--
methods. Stereotaxic Techniques.
Series: Contemporary issues in
neurological surgery
LC Classification: RD594 .C654 1992
Dewey Class No.: 617.4/8059 20

Computing techniques for robots / edited by
I. Aleksander.
Published/Created: New York:
Chapman and Hall, 1985.
Related Authors: Aleksander, Igor.
Description: 276 p.: ill.; 23 cm.
ISBN: 0412010917

Notes: Includes bibliographies.
Subjects: Robotics. Computer-aided
design.
LC Classification: TJ211 .C62186 1985
Dewey Class No.: 629.8/92 19

Concise international encyclopedia of
robotics: applications and automation /
Richard C. Dorf, editor-in-chief;
Shimon Y. Nof, consulting editor.
Published/Created: New York: Wiley,
c1990.
Related Authors: Dorf, Richard C. Nof,
Shimon Y., 1946-
Description: xix, 1190 p.: ill.; 26 cm.
ISBN: 0471516988
Notes: "A Wiley-Interscience
publication." Rev. ed. of: International
encyclopedia of robotics. c1988.
Includes bibliographical references and
index.
Subjects: Robotics--Dictionaries.
LC Classification: TJ210.4 .C66 1990
Dewey Class No.: 629.8/92/03 20

Concurrent computations: algorithms,
architecture, and technology / edited by
Stuart K. Tewksbury, Bradley W.
Dickinson, and Stuart C. Schwartz.
Published/Created: New York: Plenum
Press, c1988.
Related Authors: Tewksbury, Stuart K.
Dickinson, Bradley W. Schwartz, Stuart
C. (Stuart Carl), 1939- Princeton
University. Dept. of Electrical
Engineering and Computer Science.
Robotics Systems Research Laboratory
(AT & T Bell Laboratories)
Description: xi, 726 p.: ill.; 26 cm.
ISBN: 030642939X
Notes: "Proceedings of the 1987
Princeton Workshop on Algorithm,
Architecture, and Technology Issues for
Models of Concurrent Computation,
held September 30-October 1, 1987, at
Princeton University, Princeton, New
Jersey, sponsored by Princeton
University Departments [i.e.
Department] of Electrical Engineering
and Computer Science, and AT&T Bell

Laboratories (Computer and Robotics Systems Research Laboratory)"--T.p. verso. Includes bibliographies and index.
Subjects: Parallel processing (Electronic computers)--Congresses.
LC Classification: QA76.5 .P728 1987
Dewey Class No.: 004/.35 19

Conference proceedings, April 26-30, 1987, Chicago, Illinois / Robots 11, 17th International Symposium on Industrial Robots; sponsored by Society of Manufacturing Engineers, Robotics International of SME.
Published/Created: Dearborn, Mich.: Society of Manufacturing Engineers, Robotics International of SME, c1987.
Related Authors: Robotics International of SME. International Symposium on Industrial Robots (17th: 1987: Chicago, Ill.)
Description: 1 v. (various pagings): ill.; 28 cm.
ISBN: 0872632733
Notes: Spine Robots 11/ISIR 17. Includes bibliographies and index.
Subjects: Robots, Industrial--Congresses.
LC Classification: TS191.8 .R637 1987
Dewey Class No.: 670.42/72 20

Connell, Jonathan H.
Minimalist mobile robotics: a colony-style architecture for an artificial creature / Jonathan H. Connell.
Published/Created: Boston: Academic Press, c1990.
Description: xvii, 175 p.: ill.; 24 cm.
ISBN: 012185230X (alk. paper)
Notes: Revision of author's thesis (Ph. D.--Massachusetts Institute of Technology, 1989). Includes bibliographical references (p. 165-170) and index.
Subjects: Mobile robots.
Series: Perspectives in artificial intelligence; v. 5
LC Classification: TJ211.415 .C66 1990

Dewey Class No.: 629.8/92 20

Conrad, James M.
Stiquito for beginners: an introduction to robotics / James M. Conrad, Jonathan W. Mills.
Published/Created: Los Alamitos, Calif.: IEEE Computer Society, c1999.
Related Authors: Mills, Jonathan W. (Jonathan Wayne)
Description: ix, 177 p.: ill.; 26 cm.
ISBN: 0818675144 (pbk.)
Notes: Includes one Stiquito robot kit. Includes bibliographical references and index.
Subjects: Robotics.
LC Classification: TJ211 .C636 1999
Dewey Class No.: 629.8/92 21

Control in robotics and automation: sensor-based integration / edited by B.K. Ghosh, Ning Xi, T.J. Tarn.
Published/Created: San Diego: Academic Press, 1999.
Related Authors: Ghosh, B. K. (Bhaskar Kumar), 1936- Xi, Ning. Tarn, Tzyh-Jong, 1937-
Description: xii, 428 p.: ill.; 27 cm.
ISBN: 0122818458
Notes: Includes bibliographical references and index.
Subjects: Robots--Control systems. Automatic control.
Series: Academic Press series in engineering
LC Classification: TJ211.35 .C65 1999
Dewey Class No.: 629.8/92 21

Control problems in robotics and automation / B. Siciliano and K.P. Valavanis (eds.).
Published/Created: London; New York: Springer, c1998.
Related Authors: Siciliano, Bruno, 1959- Valavanis, K. (Kimon)
Description: xxi, 295 p.: ill.; 24 cm.
ISBN: 3540762205 (pbk.: alk. paper)
Notes: Includes bibliographical references (p.).
Subjects: Automatic control. Robots--Control systems. Automation.

Series: Lecture notes in control and information sciences; 230
LC Classification: TJ213 .C5725 1998
Dewey Class No.: 629.8 21

Cooperative intelligent robotics in space II: 12-14 November 1991, Boston, Massachusetts / William E. Stoney, chair/editor; sponsored and published by SPIE--the International Society for Optical Engineering.
Published/Created: Bellingham, Wash.: SPIE, c1992.
Related Authors: Stoney, William E. Society of Photo-optical Instrumentation Engineers.
Description: ix, 422 p.: ill.; 28 cm.
ISBN: 0819407496
Notes: Includes bibliographical references and index.
Subjects: Space stations--Automation--Congresses. Space robotics--Congresses. Artificial intelligence--Congresses.
Series: Proceedings of SPIE--the International Society for Optical Engineering; v. 1612.
LC Classification: TL797 .C628 1992
Dewey Class No.: 629.47 20

Cooperative intelligent robotics in space III: 16-18 November 1992, Boston, Massachusetts / Jon D. Erickson, chair/editor; sponsored and published by SPIE--the International Society for Optical Engineering.
Published/Created: Bellingham, Wash.: SPIE, c1992.
Related Authors: Erickson, Jon D. Society of Photo-optical Instrumentation Engineers.
Description: ix, 516 p.: ill.; 28 cm.
ISBN: 0819410306
Notes: Includes bibliographical references and index.
Subjects: Space robotics--Congresses. Space vehicles--Congresses. Artificial intelligence--Congresses.
Series: Proceedings of SPIE--the International Society for Optical

Engineering; v. 1829.
LC Classification: TL1097 .C66 1992
Dewey Class No.: 629.47 20

Cooperative intelligent robotics in space: 6-7 November 1990, Boston, Massachusetts / Rui J. deFigueriedo, William E. Stoney, chairs/editors; sponsored and published by SPIE--the International Society for Optical Engineering.
Published/Created: Bellingham, Wash., USA: SPIE, c1991.
Related Authors: DeFigueiredo, Rui J. P. Stoney, William E. Society of Photo-optical Instrumentation Engineers. Symposium on Advances in Intelligent Systems (1990: Boston, Mass.)
Description: ix, 408 p.: ill.; 28 cm.
ISBN: 0819404543
Notes: "Part of a two-conference program on Telerobotics held at SPIE's Symposium on Advances in Intelligent Systems, a part of OE/Boston '90, 4-9 November 1990"--P. viii. Includes bibliographical references and index.
Subjects: Space stations--Automation--Congresses. Space robotics--Congresses. Artificial intelligence--Congresses.
Series: Proceedings of SPIE--the International Society for Optical Engineering; v. 1387.
LC Classification: TL797 .C63 1991
Dewey Class No.: 629.47 20

Corke, Peter I., 1959-
Visual control of robots: high-performance visual servoing / Peter I. Corke.
Published/Created: Taunton, Somerset, England: Research Studies Press; New York: Wiley, c1996.
Description: xxviii, 353 p.: ill.; 24 cm.
ISBN: 0863802079 (Research Studies Press) 0471969370 (Wiley)
Notes: Includes bibliographical references (p. 303-319) and index.
Subjects: Robots--Control systems.
Series: Robotics and mechatronics

series; 2
LC Classification: TJ211.35 .C68 1996
Dewey Class No.: 629.8/92 20

Cousineau, Leslie, 1963-
Construction robots: the search for new
building technology in Japan / Leslie
Cousineau, Nobuyasu Miura.
Published/Created: Reston, VA: ACSE
Press, c1998.
Related Authors: Miura, Nobuyasu.
Description: ix, 130 p.: ill.; 23 cm.
ISBN: 0784403171
Notes: Includes bibliographical
references (p. 123-125) and index.
Subjects: Building--Japan--Automation.
Robotics--Japan.
LC Classification: TH437 .C685 1998
Dewey Class No.: 690 21

Craig, John J., 1955-
Introduction to robotics: mechanics &
control / John J. Craig.
Published/Created: Reading, Mass.:
Addison-Wesley Pub. Co., c1986
Description: viii, 303 p.: ill.; 24 cm.
ISBN: 0201103265 :
Notes: Includes bibliographies and
index.
Subjects: Robotics.
LC Classification: TJ211 .C67 1986
Dewey Class No.: 629.8/92 19

Craig, John J., 1955-
Introduction to robotics: mechanics and
control / John J. Craig.
Edition Information: 2nd ed.
Published/Created: Reading, Mass.:
Addison-Wesley, c1989.
Description: xiii, 450 p.: ill.; 24 cm.
ISBN: 0201095289 :
Notes: Includes bibliographies and
index.
Subjects: Robotics.
Series: Addison-Wesley series in
electrical and computer engineering.
Control engineering.
LC Classification: TJ211 .C67 1989
Dewey Class No.: 629.8/92 19

Critchlow, Arthur J.
Introduction to robotics / Arthur J.
Critchlow.
Published/Created: New York:
Macmillan; London: Collier Macmillan,
c1985.
Description: xxxv, 491 p.: ill.; 27 cm.
ISBN: 0023255900
Notes: Includes bibliographies and
index.
Subjects: Robotics.
LC Classification: TJ211 .C69 1985
Dewey Class No.: 629.8/92 19

Critical technologies: machine tools,
robotics, and manufacturing: hearing
before the Subcommittee on
Technology and Competitiveness of the
Committee on Science, Space, and
Technology, U.S. House of
Representatives, One Hundred Second
Congress, first session, May 9, 1991.
Published/Created: Washington: U.S.
G.P.O.: For sale by the Supt. of Docs.
Congressional Sales Office, U.S.
G.P.O., 1991.
Description: iii, 112 p.: ill.; 24 cm.
Notes: Distributed to some depository
libraries in microfiche. Shipping list no.:
91-540-P. "No. 21." Item 1025-A-1,
1025-A-2 (MF)
Subjects: Technology and state--United
States. Machine-tool industry--
Government policy--United States.
Robotics--Government policy--United
States. Manufacturing industries--
United States. Competition,
International.
LC Classification: KF27 .S39955 1991y
Govt. Doc. No.: Y 4.Sci 2:102/21

Cummings, Richard, 1931-
Make your own robots / written and
illustrated by Richard Cummings.
Published/Created: New York, NY: D.
McKay, c1981.
Description: 120 p.: ill.; 24 cm.
ISBN: 0679206868 :
Summary: Presents directions for
making a wind-powered twirling robot,

android mask, toy robot in a toy car,
scooter-like robot, writing automaton,
robot costume, and electric rover robot.
Notes: Includes index. Bibliography: p.
114.
Subjects: Robotics--Juvenile literature.
Handicraft--Juvenile literature. Robots.
Handicraft.
LC Classification: TJ211 .C85
Dewey Class No.: 629.8/92 19

Cutkosky, Mark R.
Robotic grasping and fine manipulation
/ Mark R. Cutkosky.
Published/Created: Boston: Kluwer
Academic Publishers, c1985.
Description: xv, 176 p.: ill.; 25 cm.
ISBN: 0898382009
Notes: Includes index. Bibliography: p.
165-274.
Subjects: Robotics. Manipulators
(Mechanism)
Series: Kluwer international series in
engineering and computer science;
SECS 6. Kluwer international series in
engineering and computer science.
Robotics.
LC Classification: TJ211 .C87 1985
Dewey Class No.: 670.42/7 19

CVRMed-MRCAS '97: First Joint
Conference Computer Vision, Virtual
Reality and Robotics in Medicine and
Medical Robotics and Computer-
Assisted Surgery, Grenoble, France,
March 19-22, 1997: proceedings /
Jocelyne Troccaz, Eric Grimson, Ralph
Mösges, eds.
Published/Created: Berlin; New York:
Springer, c1997.
Related Authors: Troccaz, Jocelyne.
Grimson, Eric. Mösges, Ralph.
Description: xix, 834 p.: ill. (some col.);
24 cm.
ISBN: 3540627340 (Berlin: softcover:
acid-free paper)
Notes: Includes bibliographical
references and index.
Subjects: Computer vision in medicine--
Congresses. Surgery--Data processing--

Congresses. Virtual reality in medicine-
-Congresses. Robotics in medicine--
Congresses.
Series: Lecture notes in computer
science, 0302-9743; 1205
LC Classification: R859.7.C67 J65 1997
Dewey Class No.: 610/.285/637 21

Cybersurgery: advanced technologies for
surgical practice / volume editor,
Richard M. Satava.
Published/Created: New York: Wiley-
Liss, c1998.
Related Authors: Satava, Richard M.,
1942-
Description: xiv, 201 p., [12] p. of
plates: ill. (some col.); 24 cm.
ISBN: 0471158747 (cloth: acid-free
paper)
Notes: Includes bibliographical
references and index.
Subjects: Surgery, Operative--Computer
simulation. Surgery, Operative--Data
processing. Virtual reality in medicine.
Robotics in medicine. Surgery,
Operative--methods. Computer
Simulation. Telemedicine.
Series: Protocols in general surgery
LC Classification: RD32 .C93 1998
Dewey Class No.: 617/.05 21

DaCosta, Frank.
How to build your own working robot
pet / by Frank DaCosta.
Edition Information: 1st ed.
Published/Created: Blue Ridge Summit,
Pa.: Tab Books, c1979.
Description: 238 p.: ill.; 22 cm.
ISBN: 0830697969: 083061141X (pbk.)
:
Subjects: Robotics.
LC Classification: TJ211 .D32
Dewey Class No.: 629.8/92

Danielson, Peter, 1946-
Artificial morality: virtuous robots for
virtual games / Peter Danielson.
Published/Created: London; New York:
Routledge, 1992.
Description: xiv, 240 p.; 23 cm.

ISBN: 0415034841 0415076919 (pbk.)
Notes: Includes bibliographical
references (p. 229-235) and index.
Subjects: Robotics. Artificial
intelligence. Virtue.
LC Classification: TJ211 .D36 1992
Dewey Class No.: 170/.285/63 20

Darling, David J.
    Computers of the future: intelligent
    machines and virtual reality / by David
    Darling.
    Edition Information: 1st ed.
    Published/Created: Parsippany, N.J.:
    Dillon Press, c1996.
    Description: 72 p.: col. ill.; 25 cm.
    ISBN: 0875186173 (lsb) 0382391691
    (pbk.)
    Notes: Includes bibliographical
    references (p. 69) and index.
    Subjects: Computers--Juvenile
    literature. Robotics--Juvenile literature.
    Virtual reality--Juvenile literature.
    Computers. Robotics. Robots. Virtual
    reality.
    Series: Beyond 2000
    LC Classification: QA76.23 .D35 1996
    Dewey Class No.: 004 20

Data fusion in robotics and machine
    intelligence / edited by Mongi A. Abidi,
    Ralph C. Gonzalez.
    Published/Created: Boston: Academic
    Press, c1992.
    Related Authors: Abidi, Mongi A.
    Description: xii, 546 p.: ill.; 24 cm.
    ISBN: 0120421208 (acid-free paper)
    Notes: Includes bibliographical
    references and index.
    Subjects: Robotics. Artificial
    intelligence. Multisensor data fusion.
    LC Classification: TJ211 .D38 1992
    Dewey Class No.: 629.8/92 20

Davidor, Yuval.
    Genetic algorithms and robotics: a
    heuristic strategy for optimization /
    Yuval Davidor.
    Published/Created: Singapore; Teaneck,
    NJ: World Scientific, c1991.

Description: xiv, 164 p.: ill.; 23 cm.
ISBN: 9810202172
Notes: Includes bibliographical
references (p. 153-160) and index.
Subjects: Robots--Control systems.
Combinatorial optimization.
Algorithms.
Series: World Scientific series in
robotics and automated systems; vol. 1
LC Classification: TJ211.35 .D38 1990
Dewey Class No.: 629.8/92 20

Davies, Bill, 1932-
    Practical robotics: principles and
    applications / Bill Davies.
    Published/Created: Richmond Hill,
    Ont.: WERD Technology, c1997.
    Description: viii, 337 p.: ill.; 28 cm.
    ISBN: 096818300X
    Notes: Includes bibliographical
    references and index.
    Subjects: Robotics.
    LC Classification: TJ211 .D39 1997
    Dewey Class No.: 629.8/92 21

Davies, Bill.
    The practice robotics sourcebook / Bill
    Davies.
    Published/Created: Indianapolis, IN:
    Prompt Publications, 1997.
    Description: p. cm.
    ISBN: 0790611120 (pbk.)
    LC Classification: 9707 BOOK NOT
    YET IN LC

Dean, Thomas L., 1950-
    Planning and control / Thomas L. Dean,
    Michael P. Wellman.
    Published/Created: Los Altos, Calif.: M.
    Kaufmann Publishers, c1991.
    Related Authors: Wellman, Michael P.
    Description: xv, 486 p.: ill.; 24 cm.
    ISBN: 1558602097 :
    Notes: Includes bibliographical
    references and indexes.
    Subjects: Automatic control. System
    design. Robotics. Manufacturing
    processes--Automation. Decision
    support systems.
    Series: The Morgan Kaufmann series in

representation and reasoning
LC Classification: TJ213 .D35 1991
Dewey Class No.: 629.8 20

Designing autonomous agents: theory and
practice from biology to engineering
and back / edited by Pattie Maes.
Edition Information: 1st MIT Press ed.
Published/Created: Cambridge, Mass.:
MIT Press, 1990.
Related Authors: Maes, Pattie, 1961-
Description: 194 p.: ill.; 26 cm.
ISBN: 0262631350 (pbk.)
Notes: "Reprinted from Robotics and
autonomous systems, volume 6,
numbers 1 & 2 (1990)"--T.p. verso. "A
Bradford book." Includes
bibliographical references and index.
Subjects: Self-organizing systems.
Artificial intelligence.
Series: Special issues of Robotics and
autonomous systems
LC Classification: Q325 .D47 1991
Dewey Class No.: 003/.7 20

Designing your product for robotics.
Published/Created: Warrendale, PA:
Society of Automotive Engineers,
c1982.
Related Authors: Society of Automotive
Engineers. Detroit Section.
Description: 41 p.: ill.; 29 cm.
ISBN: 0898832888 (pbk.)
Contents: Introduction / Robert W.
Matthews -- The use of vision in
designing your product for robotics /
Michael Cronin -- Robot application
data / Jerry Jellis -- Pick and place
robots / Ken Conrad -- Feasibility of
painting robots / Richard Boylan --
Robots in die casting and machine
loading / Robert Green -- Flexibility in
finishing / Kenneth Close -- Spot
welding by robots in an automated
assembly / John Mattox -- Products--
do's and don'ts / Robert Bannister --
Applications of robots / Walter Weisel.
Notes: Papers presented at a mini-
symposium held by the Detroit Section
of the Society of Automotive Engineers

on Nov. 3, 1981. "May 1982." "SP-
517."
Subjects: Robots, Industrial--
Congresses.
LC Classification: TS191.8 .D47 1982
Dewey Class No.: 629.8/92 19

Developing and applying end of arm tooling
/ Peter McCormick, editor.
Edition Information: 1st ed.
Published/Created: Dearborn, Mich.:
Robotics International of SME, c1986.
Related Authors: McCormick, Peter.
Description: 245 p.: ill.; 29 cm.
ISBN: 0872632113
Notes: Title on spine: End of arm
tooling. Includes bibliographical
references and index.
Subjects: Robots, Industrial.
LC Classification: TS191.8 .D48 1986
Dewey Class No.: 670.42/7 19

Dextrous robot hands / S.T. Venkataraman,
T. Iberall, editors.
Published/Created: New York:
Springer-Verlag, c1990.
Related Authors: Shastri, S. V.
(Subramanian V.) Iberall, Thea. IEEE
Computer Society. United States. Office
of Naval Research. Workshop on
Dextrous Robot Hands (1988:
Philadelphia, Pa.) IEEE International
Conference on Robotics and
Automation (1988: Philadelphia, Pa.)
Description: viii, 345 p.: ill.; 25 cm.
ISBN: 0387971904 (alk. paper)
Notes: "This book grew out of the
Workshop on Dextrous Robot Hands
that occurred at the 1988 IEEE
Conference on Robotics and
Automation in Philadelphia ... co-
sponsored by the IEEE Computer
Society and the Office of Naval
Research"--Pref. Includes
bibliographical references (p. [299]-
318).
Subjects: Robot hands--Congresses.
Manipulators (Mechanism)--
Congresses.
LC Classification: TJ211 .D49 1990

Dewey Class No.: 629.8/92 20

Dhillon, B. S.
  Robot reliability and safety / B.S.
  Dhillon.
  Published/Created: New York:
  Springer-Verlag, c1991.
  Description: xv, 254 p.: ill.; 24 cm.
  ISBN: 0387975357 (acid-free paper:
  New York Berlin Heidelberg)
  3540975357 (acid-free paper: Berlin
  Heidelberg New York)
  Notes: Includes bibliographical
  references and index.
  Subjects: Robotics. Robots.
  LC Classification: TJ211 .D52 1991
  Dewey Class No.: 629.8/92 20

Digeronimo, Neil.
  Robotics: markets and competitors /
  prepared by Neil Digeronimo.
  Published/Created: Cleveland, Ohio
  (200 University Circle Research Center,
  11001 Cedar Ave., Cleveland 44106):
  Predicasts, 1984.
  Related Authors: Predicasts, inc.
  Description: iii leaves, 116 p.; 30 cm.
  Notes: "December 1984." "3547."
  Subjects: Robot industry--United States.
  Market surveys--United States.
  LC Classification: HD9696.R623 U62
  1984
  Dewey Class No.: 338.4/7629892 19

D'Ignazio, Fred.
  The science of artificial intelligence /
  Fred D'Ignazio & Allen L. Wold.
  Published/Created: New York: F. Watts,
  1984.
  Related Authors: Wold, Allen L.
  Description: 87 p.: ill.; 23 cm.
  ISBN: 0531047032 (lib. bdg.)
  Summary: Defines artificial
  intelligence, compares it with human
  intelligence, and explains how it is
  being used in computer technology and
  robotics.
  Notes: Includes index.
  Subjects: Artificial intelligence--
  Juvenile literature. Electronic digital

computers--Juvenile literature.
Robotics--Juvenile literature. Artificial
intelligence. Computers. Robotics.
Series: A Computer-awareness first
book
LC Classification: Q335.4 .D53 1984
Dewey Class No.: 001.53/5 19

Distributed autonomous robotic systems / H.
  Asama ... [et al.].
  Published/Created: Tokyo; New York:
  Springer-Verlag, c1994.
  Related Authors: Asama, H. (Hajime),
  1959- International Symposium of
  Distributed Autonomous Robotic
  Systems (2nd: 1994: Wak¯o-shi, Japan)
  Description: 394 p.: ill.; 25 cm.
  ISBN: 4431701478 (Tokyo: acid-free
  paper) 3540701478 (Berlin: acid-free
  paper) 0387701478 (New York: acid-
  free paper)
  Notes: "Papers presented at the Second
  International Symposium on Distributed
  Autonomous Robotic Systems (DARS
  '94), which was held in Wako-shi,
  Saitama, Japan, in July 1994"--Pref.
  Includes bibliographical references.
  Subjects: Robotics.
  LC Classification: TJ211 .D58 1994
  Dewey Class No.: 629.8/92 20

Distributed autonomous robotic systems 3 /
  Tim Lueth ... [et al.].
  Published/Created: Berlin; New York:
  Springer-Verlag, c1998.
  Related Authors: Lueth, Tim.
  International Symposium on Distributed
  Autonomous Robotic Systems (4th:
  1998: University of Karlsruhe)
  Description: xii, 416 p.: ill.; 25 cm.
  ISBN: 3540643990
  Notes: Proceedings of the fourth
  International Symposium on Distributed
  Autonomous Robotic Systems (DARS-
  4), held at the University of Karlsruhe.
  Includes bibliographical references and
  index.
  Subjects: Robotics--Congresses.
  Autonomous robots--Congresses.
  LC Classification: TJ211 .D583 1998

Dewey Class No.: 629.8/92 21

Donald, Bruce R.
Error detection and recovery in robotics
/ Bruce R. Donald.
Published/Created: Berlin; New York:
Springer-Verlag, 1989.
Description: xxiv, 314 p.: ill.; 24 cm.
ISBN: 0387969098 (U.S.)
Notes: Revised version of thesis (Ph.
D.)--M.I.T., 1987. Bibliography: p.
[277]-285.
Subjects: Robots--Error detection and
recovery.
Series: Lecture notes in computer
science; 336
LC Classification: TJ211.417 .D66 1989
Dewey Class No.: 629.8/92 19

Donner, Marc D.
Real-time control of walking / Marc D.
Donner.
Published/Created: Boston: Birkhäuser,
1987.
Description: xv, 160 p.: ill.; 24 cm.
ISBN: 0817633324 :
Notes: Includes index. Bibliography: p.
149-155.
Subjects: Robotics. Locomotion. Real
time control.
Series: Progress in computer science;
no. 7.
LC Classification: TJ211 .D66 1987
Dewey Class No.: 629.8/92 19

Dorf, Richard C.
Robotics and automated manufacturing /
Richard C. Dorf.
Published/Created: Reston, Va.: Reston
Pub. Co., c1983.
Description: xv, 190 p.: ill.; 24 cm.
ISBN: 0835966860 :
Notes: Includes index. Bibliography: p.
183-187.
Subjects: Robots, Industrial.
LC Classification: TS191.8 .D67 1983
Dewey Class No.: 629.8/92 19

Dorigo, Marco.
Robot shaping: an experiment in
behavior engineering / Marco Dorigo
and Marco Colombetti.
Published/Created: Cambridge, Mass.:
MIT Press, c1998
Related Authors: Colombetti, Marco.
Description: xviii, 203 p.: ill.; 24 cm.
ISBN: 0262041642
Notes: "A Bradford book." Includes
bibliographical references (p. [191]-
199) and index.
Subjects: Robots--Control systems.
Machine learning. Robots--Motion.
Series: Intelligent robotics and
autonomous agents
LC Classification: TJ211.35 .D67 1998
Dewey Class No.: 629.8/92 21

Dudek, Gregory, 1958-
Computational principles of mobile
robotics / Gregory Dudek, Michael
Jenkin.
Published/Created: New York:
Cambridge University Press, 2000.
Related Authors: Jenkin, Michael
(Michael Richard MacLean), 1959-
Description: xii, 280 p. ill. 26 cm.
ISBN: 0521560217
Notes: Includes bibliographical
references (p. [257]-272) and index.
Subjects: Mobile robots.
LC Classification: TJ211.415 .D83 2000
Dewey Class No.: 629.8/92 21

Duffy, Joseph, 1937-
Analysis of mechanisms and robot
manipulators / Joseph Duffy.
Published/Created: New York: Wiley,
1980.
Description: x, 419 p.: ill.; 26 cm.
ISBN: 0470270020
Notes: "A Halsted Press book." Includes
bibliographical references and index.
Subjects: Machinery, Kinematics of.
Manipulators (Mechanism) Robotics.
LC Classification: TJ175 .D77 1980
Dewey Class No.: 629.8/92 19

Duffy, Joseph, 1937-
Statics and kinematics with applications
to robotics / Joseph Duffy.

Published/Created: Cambridge
[England]; New York, NY, USA:
Cambridge University Press, 1996.
Description: xiii, 174 p.: ill.; 24 cm.
ISBN: 0521482135 (hardcover)
Notes: Includes bibliographical
references (p. 171-172) and index.
Subjects: Manipulators (Mechanism)
Statics. Machinery, Kinematics of.
Robots--Motion.
LC Classification: TJ211 .D84 1996
Dewey Class No.: 670.42/72 20

Durrant-Whyte, Hugh F., 1961-
Integration, coordination, and control of
multi-sensor robot systems / by Hugh F.
Durrant-Whyte.
Published/Created: Boston: Kluwer
Academic Publishers, c1988.
Description: xvii, 236 p.: ill.; 25 cm.
ISBN: 0898382475
Notes: Bibliography: p. 213-236.
Subjects: Robotics.
Series: Kluwer international series in
engineering and computer science;
SECS 36. Kluwer international series in
engineering and computer science.
Robotics.
LC Classification: TJ211 .D87 1988
Dewey Class No.: 629.8/92 19

Dynamics and control of multibody/robotic
systems with space applications:
presented at the Winter Annual Meeting
of the American Society of Mechanical
Engineers, San Francisco, California,
December 10-15, 1989 / sponsored by
the Aerospace Systems Technical Panel
of the Dynamic Systems and Control
Division, ASME; edited by S.M. Joshi,
L. Silverberg, T.E. Alberts.
Published/Created: New York, N.Y.:
The Society, c1989.
Related Authors: Joshi, S. M. (Suresh
M.) Silverberg, L. (Larry) Alberts, T. E.
(Thomas E.) American Society of
Mechanical Engineers. Winter Meeting
(1989: San Francisco, Calif.) American
Society of Mechanical Engineers.
Dynamic Systems and Control Division.

Aerospace Systems Technical Panel.
Description: v, 64 p.: ill.; 28 cm.
ISBN: 0791804143
Notes: Includes bibliographical
references.
Subjects: Robots--Dynamics--
Congresses. Robots--Control systems--
Congresses. Space robotics--
Congresses.
Series: DSC; vol. 15 DSC (Series); vol.
15.
LC Classification: TJ210.3 .D96 1989
Dewey Class No.: 629.8/92 20

Education and training in robotics / edited
by T.M. Husband.
Published/Created: Bedford, UK: IFS
(Publications); New York: Springer-
Verlag, 1985.
Related Authors: Husband, T. M. (Tom
M.)
Description: 315 p.: ill.; 25 cm.
ISBN: 0387162232 (Springer-Verlag)
Notes: Includes bibliographies.
Subjects: Robotics--Study and teaching.
Robots, Industrial.
Series: International trends in
manufacturing technology
LC Classification: TJ211.26 .E38 1985
Dewey Class No.: 670.42/7 19

Electrical and chemical interactions at Mars
workshop: proceedings summary of a
workshop held at the NASA Lewis
Research Center, Cleveland, Ohio,
November 19 and 20, 1991.
Published/Created: [Washington, D.C.]:
National Aeronautics and Space
Administration, Office of Management,
Scientific and Technical Information
Program, 1992.
Related Authors: United States.
National Aeronautics and Space
Administration. Scientific and Technical
Information Program. Electrical and
Chemical Interactions at Mars
Workshop (1991: NASA Lewis
Research Center)
Description: v, 28 p.; 28 cm.
Notes: "Part I--final report." "The

Electrical and Chemical Interactions at
Mars Workshop"--Pref. Includes
bibliographical references (p. 28).
Subjects: Cosmochemistry--Congresses.
Space robotics--Congresses. Mars
(Planet)--Surface--Congresses. Mars
(Planet)--Exploration--Congresses.
Series: NASA conference publication;
10093
LC Classification: QB641 .E42 1992
Dewey Class No.: 629.43/543 21

Ellery, Alex, 1963-
An introduction to space robotics / Alex
Ellery.
Published/Created: London; New york:
Springer; Chichester: Praxis Pub.,
c2000.
Description: xviii, 663 p.: ill.; 25 cm.
ISBN: 185233164X (alk. paper)
Notes: Includes bibliographical
references (p. [627]-651) and index.
Subjects: Space robotics.
Series: Springer-Praxis books in
astronomy and space sciences
LC Classification: TL1097 .E45 2000
Dewey Class No.: 629.47 21

Engel, C. William.
The world according to Robo the Robot
/ C. William Engel.
Published/Created: Hasbrouck Heights,
N.J.: Hayden Book Co., c1985.
Description: 242 p.: ill.; 25 cm.
ISBN: 0810463318 (pbk.)
Notes: Bibliography: p. 239-242.
Subjects: Robotics. Robots.
LC Classification: TJ211 .E54 1985
Dewey Class No.: 629.8/92 19

Engelberger, Joseph F.
Robotics in practice: management and
applications of industrial robots / Joseph
F. Engelberger, associate authors,
Dennis Lock and Kenneth Willis; with a
foreword by Isaac Asimov.
Published/Created: New York, N.Y.:
AMACOM, c1980.
Related Authors: Lock, Dennis. Willis,
Kenneth.

Description: xvii, 291 p., [8] p. of
plates: ill. (some col.); 24 cm.
ISBN: 0814456456
Notes: Includes index. Bibliography: p.
[279]-283.
Subjects: Robots, Industrial.
LC Classification: T59.4 .E53 1980
Dewey Class No.: 629.8/92 19

Engelberger, Joseph F.
Robotics in service / Joseph F.
Engelberger.
Edition Information: 1st MIT Press ed.
Published/Created: Cambridge, Mass.:
MIT Press, 1989.
Description: 248 p., [12] p. of plates:
ill.; 25 cm.
ISBN: 0262050420
Notes: Includes index. Bibliography: p.
235-239.
Subjects: Robotics. Robots.
LC Classification: TJ211 .E55 1989
Dewey Class No.: 629.8/95 20

Engineering systems with intelligence:
concepts, tools, and applications / edited
by Spyros G. Tzafestas.
Published/Created: Dordrecht; Boston:
Kluwer Academic Publishers, c1991.
Related Authors: Tzafestas, S. G., 1939-
European Robotics and Intelligent
Systems Conference (1991: Kerkyra,
Greece)
Description: xiv, 688 p.: ill.; 25 cm.
ISBN: 0792315006 (alk. paper)
Notes: Papers presented at the European
Robotics and Intelligent Systems
Conference (EURISCON '91) held in
Corfu, Greece, June 23-28, 1991.
Includes index.
Subjects: Intelligent control systems--
Congresses. Artificial intelligence--
Congresses. Expert systems (Computer
science)--Congresses.
Series: International series on
microprocessor-based and intelligent
systems engineering; v. 9
LC Classification: TJ217.5 .E54 1992
Dewey Class No.: 629.8 20

Essays on mathematical robotics / John Baillieul, Shankar S. Sastry, Hector J. Sussmann, editors.
Published/Created: New York: Springer, c1998.
Related Authors: Baillieul, J. (John) Sastry, Shankar. Sussmann, Hector J., 1946-
Description: ix, 372 p.: ill.; 25 cm.
ISBN: 0387985964 (hardcover: alk. paper)
Notes: Includes bibliographical references and index.
Subjects: Robotics--Mathematics. Robots--Control systems.
Series: The IMA volumes in mathematics and its applications; v. 104
LC Classification: TJ211 .E84 1998
Dewey Class No.: 629.8/92/0151 21

ETFA '94, 1994 IEEE Symposium on Emerging Technologies & Factory Automation (SEIKEN Symposium): novel disciplines for the next century: proceedings, November 6-10, 1994, Institute of Industrial Science, University of Tokyo, Tokyo, Japan / sponsored by Institute of Industrial Science (SEIKEN), IEEE Industrial Electronics Society (IEEE IES).
Published/Created: [New York]: Institute of Electrical and Electronics Engineers; Piscataway, NJ: Copies are available from IEEE Service Center, c1994.
Related Authors: T̄okȳo Daigaku. Seisan Gijutsu Kenkȳujo. IEEE Industrial Electronics Society.
Description: 477 p.: ill.; 30 cm.
ISBN: 0780321146 (softbound) 0780321154 (microfiche)
Notes: "IEEE catalog number 94TH8000"--T.p. verso. Includes bibliographical references and index.
Subjects: Automation--Congresses. Artificial intelligence--Congresses. Robotics--Congresses.
LC Classification: T59.5 .I28 1994
Dewey Class No.: 670.42/7 20

Evolutionary robotics: First European Workshop, EvoRobot'98: Paris, France, April 16-17, 1998: proceedings / Philip Husbands, Jean-Arcady Meyer (eds.).
Published/Created: Berlin; New York: Springer, c1998.
Related Authors: Husbands, Phil. Meyer, Jean-Arcady.
Description: viii, 247 p.: ill.; 24 cm.
ISBN: 3540649573 (acid-free paper)
Notes: Includes bibliographical references and index.
Subjects: Evolutionary robotics--Congresses.
Series: Lecture notes in computer science, 0302-9743; 1468
LC Classification: TJ211.37 .E96 1998
Dewey Class No.: 629.8/92 21

Evolutionary robotics: International Symposium, ER 2001, Tokyo, Japan, October 18-19, 2001: proceedings / Takashi Gomi (ed.)
Published/Created: New York: Springer, 2001.
Related Authors: Gomi, Takashi, 1940-
Description: p. cm.
ISBN: 3540427376 (softcover: alk. paper)
Notes: Includes bibliographical references and index.
Subjects: Evolutionary robotics--Congresses.
Series: Lecture notes in computer science; 2217
LC Classification: TJ211.37 .E7 2001
Dewey Class No.: 629.8/92 21

Experimental robotics VI / Peter I. Corke and James Trevelyan, (eds.).
Published/Created: London; New York: Springer, c2000.
Related Authors: Corke, Peter I., 1959- Trevelyan, James P.
Description: xix, 528 p.: ill.; 24 cm.
ISBN: 1852332107 (alk. paper)
Notes: Includes bibliographical references and index.
Subjects: Robotics--Congresses.
Series: Lecture notes in control and

information sciences; 250
LC Classification: TJ210.3 .I5865 2000
Dewey Class No.: 629.8/92 21

Experimental robotics: the ... international
symposium.
Published/Created: London; New York:
Springer-Verlag, c1990-
Description: v.: ill.; 25 cm. 1st (1989)-
Notes: SERBIB/SERLOC merged
record
Subjects: Robotics--Congresses.
Series: Lecture notes in control and
information sciences
LC Classification: TJ210.3 .E96
Dewey Class No.: 629.8/92 20

Expert systems and robotics / edited by
Timothy Jordanides, Bruce Torby.
Published/Created: Berlin; New York:
Springer-Verlag, c1991.
Related Authors: Jordanides, Timothy.
Torby, Bruce J. North Atlantic Treaty
Organization. Scientific Affairs
Division.
Description: xii, 744 p.: ill.; 25 cm.
ISBN: 3540537317 (Berlin: alk. paper)
0387537317 (New York: alk. paper)
Notes: "Proceedings of the NATO
Advanced Study Institute on Expert
Systems and Robotics, held in Corfu,
Greece, July 15-27, 1990"--T.p. verso.
"Published in cooperation with NATO
Scientific Affairs Division." Includes
bibliographical references.
Subjects: Expert systems (Computer
science)--Periodicals. Robotics--
Periodicals.
Series: NATO ASI series. Series F,
Computer and systems sciences; no. 71.
LC Classification: QA76.76.E95 N375
1990
Dewey Class No.: 006.3/3 20

Expert systems: an overview.
Published/Created: Sacramento, CA:
Business/Technology Books, c1984.
Description: ix, 64 p.: ill.; 28 cm.
ISBN: 0899341780
Notes: Bibliography: p. 63-64.

Subjects: Expert systems (Computer
science)
Series: Robotics and artificial
intelligence applications series; v. 2
LC Classification: QA76.9.E96 E957
1984
Dewey Class No.: 001.53/5 19

Exploratory Workshop on the Social
Impacts of Robotics: summary and
issues, a background paper.
Published/Created: Washington, D.C.:
Congress of the U.S., Office of
Technology Assessment: For sale by the
Supt. of Docs., U.S. G.P.O., 1982.
Related Authors: United States.
Congress. Office of Technology
Assessment.
Description: vii, 136 p.: ill.; 26 cm.
Notes: "February 1982"--P. 4 of cover.
"OTA-BP-CIT-11"--P. 4 of cover S/N
052-003-00865-3 Item 1070-M Includes
bibliographical references.
Subjects: Robotics--Social aspects--
United States--Congresses. Automation-
-Social aspects--United States--
Congresses. Technological
unemployment--United States--
Congresses.
Series: OTA background papers
LC Classification: HD9696.R623 U63
1981
Dewey Class No.: 306/.46 19
Govt. Doc. No.: Y 3.T 22/2:R 57/sum.

Exploring the Moon and Mars: choices for
the nation.
Published/Created: [Washington]:
Congress of the U. S., Office of
Technology Assessment: For sale by the
Supt. of Docs., U.S. Govt. Print. Off.,
[1991]
Related Authors: United States.
Congress. Office of Technology
Assessment.
Description: ix, 104 p.: ill.; 26 cm.
Notes: "July 1991." Includes
bibliographical references.
Subjects: Robotics. Moon--Exploration-
-Automation. Moon--Research--United

States. Mars (Planet)--Exploration--
Automation. Mars (Planet)--Research--
United States.
LC Classification: QB581 .E86 1991
Dewey Class No.: 629.45/4/0973 20

Faraz, Ali.
Engineering approaches to mechanical
and robotic design for minimally
invasive surgery (MIS) / Ali Faraz,
Shahram Payandeh.
Published/Created: Boston: Kluwer
Academic, c2000.
Related Authors: Payandeh, Shahram,
1957-
Description: xvii, 183 p.: ill.; 24 cm.
ISBN: 0792377923
Notes: Includes bibliographical
references (p. [169]-177) and index.
Subjects: Surgical instruments and
apparatus--Design and construction.
Endoscopic surgery. Robotics in
medicine. Biomedical engineering.
Series: Kluwer international series in
engineering and computer science
LC Classification: RD71 .F36 2000
Dewey Class No.: 610/.28 21

Fast, cheap & out of control / Fourth Floor
Productions, Inc. presents American
Playhouse Theatrical Films; directed
and produced by Errol Morris.
Published/Created: United States: Sony
Pictures Classics, 1997; United States:
WETA-TV, 1999-06-07.
Related Authors: Morris, Errol,
direction, production. Hoover, Dave,
cast. Mendonca, George, cast. Mendez,
Ray, cast. Brooks, Rodney cast. LC Off-
Air Taping Collection (Library of
Congress) Copyright Collection
(Library of Congress)
Description: 2 videocassettes of 2 (80
min.) sd., col. and b&w; 3/4 in. viewing
copy. 10 reels of 10 on 5 (7200 ft.): sd.,
col. and b&w; 35 mm. ref print.
Summary: What do an elderlytopiary
gardener, a retired lion tamer, a man
fascinated by mole rats, and a cutting-
edge robotics designer have in

common? Both nothing and everything
in this unconventional documentary
directed by Erroll Morris. Fast, Cheap &
Out of Control (referring to the robot
specialist's strange philosophy of robot
design structure, not Erroll Morris's
documentary techniques!) interplays,
overlaps, and interrelates these four
separate and highly specialized
documentary subjects in order to in truth
study all of humanity, raising questions
about the future of mankind.
Notes: Copyright: Fourth Floor
Productions, Inc. DCR 1997; PUB
3Oct97; REG 22Sep98; PA900-068.
Viewing copy (VBP 5013-5014) is
American Playhouse reissue copy.
Summary taken from Internet Movie
Database, 1998. Sources used: copyright
data base; copyright data sheet; Internet
Movie Database, 1998; Baseline
11/4/99. Cast: Dave Hoover, George
Mendonca, Ray Mendez, Rodney
Brooks. Credits: Executive producer,
Lindsay Law; co-producers, Mark
Lipson Julia sheehan, Kathy Trustman;
original music, Caleb Sampson;
directors of photography, Robert
Richardson, Peter Sova; film editors,
Shondra Merrill, Karne Schmeer.
Series: American playhouse (Television
program)
LC Classification: VBP 5013-5014
(viewing copy) CGC 7417-7421 (ref
print)

Fatikow, S. (Sergej), 1960-
Microsystem technology and
microrobotics / S. Fatikow, U.
Rembold.
Published/Created: Berlin; New York:
Springer, c1997.
Related Authors: Rembold, Ulrich.
Description: xii, 408 p.: ill.; 24 cm.
ISBN: 3540606580 (hc: alk. paper)
Notes: Includes bibliographical
references (p. [366]-401) and index.
Subjects: Electromechanical devices--
Design and construction.
Microfabrication. Robotics.

LC Classification: TJ163 .F38 1997
Dewey Class No.: 621 21

Featherstone, Roy.
Robot dynamics algorithms / by Roy
Featherstone.
Published/Created: Boston: Kluwer,
c1987.
Description: x, 211 p.; 25 cm.
ISBN: 0898382300
Notes: Based on the author's thesis--
Edinburgh University, 1984. Includes
index. Bibliography: p. [197]-205.
Subjects: Robots--Dynamics. Dynamics,
Rigid. Recursive functions.
Series: Kluwer international series in
engineering and computer science;
SECS 22. Kluwer international series in
engineering and computer science.
Robotics.
LC Classification: TJ211.4 .F43 1987
Dewey Class No.: 629.8/92 19

Field and service robotics / Alexander
Zelinsky (ed.).
Published/Created: Berlin; New York:
Springer, c1998.
Related Authors: Zelinsky, Alexander,
1960- International Conference on Field
and Service Robotics (1997: Australian
National University)
Description: xi, 547 p.: ill.; 28 cm.
ISBN: 1852330392 (casebound: alk.
paper)
Notes: Papers from the International
Conference on Field and Service
Robotics, held Dec. 1997 at Australian
National University. Includes
bibliographical references.
Subjects: Robotics. Robots, Industrial.
LC Classification: TJ211 .F54 1998
Dewey Class No.: 629.8/92 21

Fifth Canadian CAD/CAM and Robotics
Conference, June 17-19, 1986, Toronto,
Ontario, Canada / sponsored by
Computer and Automated Systems
Association of SME ... [et al.].
Edition Information: 1st ed.
Published/Created: Dearborn, Mich.:

Society of Manufacturing Engineers,
c1986.
Related Authors: Computer and
Automated Systems Association of
SME.
Description: 1 v. (various pagings): ill.;
28 cm.
ISBN: 0872632407 (pbk.)
Notes: Includes bibliographies.
Subjects: CAD/CAM systems--
Congresses. Robots, Industrial--
Congresses.
LC Classification: TS155.6 .C375 1986
Dewey Class No.: 670.28/5 19

Fifth European Space Mechanisms &
Tribology Symposium: an international
symposium organised by the European
Space Agency, jointly sponsored by
NIVR and CNES and held at ESTEC,
Noordwijk, the Netherlands on 28-30
October 1992 / [compiled by W.R.
Burke].
Published/Created: Paris: European
Space Agency, c1993.
Related Authors: Burke, W. R.
European Space Agency. Netherlands
Instituut voor Vliegtuigontwikkeling en
Ruimtevaart. Centre national d'études
spatiales (France)
Description: xxxvii, 397 p.: ill.; 30 cm.
ISBN: 9290921781
Notes: Includes bibliographical
references and index.
Subjects: Space vehicles--Lubrication--
Congresses. Tribology--Congresses.
Space robotics--Congresses.
Series: ESA SP, 0379-6566; 334
LC Classification: TL917 .E93 1992
Dewey Class No.: 629.47 20

Fifth world conference on robotics research.
Published/Created: Dearborn, MI:
Society of Manufacturing Engineers,
1994.
Description: p. cm.
ISBN: 0872634558 (pbk.)
LC Classification: 9409 BOOK NOT
YET IN LC

First International Symposium on
Mathematical Models in Automation
and Robotics, September, 1-3, 1994,
Miedzyzdroje, Poland / organized by
Institute of Control Engineering,
Technical University of Szczecin;
sponsored by Committee of Automation
and Robotics, Polish Academy of
Sciences, Warsaw, Committee of
Metrology and Instrumentation, Polish
Academy of Sciences, Warsaw,
Commission of Cybernetics, Polish
Academy of Sciences, Pozna´n Branch.
Edition Information: Wyd. 1.
Published/Created: Szczecin: Technical
University of Szczecin Press, [1994]
Related Authors: Politechnika
Szczeci´nska. Instytut Automatyki
Przemyslowej.
Description: 429 p.: ill.; 30 cm.
ISBN: 8386359803
Notes: Includes bibliographical
references and index.
Subjects: Automatic control--
Mathematical models--Congresses.
Robotics--Mathematical models--
Congresses.
LC Classification: TJ212.2 .I618 1994

Fjermedal, Grant.
The tomorrow makers: a brave new
world of living-brain machines / Grant
Fjermedal.
Published/Created: Redmond, Wash.:
Tempus Books of Microsoft Press;
[New York]: Distributed to the book
trade in the U.S. by Harper $ Row,
[1988], c1986.
Description: xi, 272 p.; 23 cm.
ISBN: 1556151136 (pbk.) :
Notes: Includes index.
Subjects: Robotics--Popular works.
Artificial intelligence--Popular works.
LC Classification: TJ211.15 .F54 1988
Dewey Class No.: 629.8/92 19

Fjermedal, Grant.
The tomorrow makers: a brave new
world of living-brain machines / Grant
Fjermedal.

Published/Created: New York:
Macmillan, c1986.
Description: xi, 272 p.; 24 cm.
ISBN: 0025385607
Notes: Includes index.
Subjects: Robotics--Popular works.
Artificial intelligence--Popular works.
LC Classification: TJ211.15 .F54 1986
Dewey Class No.: 629.8/92 19

Flexible manufacturing systems / Factory
Systems Planning Service.
Published/Created: Boston, MA (89
Broad St., Boston 02110): Yankee
Group, c1985.
Related Authors: Factory Systems
Planning Service.
Description: vii, 187 p.: ill.; 28 cm.
Notes: "December 1984." "Report no. 2,
1984"--Spine. Bibliography: p. 185-187.
Subjects: Automatic control equipment
industry--United States. Robotics
industry--United States. Machine-tool
industry--United States. Flexible
manufacturing systems. Computer
integrated manufacturing systems.
Robots, Industrial. Market surveys--
United States.
Series: Industry research report
LC Classification: HD9696.A963 U65
1985
Dewey Class No.: 670.42/7 19

Flexible manufacturing systems: recent
developments / edited by A. Raouf and
M. Ben-Daya.
Published/Created: Amsterdam; New
York: Elsevier, 1995.
Related Authors: Raouf, A. (Abdul),
1929- Ben-Daya, M. (Mohamed)
International Conference on Production
Research (7th: 1983: Windsor, Ont.)
Description: vii, 316 p.: ill.; 25 cm.
ISBN: 0444897984
Notes: "This volume contains new and
updated material from Flexible
manufacturing: recent developments in
FMS, robotics, CAD/CAM, CIM which
was a selection of papers presented at
the VIIth International Conference on

Production Research, held in Windsor, Ontario during 1983"--Pref. Includes bibliographical references.
Subjects: Flexible manufacturing systems.
Series: Manufacturing research and technology; 23
LC Classification: TS155.65 .F59 1995
Dewey Class No.: 670.42/7 20

Flexible manufacturing: recent developments in FMS, robotics, CAD/CAM, CIM / edited by A. Raouf, S.I. Ahmad.
Published/Created: Amsterdam; New York: Elsevier; New York, NY: Distributors for the U.S. and Canada, Elsevier Science Pub. Co., 1985.
Related Authors: Raouf, A. (Abdul), 1929- Ahmad, S. I. (Syed Imtiaz)
International Conference on Production Research (7th: 1983: Windsor, Ont.)
Description: viii, 255 p.: ill.; 25 cm.
ISBN: 0444425047 (U.S.) :
Notes: Selected papers presented at the 7th International Conference on Production Research, Windsor, Ont., Canada in 1983. Includes bibliographies and indexes.
Subjects: Flexible manufacturing systems--Congresses. Robots, Industrial--Congresses. CAD/CAM systems--Congresses. Computer integrated manufacturing systems--Congresses.
Series: Manufacturing research and technology; 1
LC Classification: TS176 .F59 1985
Dewey Class No.: 670.42 19

Fodor, George A., 1954-
Ontologically controlled autonomous systems: principles, operations, and architecture / by George A. Fodor.
Published/Created: Boston, Mass.: Kluwer Academic Publishers, c1998.
Description: xi, 245 p.: ill.; 25 cm.
ISBN: 0792380355 (acid-free paper)
Notes: Includes bibliographical references and index.

Subjects: Programmable controllers. Robotics.
LC Classification: TJ223.P76 F63 1998
Dewey Class No.: 629.8/9 21

Fourth European Conference on Artificial Life / edited by Phil Husbands and Inman Harvey.
Published/Created: Cambridge, Mass.: MIT Press, c1997.
Related Authors: Husbands, Phil. Harvey, Inman.
Description: ix, 583 p.: ill.; 28 cm.
ISBN: 0262581574
Notes: Conference held July 28-31, 1997 in Brighton, UK. "A Bradford book." Includes bibliographical references and index.
Subjects: Biological systems--Computer simulation--Congresses. Biological systems--Simulation methods--Congresses. Artificial intelligence--Congresses. Robotics--Congresses.
Series: Complex adaptive systems
LC Classification: QH324.2 .E87 1997
Dewey Class No.: 570/.1/1 21

Fourth World Conference on Robotics Research, September 17-19, 1991, Pittsburgh, Pennsylvania.
Published/Created: Dearborn, Mich. (1 SME Dr., Dearborn): Society of Manufacturing Engineers, c1991.
Related Authors: Society of Manufacturing Engineers.
Description: 3 v.: ill.; 28 cm.
Notes: Cover title. "Creative manufacturing engineering program." Includes bibliographical references.
Subjects: Robotics Research Congresses.
LC Classification: TJ210.3 .W67 1991a
Dewey Class No.: 629.8/92 20

Fourth World Conference on Robotics Research: conference proceedings, September 17-19, 1991, Pittsburgh, Pennsylvania / sponsored by the Robotics International of the Society of Manufacturing [Engineers].

Published/Created: Dearborn, Mich.:
The Society, c1991.
Related Authors: Robotics International
of SME.
Description: 1 v. (various pagings): ill.;
28 cm.
ISBN: 0872634078
Notes: Includes bibliographical
references and index.
Subjects: Robotics--Research--
Congresses.
LC Classification: TJ210.3 .W67 1991
Dewey Class No.: 629.8/92 20

Fraser, Anthony R., 1962-
Perturbation techniques for flexible
manipulators / by Anthony R. Fraser
and Ron W. Daniel.
Published/Created: Boston: Kluwer
Academic Publishers, c1991.
Related Authors: Daniel, Ron W.
Description: xvi, 275 p.: ill.; 25 cm.
ISBN: 0792391624 (acid-free paper)
Notes: Includes bibliographical
references (p. 261-270) and index.
Subjects: Manipulators (Mechanism)
Robots--Control systems. Perturbation
(Mathematics)
Series: Kluwer international series in
engineering and computer science;
SECS 138. Kluwer international series
in engineering and computer science.
Robotics.
LC Classification: TJ211 .F72 1991
Dewey Class No.: 629.8/92 20

From animals to animats 2: proceedings of
the Second International Conference on
Simulation of Adaptive Behavior /
edited by Jean-Arcady Meyer, Herbert
L. Roitblat, and Stewart W. Wilson.
Published/Created: Cambridge, Mass.:
MIT Press, c1993.
Related Authors: Meyer, Jean-Arcady.
Roitblat, H. L. Wilson, Stewart W.
Description: x, 523 p.: ill.; 28 cm.
ISBN: 0262631490
Notes: "A Bradford book." Includes
bibliographical references and index.
Subjects: Animal behavior--Simulation

methods--Congresses. Animals--
Adaptation--Simulation methods--
Congresses. Robotics--Congresses.
Artificial intelligence--Congresses.
Series: Complex adaptive systems
LC Classification: QL751.65.S55 I58
1992
Dewey Class No.: 591.51 20

From animals to animats 3: proceedings of
the Third International Conference on
Simulation of Adaptive Behavior /
edited by Dave Cliff ... [et al.].
Published/Created: Cambridge, Mass.:
MIT Press, c1994.
Related Authors: Cliff, Dave.
Description: x, 508 p.: ill.; 28 cm.
ISBN: 0262531224
Notes: "A Bradford book." Includes
bibliographical references and index.
Subjects: Animal behavior--Simulation
methods--Congresses. Animals--
Adaptation--Simulation methods--
Congresses. Robotics--Congresses.
Artificial intelligence--Congresses.
Series: Complex adaptive systems
LC Classification: QL751.65.S55 I58
1994
Dewey Class No.: 591.51 20

From animals to animats 4: proceedings of
the Fourth International Conference on
Simulation of Adaptive Behavior /
edited by Pattie Maes ... [et al.].
Published/Created: Cambridge, Mass.:
MIT Press, c1996.
Related Authors: Maes, Pattie, 1961-
Description: xii, 644 p.: ill.; 28 cm.
ISBN: 0262631784
Notes: "A Bradford book." Includes
bibliographical references and index.
Subjects: Animal behavior--Simulation
methods--Congresses. Animals--
Adaptation--Simulation methods--
Congresses. Robotics--Congresses.
Artificial intelligence--Congresses.
Series: Complex adaptive systems
LC Classification: QL751.65.S55 I58
1996

Dewey Class No.: 591.5 21

From animals to animats 5: proceedings of
the Fifth International Conference on
Simulation of Adaptive Behavior /
edited by Rolf Pfeifer ... [et al.].
Published/Created: Cambridge, Mass.:
MIT Press, c1998.
Related Authors: Pfeifer, Rolf, 1947-
Description: xvi, 564 p.: ill.; 28 cm.
ISBN: 0262661446
Notes: "A Bradford book." Includes
bibliographical references and index.
Subjects: Animal behavior--Simulation
methods--Congresses. Animals--
Adaptation--Simulation methods--
Congresses. Robotics--Congresses.
Artificial intelligence--Biological
applications Congresses.
Series: Complex adaptive systems,
1089-4365
LC Classification: QL751.65.S55 I58
1998
Dewey Class No.: 591.5/01/1 21

From animals to animats 6: proceedings of
the Sixth International Conference on
Simulation of Adaptive Behavior /
edited by Jean-Arcady Meyer... [et al.].
Published/Created: Cambridge, Mass.:
MIT Press, c2000.
Related Authors: Pfeifer, Rolf, 1947-
Description: xii, 540 p.: ill.; 28 cm.
ISBN: 0262632004
Notes: "A Bradford book." Includes
bibliographical references and index.
Subjects: Animal behavior--Simulation
methods--Congresses. Animals--
Adaptation--Simulation methods--
Congresses. Robotics--Congresses.
Artificial intelligence--Biological
applications Congresses.
Series: Complex adaptive systems,
1089-4365
LC Classification: QL751.65.S55 I58
2000
Dewey Class No.: 591.5/01/13 21

Frude, Neil.
The robot heritage / Neil Frude.

Published/Created: London: Century
Pub., 1984.
Description: 252 p., [8] p. of plates: ill.;
21 cm.
ISBN: 0712609180: 0712609210 (pbk.)
Notes: Includes index. Bibliography: p.
[240]-242. Filmography: p. [243].
Subjects: Robotics.
LC Classification: TJ211 .F78 1984
Dewey Class No.: 303.4/83 19

Fu, K. S. (King Sun), 1930-
Robotics: control, sensing, vision, and
intelligence / K.S. Fu, R.C. Gonzalez,
C.S.G. Lee.
Published/Created: New York:
McGraw-Hill, c1987.
Related Authors: Gonzalez, Rafael C.
Lee, C. S. G. (C. S. George)
Description: xiii, 580 p.: ill.; 25 cm.
ISBN: 0070226253: 0070226261
(solutions manual)
Notes: Includes index. Bibliography: p.
556-570.
Subjects: Robotics.
Series: CAD/CAM, robotics, and
computer vision
LC Classification: TJ211 .F82 1987
Dewey Class No.: 629.8/92 19

Fukuda, T. (Toshio), 1948-
Cellular robotics and micro robotic
systems / T. Fukuda & T. Ueyama.
Published/Created: Singapore; River
Edge, N.J.: World Scientific, c1994.
Related Authors: Ueyama, T.
(Tsuyoshi)
Description: xv, 267 p.: ill.; 23 cm.
ISBN: 981021457X
Notes: Includes bibliographical
references (p. 257-262) and index.
Subjects: Robotics. Automatic control.
Microelectronics.
Series: World Scientific series in
robotics and automated systems; vol. 10
LC Classification: TJ211 .F84 1994
Dewey Class No.: 629.8/92 20

Fuller, James L., 1941-
Robotics: introduction, programming,

and projects / James L. Fuller.
Edition Information: 2nd ed.
Published/Created: Upper Saddle River,
N.J.: Prentice Hall, c1999.
Description: xv, 489 p.: ill.; 24 cm.
ISBN: 0130955434 (case)
Notes: Includes bibliographical
references (p. 483-484) and index.
Subjects: Robotics.
LC Classification: TJ211 .F85 1999
Dewey Class No.: 670.42/72 21

Fuller, James L., 1941-
Robotics: introduction, programming,
and projects / James L. Fuller.
Published/Created: New York: Merrill,
c1991.
Description: xvii, 465 p.: ill.; 24 cm.
ISBN: 067521078X
Notes: Includes bibliographical
references (p. 457-459) and index.
Subjects: Robotics.
LC Classification: TJ211 .F85 1991
Dewey Class No.: 629.8/92 20

Fuzzy logic control: advances in
applications / edited by Henk B.
Verbruggen, Robert Babuska.
Published/Created: Singapore; River
Edge, NJ: World Scientific, 1999.
Related Authors: Verbruggen, H. B.
Babuska, Robert.
Description: p. cm.
ISBN: 9810238258 (alk. paper)
Subjects: Intelligent control systems.
Fuzzy logic.
Series: World Scientific series in
robotics and intelligent systems; vol. 23
LC Classification: TJ217.5 .F885 1999
Dewey Class No.: 629.8 21

Garoogian, Andrew.
Robotics, 1960-1983: an annotated
bibliography / by Andrew Garoogian.
Published/Created: Brooklyn, N.Y.:
CompuBibs, 1984.
Description: i, 119 p.; 28 cm.
ISBN: 0914791036 (pbk.)
Notes: Includes bibliographical
references.

Subjects: Robotics--Bibliography.
Robots--Bibliography.
Series: CompuBibs (Series); #1.
LC Classification: Z5853.R58 G37
1984 TJ211
Dewey Class No.: 016.6298/92 19

Ge, S. S. (Shuzhi S.)
Adaptive neural network control of
robotic manipulators / S.S. Ge, T.H.
Lee, C.J. Harris.
Published/Created: Singapore; River
Edge, NJ: World Scientific, c1998.
Related Authors: Lee, Tong Heng,
1958- Harris, C. J. (Christopher John)
Description: xiv, 381 p.: ill.; 23 cm.
ISBN: 981023452X
Notes: Includes bibliographical
references (p. 327-344) and index.
Subjects: Robots--Control systems.
Neural networks (Computer science)
Adaptive control systems.
Series: World Scientific series in
robotics and intelligent systems; vol. 19
LC Classification: TJ211.35 .G4 1998
Dewey Class No.: 670.42/72 21

General robotics and robotics in medicine:
The 3rd French - Israeli Symposium on
Robotics, 22-23 May 1995, Hasharon
Hotel, Hertzelia, Israel: Scientific
Program and Symposium Proceedings.
Published/Created: Jerusalem: Ministry
of Science & the Arts, 1995.
Description: 172 p.
LC Classification: IN PROCESS

Geometric computing with Clifford algebra:
theoretical foundations and applications
in computer vision and robotics / Gerald
Sommer (ed.).
Published/Created: Berlin; New York:
Springer, 2001.
Related Authors: Sommer, Gerald.
Description: xviii, 551 p.: ill.; 25 cm.
ISBN: 3540411984 (alk. paper)
Notes: Includes bibliographical
references (p. [531]-542) and indexes.
Subjects: Clifford algebras. Clifford
algebras--Industrial applications.

LC Classification: QA199 .G46 2001
Dewey Class No.: 512/.57 21

Geometric modeling: algorithms and new
trends / edited by Gerald E. Farin.
Published/Created: Philadelphia, Pa.:
Society for Industrial and Applied
Mathematics, c1987.
Related Authors: Farin, Gerald E.
Society for Industrial and Applied
Mathematics. SIAM Conference on
Geometric Modeling and Robotics
(1985: Albany, N.Y.)
Description: xi, 399 p., [8] p. of plates:
ill. (some col.); 27 cm.
ISBN: 0898712068
Notes: Papers presented at the SIAM
Conference on Geometric Modeling and
Robotics, held at Albany, New York,
July 15-19, 1985. Includes
bibliographies and index.
Subjects: Geometry--Data processing--
Congresses. Mathematical models--
Congresses. CAD/CAM systems--
Congresses. Computer graphics--
Congresses.
LC Classification: QA447 .G46 1987
Dewey Class No.: 516/.00724 19

Geometric reasoning / edited by Deepak
Kapur and Joseph L. Mundy.
Edition Information: 1st MIT Press ed.
Published/Created: Cambridge, Mass.:
MIT Press, 1989, c1988.
Related Authors: Kapur, Deepak.
Mundy, Joseph L. International
Workshop on Geometric Reasoning
(1986: Oxford University)
Description: 512 p.: ill.; 23 cm.
ISBN: 0262610582
Notes: Papers presented at an
International Workshop on Geometric
Reasoning, held at Oxford University,
June 30 to July 3, 1986. Reprinted from
Artificial intelligence, v, 37, no. 1-3
(1988). "A Bradford book." Includes
bibliographical references.
Subjects: Artificial intelligence--
Congresses. Automatic theorem
proving--Congresses. Computer vision--

Congresses. Robotics--Congresses.
LC Classification: Q334 .G46 1989
Dewey Class No.: 006.3 20

Geometric reasoning for perception and
action: workshop, Grenoble, France,
September 16-17, 1991: selected papers
/ Christian Laugier (ed.).
Published/Created: Berlin; New York:
Springer-Verlag, c1993.
Related Authors: Laugier, Christian.
Description: viii, 281 p.: ill.; 24 cm.
ISBN: 3540571329 (Berlin: acid-free
paper) 0387571329 (New York)
Notes: Includes bibliographical
references.
Subjects: Artificial intelligence--
Congresses. Computer vision--
Congresses. Robotics--Congresses.
Automatic theorem proving--
Congresses.
Series: Lecture notes in computer
science; 708
LC Classification: Q334 .G465 1993
Dewey Class No.: 006.3/7/01516 20

Geometry and robotics: workshop,
Toulouse, France, May 26-28, 1988:
proceedings / J.-D. Boissonnat, J.-P.
Laumond (eds.).
Published/Created: Berlin; New York:
Springer-Verlag, c1989.
Related Authors: Boissonnat, J.-D.
(Jean-Daniel), 1953- Laumond, J.-P.
(Jean-Paul) Institut national de
recherche en informatique et en
automatique (France)
Description: vi, 413 p.: ill. (some col.);
24 cm.
ISBN: 0387516832 (U.S.) :
Notes: "The workshop has been jointly
organized by INRIA and LAAS-
CNRS"--Pref. Includes bibliographical
references.
Subjects: Robotics--Congresses.
Geometry--Congresses.
Series: Lecture notes in computer
science; 391
LC Classification: TJ210.3 .G46 1989

Dewey Class No.: 629.8/92 20

Gevarter, William B.
Artificial intelligence & robotics: five overviews.
Published/Created: Orinda, CA: Business Technology Books, c1984.
Description: 618 p. in various pagings: ill.; 29 cm.
ISBN: 0899342272
Notes: Includes bibliographies.
Subjects: Artificial intelligence. Robotics. Expert systems (Computer science) Computer vision. Computational linguistics.
Series: Robotics and artificial intelligence applications series; v. 1-5c
LC Classification: Q335 .G473 1984
Dewey Class No.: 006.3 19

Gevarter, William B.
Artificial intelligence: an overview.
Published/Created: Sacramento, CA: Business/Technology Books, c1984.
Description: 3 v.: ill.; 28 cm.
ISBN: 0899341942 (pbk.)
Notes: Includes bibliographies.
Subjects: Artificial intelligence. Expert systems (Computer science)
Series: Robotics and artificial intelligence applications series; v. 5A-5C
LC Classification: Q335 .G475 1984
Dewey Class No.: 001.53/5 19

Gevarter, William B.
Intelligent machines: an introductory perspective of artificial intelligence and robotics / William B. Gevarter.
Published/Created: Englewood Cliffs, N.J.: Prentice Hall, c1985.
Description: xvi, 282 p.: ill.; 24 cm.
ISBN: 0134688104 :
Notes: Includes bibliographies and indexes.
Subjects: Artificial intelligence. Robotics.
LC Classification: Q335 .G483 1985
Dewey Class No.: 001.53/5 19

Gill, Mark A. C.
Obstacle avoidance in multi-robot systems: experiments in parallel genetic algorithms / Mark A.C. Gill & Albert Y. Zomaya.
Published/Created: Singapore; River Edge, N.J.: World Scientific, c1998.
Related Authors: Zomaya, Albert Y.
Description: xii, 185 p.: ill.; 23 cm.
ISBN: 9810234236
Notes: Includes bibliographical references (p. 161-176) and index.
Subjects: Robots--Control systems. Genetic algorithms. Parallel processing (Electronic computers) Intelligent control systems.
Series: World Scientific series in robotics and intelligent systems; vol. 20
LC Classification: TJ211.35 .G55 1998
Dewey Class No.: 629.8/925275 21

Goddard Conference on Space Applications of Artificial Intelligence: proceedings of a workshop held at NASA Goddard Space Flight Center, Greenbelt, Maryland ...
Published/Created: Washington, D.C.: National Aeronautics and Space Administration, Office of Management, Scientific and Technical Information Division, -1993 [i.e. 1994]
Related Authors: Goddard Space Flight Center. United States. National Aeronautics and Space Administration. Scientific and Technical Information Division. United States. National Aeronautics and Space Administration. Scientific and Technical Information Program. United States. National Aeronautics and Space Administration. Scientific and Technical Information Branch.
Description: 7 v.: ill.; 28 cm. Began with 1988. -9th (1994)
Notes: Subtitle varies. Description based on: 1989. Vols. for 1988-1991 issued by: NASA, Scientific and Technical Information Division; 1992-1993 by: NASA, Scientific and Technical Information Program; 1994

by: NASA, Scientific and Technical
Information Branch. SERBIB/SERLOC
merged record
Subjects: Astronautics--Data
processing--Congresses. Artificial
intelligence--Congresses. Space flight--
Automation--Congresses.
Series: NASA conference publication
LC Classification: TL787 .C6a
Dewey Class No.: 629.4/0285/63 20
Govt. Doc. No.: NAS 1.55:

Gorinevskii, D. M.
Force control of robotics systems /
Dimitry M. Gorinevsky, Alexander M.
Formalsky, Anatoly Yu. Schneider.
Published/Created: Boca Raton: CRC
press, c1997.
Related Authors: Formal'skii, A. M.
(Aleksandr Moiseevich) Shneider, A.
IU.
Description: xiii, 350 p.: ill.; 25 cm.
Notes: Includes bibliographical
references (p. 317-342) and index.
Subjects: Robots--Control systems.
LC Classification: TJ211.35 .G6713
1997
Dewey Class No.: 670.42/72 21

Government and industry cooperation to
promote economic conversion: hearing
before the Subcommittee on
Investigations and Oversight of the
Committee on Science, Space, and
Technology, U.S. House of
Representatives, One Hundred Third
Congress, first session, September 16,
1993.
Published/Created: Washington: U.S.
G.P.O.: For sale by the U.S. G.P.O.,
Supt. of Docs., Congressional Sales
Office, 1994.
Description: iii, 92 p.: ill.; 24 cm.
ISBN: 0160433878
Notes: Distributed to some depository
libraries in microfiche. Shipping list no.:
94-0015-P. "No. 68."
Subjects: University of Texas at
Arlington. Automation & Robotics
Research Institute. Economic

conversion--Texas. Manufacturing
processes--Texas. Economic
conversion--United States. Industrial
policy--United States. Technology
transfer--United States.
LC Classification: KF27 .S3975 1993c
Govt. Doc. No.: Y 4.SCI 2:103/68

Graham, Ian, 1953-
Artificial intelligence / Ian Graham.
Published/Created: Chicago, Ill.:
Heinemann Library, c2002.
Description: p. cm.
ISBN: 1403403236
Summary: Explores the different types
of artificial intelligence, how they are
developed, how they are used now and
how they may be used in the future, and
controversies surrounding this new
technology.
Contents: Robots to the rescue --
Robotics, A.I. and A. life -- Robots --
Human machines -- Intelligent machines
-- Expert systems -- The web as an
artificial intelligence -- A-life -- The
ethics of A.I.
Notes: Includes bibliographical
references and index.
Subjects: Artificial intelligence--
Juvenile literature. Artificial
intelligence. Robotics.
Series: Science at the edge
LC Classification: Q335.4 .G73 2002
Dewey Class No.: 006.3 21

Graphics and robotics / Wolfgang Strasser,
Friedrich Wahl, editors.
Published/Created: Berlin; New York:
Springer, c1995.
Related Authors: Strasser, Wolfgang,
1941- Wahl, Friedrich M., 1948-
Description: viii, 247 p.: ill. (some col.);
25 cm.
ISBN: 3540583580 (Berlin: acid-free
paper): 0387583580 (New York: acid-
free paper)
Notes: Includes bibliographical
references.
Subjects: Robotics. Computer graphics.
LC Classification: TJ211 .G72 1995

Dewey Class No.: 629.8/9266 20

Gray, Chris Hables.
    Cyborg citizen: politics in the
    posthuman age / by Chris Hables Gray.
    Published/Created: New York:
    Routledge, 2000.
    Description: p. cm.
    ISBN: 0415919789 (hb: alk. paper)
    0415919797 (pb: alk. paper)
    Notes: Includes bibliographical
    references and index.
    Subjects: Human-machine systems.
    Robotics. Cyborgs.
    LC Classification: TA167 .G75 2000
    Dewey Class No.: 303.48/3 21

Greene, Carol.
    Robots / by Carol Greene.
    Published/Created: Chicago: Childrens
    Press, c1983.
    Description: 45 p.: ill. (some col.); 22
    cm.
    ISBN: 0516016849 (lib. bdg.)
    Summary: Briefly discusses the history
    and uses of automatically operating
    devices from mechanical toys through
    dishwashers and clock radios to cyborgs
    and manipulators.
    Notes: Includes index.
    Subjects: Robots--Juvenile literature.
    Robotics. Robots.
    LC Classification: TJ211 .G73 1983
    Dewey Class No.: 629.8/92 19

Guest, Robert H.
    Robotics, the human dimension / by
    Robert H. Guest.
    Edition Information: 1st ed.
    Published/Created: New York:
    Pergamon Press, c1984.
    Description: 42 p.; 28 cm.
    ISBN: 0080315771 (pbk.) :
    Notes: Bibliography: p. 31-33.
    Subjects: Automation--Social aspects.
    Robots, Industrial. Technological
    unemployment.
    Series: Work in America Institute
    studies in productivity; 36
    LC Classification: HC79.A9 G83 1984

Dewey Class No.: 303.4/83 19

Gupta, Krishna C.
    Mechanics and control of robots: with
    38 illustrations / Krishna C. Gupta.
    Published/Created: New York: Springer,
    c1997.
    Description: xii, 178 p.: ill.; 24 cm.
    ISBN: 0387949232 (alk. paper)
    Notes: Includes bibliographical
    references (p. [161]-171) and index.
    Subjects: Robotics. Robots--Control
    systems. Automatic machinery.
    Series: Mechanical engineering series
    (Berlin, Germany)
    LC Classification: TJ211 .G86 1997
    Dewey Class No.: 629.8/92 21

Hall, Ernest L.
    Robotics, a user-friendly introduction /
    Ernest L. Hall, Bettie C. Hall.
    Published/Created: New York: Holt,
    Rinehart, and Winston, c1985.
    Related Authors: Hall, Bettie C.
    Description: xvi, 254 p.: ill.; 24 cm.
    ISBN: 0030697182 (pbk.)
    Notes: Includes index. Bibliography: p.
    229-234.
    Subjects: Robotics.
    LC Classification: TJ211 .H24 1985
    Dewey Class No.: 629.8/92 19

Hamlin, Gregory J.
    Tetrobot: a modular approach to
    reconfigurable parallel robotics / by
    Gregory J. Hamlin and Arthur C.
    Sanderson.
    Published/Created: Boston: Kluwer
    Academic Publishers, c1998.
    Related Authors: Sanderson, A. C.
    (Arthur C.)
    Description: vii, 182 p.: ill.; 24 cm.
    ISBN: 0792380258 (alk. paper)
    Notes: Includes bibliographical
    references (p. [163]-167) and index.
    Subjects: Robotics.
    Series: Kluwer international series in
    engineering and computer science;
    SECS 423. Kluwer international series
    in engineering and computer science.

Robotics.
LC Classification: TJ211 .H26 1998
Dewey Class No.: 629.8/92 21

Handbook of clinical automation, robotics,
and optimization / edited by Gerald J.
Kost with the collaboration of Judith
Welsh.
Published/Created: New York: Wiley,
c1996.
Related Authors: Kost, Gerald J. Welsh,
Judith, R.N.
Description: xvi, 952 p., [2] leaves of
plates: ill. (some col.); 24 cm.
ISBN: 0471031798 (cloth: acid-free
paper)
Notes: "A Wiley-Interscience
publication." Includes bibliographical
references and index.
Subjects: Diagnosis--Data processing.
Computers in medicine. Robotics in
medicine. Expert systems (Computer
science)
Series: Wiley-Interscience series on
laboratory automation
LC Classification: RC78.7.D35 H36
1996
Dewey Class No.: 610/.285 20

Handbook of industrial robotics / edited by
Shimon Y. Nof.
Edition Information: 2nd ed.
Published/Created: New York: John
Wiley, c1999.
Related Authors: Nof, Shimon Y., 1946-
Description: xxii, 1348 p.: ill.; 25 cm. +
1 computer laser optical disc (4 3/4 in.)
ISBN: 0471177830 (alk. paper)
Notes: Includes bibliographical
references and index. System
requirements for accompanying
computer disc: Windows 95 or better.
Subjects: Robots, Industrial--
Handbooks, manuals, etc.
LC Classification: TS191.8 .H36 1999
Dewey Class No.: 670.42/72 21

Handbook of industrial robotics / Shimon Y.
Nof, editor.; with a foreward by Isaac
Asimov.

Published/Created: Malabar, Fla.:
Krieger Pub., 1992.
Related Authors: Nof, Shimon Y., 1946-
Description: xvii, 1358 p.: ill.; 25 cm.
ISBN: 0894647229
Notes: Originally published: New York:
J. Wiley, c1985. Includes
bibliographical references and index.
Subjects: Robots, Industrial--
Handbooks, manuals, etc.
LC Classification: TS191.8 .H36 1992
Dewey Class No.: 670.42/72 20

Handbook of industrial robotics / Shimon Y.
Nof, editor; with a foreword by Isaac
Asimov.
Published/Created: New York: J. Wiley,
c1985.
Related Authors: Nof, Shimon Y., 1946-
Description: xvii, 1358 p.: ill.; 25 cm.
ISBN: 0471896845 :
Notes: Includes bibliographies and
index.
Subjects: Robots, Industrial--
Handbooks, manuals, etc.
LC Classification: TS191.8 .H36 1985
Dewey Class No.: 629.8/92 19

Harrar, George, 1949-
Radical robots: can you be replaced? /
George Harrar.
Published/Created: New York:
Published by Simon & Schuster in
association with WGBH Boston, c1990.
Description: 48 p.: col. ill.; 26 cm.
ISBN: 0671694200: 0671694219 (pbk.)
Summary: Examines the design,
construction, and applications of robots,
discussing what they can and cannot do
and the extent to which they can
develop their own intelligence.
Notes: Includes index.
Subjects: Robots--Juvenile literature.
Robots. Robotics.
Series: A Novabook
LC Classification: TJ211.2 .H37 1990
Dewey Class No.: 629.8/92 20

Harris, C. J. (Christopher John)
Intelligent control: aspects of fuzzy

logic and neural nets / C.J. Harris, C.G. Moore & M. Brown.
Published/Created: Singapore; River Edge, NJ: World Scientific, 1993 (1994 printing)
Related Authors: Moore, C. G. (Chris G.) Brown, M. (Martin)
Description: xvii, 380 p.: ill.; 23 cm.
ISBN: 9810210426
Notes: Includes bibliographical references and index.
Subjects: Intelligent control systems. Fuzzy logic. Neural networks (Computer science)
Series: World Scientific series in robotics and automated systems; vol. 6
LC Classification: TJ217.5 .H37 1993
Dewey Class No.: 629.8/9 20

Hart, Anne, 1941-
Robotics / Anne Cardoza and Suzee J. Vlk.
Edition Information: 1st ed.
Published/Created: Blue Ridge Summit, Pa.: Tab Books, c1985.
Related Authors: Vlk, Suzee.
Description: ix, 149 p.: ill.; 25 cm.
ISBN: 0830608583: 0830618589 (pbk.) :
Notes: Includes index. "No. 1858."
Bibliography: p. 143-146.
Subjects: Robotics.
LC Classification: TJ211 .C26 1985
Dewey Class No.: 629.8/92 19

Hart, Anne, 1941-
The robotics careers handbook / Anne Cardoza and Suzee J. Vlk.
Published/Created: New York: Arco Pub., c1985.
Related Authors: Vlk, Suzee.
Description: vi, 154 p.: ill.; 27 cm.
ISBN: 0668062894: 0668063009 (pbk.) :
Notes: Includes index. Bibliography: p. 151.
Subjects: Robotics--Vocational guidance.
LC Classification: TJ211.25 .C37 1985

Dewey Class No.: 629.8/92 19

Hausser, Roland R.
Foundations of computational linguistics: man-machine communication in natural language / Roland Hausser.
Published/Created: Berlin; New York: Springer, c1999.
Description: xii, 534 p.; 25 cm.
ISBN: 3540660151 (hardcover: alk. paper)
Notes: Includes bibliographical references (p. [499]-536) and indexes.
Subjects: Computational linguistics. Robotics.
LC Classification: P98 .H35 1999
Dewey Class No.: 410/.285 21

Heath, Larry.
Fundamentals of robotics: theory and applications / Larry Heath.
Published/Created: Reston, Va.: Reston Pub. Co., c1985.
Description: xvi, 412 p.: ill.; 25 cm.
ISBN: 0835921891 :
Notes: Includes index.
Subjects: Robotics.
LC Classification: TJ211 .H34 1985
Dewey Class No.: 629.8/92 19

Heiserman, David L., 1940-
Build your own working robot / by David L. Heiserman.
Edition Information: 1st ed.
Published/Created: Blue Ridge Summit, Pa.: G/L Tab Books, 1976.
Description: 234 p.: ill.; 22 cm.
ISBN: 0830668411: 0830658416
Notes: Includes index.
Subjects: Robotics.
LC Classification: TJ211 .H35
Dewey Class No.: 629.8/92

Heiserman, David L., 1940-
Build your own working robot: the second generation / David L. Heiserman.
Edition Information: 1st ed.
Published/Created: Blue Ridge Summit,

PA: TAB Books, c1987.
Description: x, 130 p.: ill.; 22 cm.
ISBN: 0830611819: 0830627812 (pbk.)
:
Notes: "No. 2781." Includes index.
Subjects: Robotics.
LC Classification: TJ211 .H35 1987
Dewey Class No.: 629.8/92 19

Heiserman, David L., 1940-
How to build your own self-
programming robot / by David L.
Heiserman.
Edition Information: 1st ed.
Published/Created: Blue Ridge Summit,
Pa.: Tab Books, c1979.
Description: 237 p.: ill.; 22 cm.
ISBN: 0830697608: 0830612416 (pbk).
:
Notes: "No. 1241." Includes index.
Subjects: Robotics.
LC Classification: TJ211 .H36
Dewey Class No.: 629.8/92

Heiserman, David L., 1940-
How to design and build your own
custom robot / by David L. Heiserman.
Edition Information: 1st ed.
Published/Created: Blue Ridge Summit,
Pa.: Tab Books, 1981.
Description: 462 p.: ill.; 22 cm.
ISBN: 0830696296 :
Notes: "Tab book #1341." Includes
index.
Subjects: Robotics.
LC Classification: TJ211 .H37
Dewey Class No.: 629.8/92 19

Henson, Hilary.
Robots / Hilary Henson; [editor,
Vanessa Clarke].
Edition Information: A Warwick Press
library ed.
Published/Created: New York: Warwick
Press, 1982, c1981.
Related Authors: Clarke, Vanessa.
Description: 77 p.: ill. (some col.); 28
cm.
ISBN: 0531091899 (lib. bdg.)
Notes: Includes index.

Subjects: Robotics. Robots, Industrial.
LC Classification: TJ211 .H385 1982
Dewey Class No.: 629.8/92 19

High technology industries--profiles and
outlooks. The robotics industry.
Published/Created: [Washington,
D.C.?]: U.S. Dept. of Commerce,
International Trade Administration: For
sale by the Supt. of Docs., U.S. G.P.O.,
[1983]
Related Authors: United States.
International Trade Administration.
Description: 54 p.: ill.; 28 cm.
Notes: "April 1983." S/N 003-009-
00363-6 Item 231-B-1
Subjects: Robot industry--United States-
-Congresses. Robot industry--
Government policy--United States
Congresses.
LC Classification: HD9696.R623 U633
1983
Dewey Class No.: 338.4/7629892 19
Govt. Doc. No.: C 61.2:T 22/3

High technology international trade and
competition: robotics, computers,
telecommunications, semiconductors:
based on research by International
Trade Administration, U.S. Department
of Commerce, Washington, DC and
Office of Technology Assessment,
Congress of the United States,
Washington, DC / edited by J.K. Paul.
Published/Created: Park Ridge, N.J.:
Noyes Publications, c1984.
Related Authors: Paul, J. K. United
States. International Trade
Administration. United States.
Congress. Office of Technology
Assessment.
Description: xiii, 394 p.: ill.; 25 cm.
ISBN: 081550988X :
Notes: Includes bibliographical
references and index.
Subjects: High technology industries--
United States--Congresses.
Competition, International--Congresses.
LC Classification: HD9696.A2 H53
1984

Dewey Class No.:
382/.456213817/0973 19

Highly redundant sensing in robotic systems
/ edited by Julius T. Tou, Jens G.
Balchen.
Published/Created: Berlin; New York:
Springer-Verlag, c1990.
Related Authors: Tou, Julius T., 1926-
Balchen, Jens G. North Atlantic Treaty
Organization. Scientific Affairs
Division.
Description: x, 320 p.: ill.; 25 cm.
ISBN: 0387520465 (U.S.: alk. paper)
Notes: "Published in cooperation with
NATO Scientific Affairs Division."
"Proceedings of the NATO Advanced
Research Workshop on Highly
Redundant Sensing in Robotic Systems
held in Il Ciocco, Italy, May 16-20,
1988"--T.p. verso. Includes
bibliographical references.
Subjects: Robotics--Congresses.
Intelligent control systems--Congresses.
Series: NATO ASI series. Series F,
Computer and systems sciences; no. 58.
LC Classification: TJ210.3 .N36 1988
Dewey Class No.: 629.8/92 20

Hirose, Shigeo, 1947-
Biologically inspired robots: snake-like
locomotors and manipulators / Shigeo
Hirose; translated by Peter Cave and
Charles Goulden.
Published/Created: Oxford; New York:
Oxford University Press, 1993.
Description: xiv, 220 p.: ill.; 25 cm.
ISBN: 0198562616 :
Notes: Includes bibliographical
references (p. 211-216) and index.
Subjects: Robotics. Biomechanics.
LC Classification: TJ211 .H52 1993
Dewey Class No.: 629.8/92 20

Hodges, Bernard.
Industrial robotics / Bernard Hodges
and Paul Hallam.
Published/Created: Oxford: Heinemann
Newnes, 1990.
Related Authors: Hallam, Paul.

Description: ix, 193 p.: ill.; 25 cm.
ISBN: 0434907820
Notes: Includes bibliographical
references (p. 185) and index.
Subjects: Robots, Industrial. Robotics.
LC Classification: TS191.8 .H64 1990
Dewey Class No.: 670.42/72 20

Hodges, Bernard.
Industrial robotics / Bernard Hodges.
Edition Information: 2nd ed.
Published/Created: Oxford; Boston:
Newnes, 1992.
Description: xi, 241 p.: ill.; 25 cm.
ISBN: 0750607815
Notes: Includes index.
Subjects: Robots, Industrial. Robotics.
LC Classification: TS191.8 .H64 1992
Dewey Class No.: 670.42/72 20

Hoekstra, Robert L.
Robotics and automated systems /
Robert L. Hoekstra.
Published/Created: Cincinnati: South-
Western Pub. Co., c1986.
Description: ix, 677 p.: ill.; 24 cm.
ISBN: 0538336501
Notes: Includes index.
Subjects: Robotics. Automatic control.
LC Classification: TJ211 .H6 1986
Dewey Class No.: 629.8 19

Holland, John M.
Basic robotics concepts / by John M.
Holland; [edited by Arlet Pryor;
illustrated by T.R. Emrick].
Edition Information: 1st ed.
Published/Created: Indianapolis, Ind.:
H.W. Sams, c1983.
Description: 270 p.: ill.; 22 cm.
ISBN: 0672219522 (pbk.) :
Notes: Includes index.
Subjects: Robotics.
Series: The Blacksburg continuing
education series
LC Classification: TJ211 .H64 1983
Dewey Class No.: 629.8/92 19

Holzbock, Werner G.
Robotic technology, principles and

practice / Werner G. Holzbock; with a foreword by Jack D. Lane.
Published/Created: New York: Van Nostrand Reinhold Co., c1986.
Description: xxi, 494 p.: ill.; 24 cm.
ISBN: 0442231547
Notes: Includes bibliographies and index.
Subjects: Robotics--Handbooks, manuals, etc.
LC Classification: TJ211 .H65 1986
Dewey Class No.: 629.8/92 19

Hubbard, John D., 1944-
HERO 2000: programming and interfacing / written by John D. Hubbard, Lawrence P. Larsen; Heathkit/Zenith Educational Systems.
Published/Created: Benton Harbor, Mich.: Heath Co., c1986.
Related Authors: Larsen, Lawrence P., 1944- Heathkit/Zenith Educational Systems (Group)
Description: [496] p. in various pagings: ill.; 30 cm.
ISBN: 0871191539
Notes: Final examination kit in pocket. Includes index.
Subjects: Robotics. Robots--Programming.
LC Classification: TJ211 .H83 1986
Dewey Class No.: 629.8/92 19

Human-oriented design of advanced robotics systems (DARS'95): a postprint volume from the IFAC workshop, Vienna, Austria, 19-20 September 1995 / edited by P. Kopacek.
Edition Information: 1st ed.
Published/Created: Oxford; Tarrytown, N.Y., U.S.A.: Published for the International Federation of Automatic Control by Pergamon, 1996.
Related Authors: Kopacek, Peter. International Federation of Automatic Control. IFAC Workshop on Human-Oriented Design of Advanced Robotics Systems (1995: Vienna, Austria)
Description: vii, 199 p.: ill.; 30 cm.
ISBN: 0080426042

Notes: "IFAC Workshop on Human-Oriented Design of Advanced Robotics Systems"--P. [iii]. Includes bibliographical references and index.
Subjects: Robotics--Human factors--Congresses. Robots--Design and construction--Congresses.
LC Classification: TJ211.49 .H86 1996
Dewey Class No.: 629.8/92 20

Human-robot interaction / edited by Mansour Rahimi and Waldemar Karwowski.
Published/Created: London; Washington, DC: Taylor & Francis, 1992.
Related Authors: Rahimi, Mansour. Karwowski, Waldemar, 1953-
Description: viii, 378 p.: ill.; 25 cm.
ISBN: 0850668093
Notes: Includes bibliographical references and index.
Subjects: Robotics. Human-machine systems.
LC Classification: TJ211 .H847 1992
Dewey Class No.: 629.8/92 20

Hunt, H. Allan.
Human resource implications of robotics / by H. Allan Hunt and Timothy L. Hunt.
Published/Created: Kalamazoo, Mich.: W.E. Upjohn Institute for Employment Research, c1983.
Related Authors: Hunt, Timothy L.
Description: xiii, 207 p.; 23 cm.
ISBN: 0880990090 0880990082 (pbk.)
Notes: Bibliography: p. 181-207.
Subjects: Technological unemployment--United States. Robotics--Social aspects--United States. Robotics--Economic aspects--United States.
LC Classification: HD6331.2.U5 H86 1983
Dewey Class No.: 331.12/929892/0973 19

Hunt, V. Daniel.
Industrial robotics handbook / V. Daniel Hunt.

Published/Created: New York, N.Y.:
Industrial Press, c1983.
Description: xiii, 432 p.: ill.; 24 cm.
ISBN: 0831111488 :
Notes: Includes index. Bibliography: p.
[401]-417.
Subjects: Robot industry--United States.
Robots, Industrial. Market surveys--
United States.
LC Classification: HD9696.R623 U635
1983
Dewey Class No.: 629.8/92 19

Hunt, V. Daniel.
Robotics sourcebook / V. Daniel Hunt.
Published/Created: New York: Elsevier,
1988.
Description: xi, 321 p.: ill.; 28 cm.
ISBN: 0444012982
Notes: Bibliography: p. 311-313.
Subjects: Robotics--Handbooks,
manuals, etc.
LC Classification: TJ211 .H85 1988
Dewey Class No.: 629.8/92 19

Hunt, V. Daniel.
Smart robots: a handbook of intelligent
robotic systems / V. Daniel Hunt.
Published/Created: New York:
Chapman and Hall, 1985.
Description: xxii, 377 p.: ill.; 25 cm.
ISBN: 041200531X :
Notes: Includes index. Bibliography: p.
[363]-368.
Subjects: Robotics. Robots, Industrial.
LC Classification: TJ211 .H86 1985
Dewey Class No.: 629.8/92 19

Hunt, V. Daniel.
Understanding robotics / V. Daniel
Hunt.
Published/Created: San Diego:
Academic Press, c1990.
Description: xvii, 286 p.: ill.: 26 cm.
ISBN: 0123617758 (alk. paper)
Notes: Includes bibliographical
references (p. 279-282) and index.
Subjects: Robotics.
LC Classification: TJ211 .H863 1990

Dewey Class No.: 629.8/92 20

Hurst, W. Jeffrey (William Jeffrey), 1948-
Laboratory robotics: a guide to
planning, programming, and
applications / W. Jeffrey Hurst and
James W. Mortimer.
Published/Created: New York, NY:
VCH Publishers, c1987.
Related Authors: Mortimer, James W.
(James Winslow), 1955-
Description: xi, 129 p.: ill.; 24 cm.
ISBN: 0895733226
Notes: Includes bibliographies and
index.
Subjects: Robotics. Laboratories--
Automation.
LC Classification: TJ211 .H87 1987
Dewey Class No.: 542/.1 19

IECON 2000: 2000 26th Annual Conference
of the IEEE Industrial Electronics
Society: 2000 IEEE International
Conference on Industrial Electronics,
Control and Instrumentation: 21st
Century technologies and industrial
opportunities: 22-28 October, 2000,
Nagoya, Aichi, Japan / IEEE.
Published/Created: Piscataway, New
Jersey: IEEE, c2000.
Related Authors: Institute of Electrical
and Electronics Engineers. International
Conference on Industrial Electronics,
Control, and Instrumentation (26th:
2000: Nagoya, Japan)
Description: 4 v.: ill.; 28 cm.
ISBN: 0780364562 (softbound)
0780364570 (casebound) 0780364589
(microfiche) 0780364597 (CD-ROM)
Notes: "IEEE Catalog Number:
00CH37141 (softbound)"--verso of T.p.
"IEEE Catalog Number: 00CB37141
(casebound)"--verso of T.p. "IEEE
Catalog Number: 00CH37141C (CD-
ROM)"--verso of T.p. Includes
bibliographic references and author
index.
Subjects: Industrial electronics--
Congresses. Automation--Congresses.
Power electronics--Congresses.

Robotics--Congresses. Electronic
control--Congresses. Electronic
instruments--Congresses.

IECON '87 / 1987 International Conference
on Industrial Electronics, Control, and
Instrumentation; sponsored by IEEE
Industrial Electronics Society in
association with Society of Instrument
and Control Engineers of Japan and
SPIE--the International Society for
Optical Engineering.
Published/Created: Bellingham, Wash.,
USA: SPIE, c1987.
Related Authors: IEEE Industrial
Electronics Society. Keisoku Jid¯o
Seigyo Gakkai (Japan) Society of
Photo-optical Instrumentation
Engineers.
Description: 6 v. (viii, 1156 p.): ill.; 28
cm.
ISBN: 0892528885 (pbk.: v. 1)
Contents: [1] Industrial applications of
control and simulation: 3-4 November
1987 / Tom T. Hartley, chair/editor --
[2] Motor control and power
electronics: 4-6 November 1987 /
Martin F. Schlecht, chair/editor -- [3]
Small computer applications--hardware
and software: 3 November 1987 /
Phillip Gold, chair/editor -- [4]
Industrial applications of robotics and
machine vision: 5-6 November 1987 /
Abe Abramovich, chair/editor -- [5]
Automated design and manufacturing:
5-6 November 1987 / Hubert Y.K. Wo,
Victor K.L. Huang, chairs/editors -- [6]
Signal acquisition and processing: 3-4
November 1987 / Russell J. Niederjohn,
chair/editor.
Notes: Includes bibliographies and
indexes.
Subjects: Industrial electronics--
Congresses.
Series: Proceedings of the Society of
Photo-optical Instrumentation
Engineers; v. 853-858.
LC Classification: TK7881 .I55 1987a
Dewey Class No.: 670.42 19

IECON '94, 20th International Conference
on Industrial Electronics, Control, and
Instrumentation.
Published/Created: New York: Institute
of Electrical and Electronics Engineers,
c1994.
Related Authors: IEEE Industrial
Electronics Society. Keisoku Jid¯o
Seigyo Gakkai (Japan) Associazione
elettrotecnica ed elettronica italiana.
Description: 3 v. (xxxvii, 23, 2154 p.):
ill.; 28 cm.
ISBN: 0780313291 (softbound)
0780313283 (casebound) 0780313305
(microfiche)
Contents: v. 1. Plenary session. Power
electronics -- v. 2. Robotics, vision, and
sensors. Factory automation. Emerging
technologies -- v. 3. Special sessions.
Signal processing and control.
Notes: Held Sept. 5-9, 1994, Università
di Bologna, Italy. "Sponsored by the
Industrial Electronics Society (IES) of
IEEE in technical co-sponsorship with
the Society of Instrument and Control
Engineers of Japan (SICE), the
Electrical and Electronic Association of
Italy (AEI)"--Cover. "IEEE catalog
number 94CH3319-1"--T.p. verso.
Includes bibliographical references and
index.
Subjects: Industrial electronics--
Congresses. Automation--Congresses.
Robotics--Congresses. Electronic
control--Congresses.
LC Classification: TK7881 .I55 1994
Dewey Class No.: 670.4/2 20

IECON '98: proceedings of the 24th Annual
Conference of the IEEE Industrial
Electronics Society, Aachen, Germany,
August 31-September 4, 1998 /
sponsored by IEEE Industrial
Electronics Society; technical co-
sponsors, the European Center for
Mechatronics, Society of Instrument
and Control Engineers of Japan (SICE).
Published/Created: [New York]:
Institute of Electronic and Engineering
Institute, c1998.

Related Authors: IEEE Industrial
Electronics Society. European Center
for Mechatronics. Keisoku Jid‾o Seigyo
Gakkai (Japan)
Description: 4 v.: ill.; 29 cm.
ISBN: 0780345037 (softbound)
0780345045 (casebound) 0780345053
(microfiche)
Notes: "IEEE catalog number
98CH36200"--T.p. verso. Includes
bibliographical references.
Subjects: Industrial electronics--
Congresses. Automation--Congresses.
Power electronics--Congresses.
Robotics--Congresses. Electronic
control--Congresses. Electronic
instruments--Congresses.
LC Classification: TK7881 .I284 1998
Dewey Class No.: 670.42 21

IEEE 2nd International Workshop on
Emerging Technologies and Factory
Automation: design and operations of
intelligent factories: Palm Cove, Cairns,
Australia, September 27-29, 1993:
workshop proceedings / sponsored by
the Industrial Electronics Society (IES)
of the Institute of Electrical and
Electronics Engineers (IEEE) in
cooperation with IEEE Neural Network
Council, the Society of Instrument and
Control Engineers of Japan (SICE),
Intelligent Automation Systems.
Published/Created: Piscataway, NJ:
IEEE, c1993.
Related Authors: IEEE Industrial
Electronics Society.
Description: xiv, 283 p.: ill.; 30 cm.
ISBN: 0780309855 (softbound ed.)
0780309863 (microfiche ed.)
Notes: Cover and spine 2nd IEEE
International Workshop on Emerging
Technologies and Factory Automation.
Includes bibliographical references.
Subjects: Automation--Congresses.
Artificial intelligence--Congresses.
Robotics--Congresses.
LC Classification: T59.5 .I28 1993
Dewey Class No.: 670.42/7 20

IEEE International Conference on Robotics
and Automation: [proceedings].
Published/Created: Silver Spring, MD:
IEEE Computer Society Press, c1984.
Related Authors: IEEE Computer
Society.
Description: 1 v.: ill.; 28 cm. 1985.
ISSN: 1049-3492
Notes: Issued by: IEEE Computer
Society, the Institute of Electrical and
Electronics Engineers.
SERBIB/SERLOC merged record
Indexed entirely by: Index to IEEE
publications 0099-1368
Subjects: Robotics--Congresses. Robots,
Industrial--Congresses. Automatic
control--Congresses.
LC Classification: TJ210.3 .I58a
Dewey Class No.: 670.42/7 20

IEEE International Symposium on
Assembly and Task Planning:
proceedings, August 10-11, 1995,
Pittsburgh, Pennsylvania / sponsored by
IEEE Robotics and Automation Society.
Published/Created: Los Alamitos,
Calif.: IEEE Computer Society Press,
c1995.
Related Authors: IEEE Robotics and
Automation Society.
Description: xiii, 432 p.: ill.; 28 cm.
ISBN: 0818669950 (paper) 0818669969
(microfiche)
Notes: "IEEE catalog number
95TB8123"--T.p. verso. Includes
bibliographical references and index.
Subjects: Assembly-line methods--
Congresses. Production planning--
Congresses. Robots, Industrial--
Congresses. Flexible manufacturing
systems--Congresses.
LC Classification: TS178.4 .I34 1995
Dewey Class No.: 670.42/7 20

IEEE journal of robotics and automation.
Published/Created: New York, NY:
Institute of Electrical and Electronics
Engineers, c1985-c1988.
Related Authors: Institute of Electrical
and Electronics Engineers. IEEE

Robotics and Automation Council.
Description: 4 v.: ill.; 28 cm. Vol. RA-1,
no. 1 (Mar. 1985)-v. 4, no. 6 (Dec.
1988).
ISSN: 0882-4967
Notes: Title from cover. Published by:
the IEEE Robotics and Automation
Council on behalf of the following
societies: Industry Applications, etc.
SERBIB/SERLOC merged record
Indexed entirely by: Computer &
control abstracts 0036-8113 Mar. 1985-
Electrical & electronics abstracts 0036-
8105 Mar. 1985- Index to IEEE
publications 0099-1368 Mar. 1985-
Physics abstracts. Science abstracts.
Series A 0036-8091 Mar. 1985-
Subjects: Robotics--Periodicals.
Automatic control--Periodicals.
Automation--Periodicals.
LC Classification: TJ210.2 .I33
Dewey Class No.: 629.8/92/05 19

IEEE micro electro mechanical systems: an
investigation of micro structures,
sensors, actuators, machines and robots:
proceedings, Travemünde, Germany, 4
February-7 February 1992 / W.
Benecke, H.-C. Petzold, eds.; sponsored
by the IEEE Robotics and Automation
Society and in cooperation with the
German IEEE Section.
Published/Created: [New York]:
Institute of Electrical and Electronics
Engineers, 1992.
Related Authors: Benecke, W.
(Wolfgang) Petzold, H.-C. (Hans-
Christian) IEEE Robotics and
Automation Society. Institute of
Electrical and Electronics Engineers.
German Section. IEEE Workshop on
Micro Electro Mechanical Systems
(1992: Travemünde, Lübeck, Germany)
Description: xv, 240 p.: ill.; 30 cm.
ISBN: 0780304977 (softbound)
0780304985 (casebound) 0780304993
(microfiche)
Notes: "IEEE Micro Electro Mechanical
Systems Workshop"--Spine. "IEEE
catalog number 92CH3093-2." Includes

bibliographical references.
Subjects: Microelectromechanical
systems--Congresses. Robotics--
Congresses. Detectors--Congresses.
Actuators--Congresses.
LC Classification: TK153 .I34 1992
Dewey Class No.: 620.4 20

IEEE micro electro mechanical systems: an
investigation of micro structures,
sensors, actuators, machines, and
robots: proceedings, Nara, Japan, 30
January-2 February 1991 / sponsored by
the IEEE Robotics and Automation
Society and in cooperation with the IEE
of Japan and the ASME Dynamic
Systems and Control Division.
Published/Created: New York: Institute
of Electrical and Electronics Engineers,
c1991.
Related Authors: IEEE Robotics and
Automation Society. Denki Gakkai
(1988) American Society of Mechanical
Engineers. Dynamic Systems and
Control Division. IEEE Workshop on
Micro Electro Mechanical Systems (4th:
1991: Nara-shi, Japan)
Description: xiv, 288 p.: ill.; 29 cm.
ISBN: 087942642X (lib. bdg.)
0879426411 (softbound) 0879426438
(microfiche)
Notes: "The Fourth IEEE Workshop on
Micro Electro Mechanical Systems
(MEMS '91)"--P. iii. Cover IEEE micro
electro mechanical systems, 1991.
"IEEE catalog number: 91CH2957-9"--
Label on verso of t.p. Includes
bibliographical references and indexes.
Subjects: Microelectromechanical
systems--Congresses. Miniature objects-
-Congresses. Robotics--Congresses.
LC Classification: TK153 .I33 1991
Dewey Class No.: 621.3 20

IEEE micro electro mechanical systems: an
investigations [sic] of micro structures,
sensors, actuators, machines, and
robotic systems: proceedings, Oiso,
Japan, January 25-28, 1994 / sponsored
by the IEEE Robotics and Automation

Society in cooperation with the ASME
Dynamic Systems and Control Division
and the Micromachine Center.
Published/Created: [New York: Institute
of Electrical and Electronics Engineers],
1994.
Related Authors: IEEE Robotics and
Automation Society. American Society
of Mechanical Engineers. Dynamic
Systems and Control Division.
Micromachine Center (Japan) IEEE
Workshop on Micro Electro Mechanical
Systems (1994: ¯Oiso-machi, Japan)
Description: xiv, 360 p.: ill.; 30 cm.
ISBN: 078031834X (casebound)
0780318331 (softbound) 0780318358
(microfiche)
Notes: "IEEE Workshop on Micro
Electro Mechanical Systems"--P. iii.
"IEEE catalog number 94CH3404-1."
Title on spine: IEEE Micro Electro
Mechanical Systems Workshop.
Includes bibliographical references and
index.
Subjects: Electromechanical devices--
Congresses. Microelectronics--
Congresses. Detectors--Congresses.
Actuators--Congresses.
LC Classification: TK153 .I343 1993
Dewey Class No.: 621.3 20

IEEE micro electro mechanical systems:
proceedings, Amsterdam, the
Netherlands, January 29-February 2,
1995 / sponsored by the IEEE Robotics
and Automation Society in cooperation
with the ASME Dynamic Systems and
Control Division.
Published/Created: [New York]:
Institute of Electrical and Electronics
Engineers, 1995.
Related Authors: IEEE Robotics and
Automation Society. American Society
of Mechanical Engineers. Dynamic
Systems and Control Division. IEEE
Workshop on Micro Electro Mechanical
Systems (1995: Amsterdam, the
Netherlands)
Description: 418 p.: ill.; 30 cm.
ISBN: 0780325036 (softbound)

0780325044 (casebound) 0780325052
(microfiche)
Notes: "Micro Electro Mechanical
Systems Workshop"--Spine. "IEEE
catalog number 95CH35754." Includes
bibliographical references and index.
Subjects: Electromechanical devices--
Congresses. Microelectronics--
Congresses. Detectors--Congresses.
Actuators--Congresses.
LC Classification: TK153 .I344 1995

IEEE robotics & automation magazine /
IEEE Robotics & Automation Society.
Published/Created: New York, NY:
Institute of Electrical and Electronics
Engineers, c1994-
Related Authors: Institute of Electrical
and Electronics Engineers. IEEE
Robotics and Automation Society.
Description: v.: ill.; 28 cm. Vol. 1, no. 1
(Mar. 1994)-
ISSN: 1070-9932
Notes: Title from cover.
SERBIB/SERLOC merged record
Subjects: Robotics--Periodicals.
Automation--Periodicals.
LC Classification: TJ210.2 .I32
Dewey Class No.: 629.8/92 20

IEEE the Ninth Annual International
Workshop on Micro Electro Mechanical
Systems: an investigation of micro
structures, sensors, actuators, machines,
and systems: proceedings, San Diego,
California, USA, February 11-15, 1996 /
sponsored by the IEEE Robotics and
Automation Society.
Published/Created: [New York]:
Institute of Electrical and Electronics
Engineers, c1996.
Related Authors: IEEE Robotics and
Automation Society.
Description: xxiv, 530 p.: ill.; 30 cm.
ISBN: 0780329856 (softbound)
0780329864 (casebound) 0780329872
(microfiche)
Notes: "IEEE catalog number
96CH35856." Includes bibliographical
references.

Subjects: Electromechanical devices--
Congresses. Microelectronics--
Congresses. Detectors--Congresses.
Actuators--Congresses.
LC Classification: TK153 .I35 1996

IEEE transactions on robotics and
automation: a publication of the IEEE
Robotics and Automation Society.
Published/Created: New York, NY:
Institute of Electrical and Electronics
Engineers, c1989-
Related Authors: Institute of Electrical
and Electronics Engineers. IEEE
Robotics and Automation Society.
Description: v.: ill.; 28 cm. Vol. 5, no. 1
(Feb. 1989)-
ISSN: 1042-296X Incorrect
ISSN: 0882-4967
Notes: SERBIB/SERLOC merged
record Indexed entirely by: Applied
science & technology index 0003-6986
1991- Computer & control abstracts
0036-8113 Feb. 1989- Electrical &
electronics abstracts 0036-8105 Feb.
1989- Index to IEEE publications 0099-
1368 1989- Physics abstracts 0036-8091
Feb. 1989-
Subjects: Robotics--Periodicals.
Automatic control--Periodicals.
Automation--Periodicals.
LC Classification: TJ210.2 .I33
Dewey Class No.: 628.8/92/05 20

IEEE Workshop on Languages for
Automation: [proceedings].
Published/Created: Silver Spring, MD:
IEEE Computer Society Press, c1983-
c1988.
Related Authors: IEEE Computer
Society.
Description: v.: ill.; 28 cm. Nov. 7-9,
1983- Ceased with 1988?
ISSN: 1042-4563
Notes: Some vols. have also theme
titles. Published: Washington, D.C.,
1986-1988. Issued by: IEEE Computer
Society. SERBIB/SERLOC merged
record
Subjects: Automatic control--

Congresses. Robotics--Congresses.
Programming languages (Electronic
computers)--Congresses. Programming
languages (Electronic computers)--
Computer graphics--Congresses.
LC Classification: TJ212.2 .I326a
Dewey Class No.: 629.8/92513/05 20

Impact 2000: materials, computer hardware
& software, automation & robotics,
sensors & measuring techniques,
mechanical components, precision
mechanics & optics, communications,
electronics & opto-electronics, medical /
[editor, Bruce Battrick].
Published/Created: Noordwijk: ESA
Publications, c1999.
Related Authors: Battrick, B., 1946-
European Space Agency.
Description: 61 p.: col. ill.; 30 cm.
ISBN: 9290926384
Subjects: Technology transfer--Europe.
Aeronautics--Europe.
Series: ESA BR; 154
LC Classification: T174.3 .I45 1999
Dewey Class No.: 629.13/0094 21

Industrial robotics, machine vision, and
artificial intelligence / edited by Ken
Stonecipher.
Edition Information: 1st ed.
Published/Created: Indianapolis, Ind.,
USA: H.W. Sams & Co., 1989.
Related Authors: Stonecipher, Ken.
Description: xvi, 313 p.: ill.; 25 cm.
ISBN: 0672225026 :
Notes: Includes bibliographical
references.
Subjects: Robots, Industrial. Computer
vision. Artificial intelligence.
LC Classification: TS191.8 .I487 1989
Dewey Class No.: 670.42/72 20

Industrial robotics: technology,
programming, and applications / Mikell
P. Groover ... [et al.]
Published/Created: New York, NY:
McGraw-Hill, c1986.
Related Authors: Groover, Mikell P.,
1939-

Description: xiv, 546 p.: ill.; 25 cm.
ISBN: 007024989X :
Notes: Includes bibliographies and
index.
Subjects: Robots, Industrial.
LC Classification: TS191.8 .I49 1986
Dewey Class No.: 629.8/92 19

Industrial robots / William R. Tanner, editor.
  Edition Information: 2nd ed.
  Published/Created: Dearborn, Mich.:
  Robotics International of SME, Society
  of Manufacturing Engineers, Marketing
  Services Dept., c1981.
  Related Authors: Tanner, William R.
  Description: 2 v.: ill.; 29 cm.
  ISBN: 0872630706
  Contents: v. 1. Fundamentals -- v. 2.
  Applications.
  Notes: Includes bibliographical
  references and indexes.
  Subjects: Robots, Industrial.
  Series: Manufacturing update series
  LC Classification: TS191.8 .I5 1981
  Dewey Class No.: 629.8/92 19

Industry in Israel / [production company
  unknown].
  Published/Created: United States:
  Embassy of Israel, [1993?]
  Related Authors: Israel. Shagrirut (U.S.)
  Embassy of Israel Collection (Library of
  Congress)
  Description: 1 videocassette of 1 (VHS)
  (ca. 7 min.): sd., [col.]; 1/2 in. viewing
  copy.
  Summary: This documentary presents a
  brief look at the high quality and
  competitiveness of Israeli industry in
  today's global economy. It features
  interviews with executives from major
  multinational corporations, details
  Israeli industry's strides in the fields of
  software development and robotics and
  highlights its success in applying
  military technology to civilian uses.
  Notes: Copyright: reg. unknown. Title
  taken from videocassette label and
  Embassy of Israel home page, films
  catalog, 9/17/97. Possible release date

taken from Embassy of Israel home
page, films catalog, 9/17/97; may have
been released in U.S. before this date.
Summary taken from Embassy of Israel
home page, films catalog, 9/17/97.
Embassy of Israel films catalog no. S-3.
Sources used: Embassy of Israel home
page, films catalog, 9/17/97;
videocassette label.
Subjects: Industries--Israel.
Genre/Form: Documentary--Short.
LC Classification: VAF 3516 (viewing
copy)

Information technologies in medicine /
  edited by Metin Akay, Andy Marsh.
  Published/Created: New York: Wiley,
  c2001.
  Related Authors: Akay, Metin. Marsh,
  Andy.
  Description: 2 v.: ill.; 24 cm.
  ISBN: 0471388637 (cloth: alk. paper)
  0471414921 (v. 2)
  Notes: Includes bibliographical
  references and indexes.
  Subjects: Virtual reality in medicine.
  Computer vision in medicine. Robotics
  in medicine.
  LC Classification: R859.7.C67 V575
  2001
  Dewey Class No.: 610/.285/6 21

Information technology in manufacturing
  processes: case studies in technological
  change / edited by Graham Winch.
  Published/Created: London:
  Rossendale, c1983.
  Related Authors: Winch, Graham.
  Description: 148 p.: ill.; 24 cm.
  ISBN: 0946138036 (pbk.)
  Contents: Management and
  manufacturing innovation / John
  Bessant -- Information technology as a
  technological fix / Erik Arnold --
  Robotics in manufacturing organisations
  / James Fleck -- Organisation design for
  CAD/CAM / Graham Winch --
  Technological imperatives and strategic
  choice / David Buchanan --
  Computerised machine tools, manpower

training, and skill polarisation / Ian Nicholas ... [et al.] -- Technical training and technical knowledge in an Irish electronics factory / Peter Murray and James Wickham -- Training for new technology / Sheila Rothwell and David Davidson -- New Technology in print / Cynthia Cockburn.
Notes: Includes bibliographies.
Subjects: Manufacturing processes--Data processing--Case studies.
LC Classification: TS183 .I54 1983

Inspector Gadget / an Avnet Kerner, Roger Birnbaum production; directed by David Kellogg; written by Andy Heyward and Jean Chaplopin.
Published/Created: 1999.
Related Authors: Copyright Collection (Library of Congress)
Description: 8 reels of 8 on 4 sd., col.; 35 mm. viewing print. 1 compact discs in 1 can: digital sd.; 4 3/4 in. trk.
Summary: A security guard's dreams come true when he is selected to be transformed into a cybernetic police office.
Notes: Copyright: Disney Enterprises, Inc. Appl. au.: Dr. Artemus Bradford Robotics Pictures, Inc. NM: all other cinematographic material. DCR 1999; PUB 23Jul99; REG 16Aug99; PA943-899. Digital sound track (DTS) on 1 compact disc stored in 1 film can (CGC 9202), serial no. 1958. Sources used: copyright database; Internet Movie Database, Ltd., 1990-2001; Television Programming Source books series, (1999-2000). Part summary taken from Internet database.
LC Classification: CGC 9198-9201 (viewing print) CGC 9202 (trk)

Integrated micro-motion systems: micromachining, control, and applications: a collection of contributions based on lectures presented at the Third Toyota Conference, Aichi, Japan, 22-25 October 1989 / edited by Fumio

Harashima.
Published/Created: Amsterdam; New York: Elsevier; New York, N.Y., U.S.A.: Distributors for the U.S. and Canada, Elsevier Science Pub. Co., c1990.
Related Authors: Harashima, Fumio.
Description: xii, 462 p.: ill.; 25 cm.
ISBN: 0444884173
Notes: Includes bibliographical references.
Subjects: Robotics--Congresses. Micromachining--Congresses. Miniature electronic equipment--Congresses.
LC Classification: TJ210.3 .T69 1990
Dewey Class No.: 670.42/7 20

Integration of assistive technology in the information age: ICORR '2001, 7th International Conference on Rehabilitation Robotics / edited by Mounir Mokhtari.
Published/Created: Amsterdam; Washington, D.C.: IOS Press, c2001.
Description: 354 p.; 25 cm.
ISBN: 1586031716 (IOS Press)
Series: Assistive technology research series; v. 9
LC Classification: *

Intelligent assembly systems / editors, M.H. Lee, J.J. Rowland.
Published/Created: Singapore; New Jersey: World Scientific, c1995.
Related Authors: Lee, M. H. Rowland, J. J.
Description: xxii, 239 p.: ill.; 23 cm.
ISBN: 981022494X
Notes: Includes bibliographical references and index.
Subjects: Robots, Industrial. Assembly-line methods--Automation.
Series: World Scientific series in robotics and intelligent systems; vol. 12
LC Classification: TS191.8 .I5564 1995
Dewey Class No.: 670.42/72 20

Intelligent autonomous systems, IAS--3: an international conference, Pittsburgh,

Pennsylvania, February 15-18, 1993 / edited by F.C.A. Groen, S. Hirose, Charles E. Thorpe.
Published/Created: Washington: IOS Press, c1993.
Related Authors: Groen, F. C. A. Hirose, Shigeo, 1947- Thorpe, Charles E.
Description: xv, 743 p.: ill.; 27 cm.
ISBN: 9051991223
Notes: Includes bibliographical references and indexes.
Subjects: Intelligent control systems--Congresses. Robotics--Congresses.
LC Classification: TJ217.5 .I52 1993
Dewey Class No.: 629.8/9263 20

Intelligent machines: myths and realities / edited by Clarence W. de Silva.
Published/Created: Boca Raton, FL: CRC Press, 2000.
Related Authors: De Silva, Clarence W.
Description: 326 p.: ill.; 24 cm.
ISBN: 0849303303 (alk.)
Notes: Includes bibliographical references and index.
Subjects: Intelligent control systems. Robotics. Artificial intelligence.
LC Classification: TJ217.5. I5445 2000
Dewey Class No.: 006.3 21

Intelligent motion control: proceedings of the IEEE International Workshop on Intelligent Motion Control, Bogaziçi University, Istanbul, Turkey, 20-22 August 1990 / cosponsored by IEEE Industrial Electronics Society; in cooperation with IEEE Robotics and Automation Society; hosted by Bogaziçi University; editor/chair, Okyay Kaynak.
Published/Created: [New York, N.Y.]: IEEE; [Piscataway, NJ, U.S.A.: Order copies from IEEE Service Center, Publications Sales Dept., 1990]
Related Authors: Kaynak, Okyay. IEEE Industrial Electronics Society. IEEE Robotics and Automation Society.
Description: 2 v.: ill.; 28 cm.
Notes: Spine Proceedings of the IEEE International Workshop on Intelligent

Motion Control. "Catalog number: 90 TH0272-5"--V. 1-2, cover. Includes bibliographical references and index.
Subjects: Intelligent control systems--Congresses. Motion control devices--Congresses.
LC Classification: TJ217.5 .I34 1990
Dewey Class No.: 629.8/92 20

Intelligent robotic systems / edited by Spyros G. Tzafestas.
Published/Created: New York: M. Dekker, c1991.
Related Authors: Tzafestas, S. G., 1939-
Description: x, 720 p.: ill.; 23 cm.
ISBN: 082478135X (alk. paper)
Notes: Includes bibliographical references and index.
Subjects: Robotics.
Series: Electrical engineering and electronics
LC Classification: TJ211 .I4825 1991
Dewey Class No.: 629.8/92 20

Intelligent robotics: proceedings of the International Symposium on Intelligent Robotics, January 2-5, 1991, Bangalore, India / editors, M. Vidyasagar, Mohan Trivedi.
Published/Created: New Delhi; New York: Tata McGraw-Hill Pub. Co., c1991.
Related Authors: Vidyasagar, M. (Mathukumalli), 1947- Trivedi, Mohan M. Centre for Artificial Intelligence and Robotics (Bangalore, India) India. Ministry of Defence. Defence Research and Development Organisation.
Description: xii, 736 p.: ill.; 25 cm.
ISBN: 0074604678 :
Notes: Organized by the Centre for Artificial Intelligence and Robotics, Bangalore, and sponsored by the Defence Research and Development Organisation, Ministry of Defence, Govt. of India. Also issued without series statement. Includes index. Includes bibliographical references.
Subjects: Robotics--Congresses.
Series: Proceedings of SPIE--the

International Society for Optical
Engineering; v. 1571
LC Classification: TJ210.3 .I587 1991
Dewey Class No.: 629.8/92 20

Intelligent robots and computer vision XVII:
algorithms, techniques, and active
vision: 2-3 November, 1998, Boston,
Massachusetts / David P. Casasent,
chair/editor; sponsored and published by
SPIE--the International Society for
Optical Engineering; endorsed by SME-
-the Society of Manufacturing
Engineers.
Published/Created: Bellingham,
Washington: SPIE, c1998.
Related Authors: Casasent, David Paul.
Society of Photo-optical
Instrumentation Engineers. Society of
Manufacturing Engineers.
Description: x, 548 p.: ill.; 28 cm.
ISBN: 081942983X
Notes: Includes bibliographical
references and author index.
Subjects: Robotics--Congresses.
Artificial intelligence--Congresses.
Computer vision--Congresses. Image
processing--Congresses.
Series: Proceedings of SPIE--the
International Society for Optical
Engineering; v. 3522.

Intelligent robots and computer vision XX:
algorithms, techniques, and active
vision: 29-31 October, 2001, Newton,
[Massachusetts] USA / David P.
Casasent, Ernest L. Hall, chairs/editors;
sponsored and published by SPIE--the
International Society for Optical
Engineering.
Published/Created: Bellingham,
Washington: SPIE, c2001.
Related Authors: Casasent, David Paul.
Hall, Ernest L. Society of Photo-optical
Instrumentation Engineers.
Description: x, 582 p.: ill.; 28 cm.
ISBN: 081944300X
Notes: Includes bibliographic references
and author index.
Subjects: Robotics--Congresses.

Artificial intelligence--Congresses.
Computer vision--Congresses. Image
processing--Congresses.
Series: Proceedings of SPIE--the
International Society for Optical
Engineering; v. 4572.

Intelligent robots and computer vision.
Published/Created: Bellingham, Wash.:
SPIE--the International Society for
Optical Engineering,
Related Authors: Society of Photo-
optical Instrumentation Engineers.
Description: v.: ill.; 28 cm. Vols. for
1989- also have distinctive subtitles
indicating contents of vol. Issues for 6-
10 Nov. 1989- also called 8-; vol. no.
remains same for two issues a year.
Notes: Description based on: Nov. 5-8,
1984. SERBIB/SERLOC merged record
Subjects: Robotics--Congresses.
Artificial intelligence--Congresses.
Computer vision--Congresses.
Series: Proceedings of SPIE--the
International Society for Optical
Engineering
LC Classification: TJ210.3 .I53
Dewey Class No.: 629.28/7 20

Intelligent robots and systems: selections of
the International Conference on
Intelligent Robots and Systems 1994,
IROS 94, Munich, Germany, 12-16
September 1994 / edited by Volker
Graefe.
Published/Created: Amsterdam; New
York: Elsevier, 1995.
Related Authors: Graefe, Volker, 1938-
Description: x, 733 p.: ill.; 25 cm.
ISBN: 044482250X (acid-free paper)
Notes: Includes bibliographical
references.
Subjects: Robotics--Congresses.
Intelligent control systems--Congresses.
LC Classification: TJ210.3 .I447 1994a
Dewey Class No.: 629.8/92 20

Intelligent robots: Third International
Conference on Robot Vision and
Sensory Controls, RoViSeC3: 7-10

November, 1983, Cambridge, Massachusetts / David P. Casasent, Ernest L. Hall, chairmen/editors; sponsored by SPIE--the International Society for Optical Engineering; organizational assistance in Europe, IFS (Conferences Ltd.; cosponsors, the British Robot Association ... [et al.]; cooperating organizations, Robotics Institute/Carnegie-Mellon University ... [et al.].
Published/Created: Bellingham, Wash., USA: The Society, c1984.
Related Authors: Casasent, David Paul. Hall, Ernest L. Society of Photo-optical Instrumentation Engineers. IFS (Conferences) Ltd.
Description: 2 v.: ill.; 28 cm.
ISBN: 0892524847 (pbk.: set)
Notes: Includes bibliographies and index.
Subjects: Robots, Industrial--Congresses. Robot vision--Congresses.
Series: Proceedings of SPIE--the International Society for Optical Engineering; v. 449
LC Classification: TS191.8 .I56 1983
Dewey Class No.: 629.8/92 19

Intelligent unmanned ground vehicles: autonomous navigation research at Carnegie Mellon / edited by Martial H. Hebert, Charles Thorpe, Anthony Stentz.
Published/Created: Boston: Kluwer Academic Publishers, c1997.
Related Authors: Hebert, Martial. Thorpe, Charles E. Stentz, Anthony.
Description: 308 p.: ill.; 24 cm.
ISBN: 0792398335 (acid-free paper)
Notes: Includes bibliographical references and index.
Subjects: Carnegie-Mellon University--Research. Motor vehicles--Automatic control. All terrain vehicles--Automatic control. Intelligent control systems. Mobile robots.
Series: Kluwer international series in engineering and computer science; SECS 388. Kluwer international series

in engineering and computer science. Robotics.
LC Classification: TL152.8 .I57 1997
Dewey Class No.: 629.2/9 21

International Conference on Robotics: [proceedings].
Published/Created: Silver Spring, MD: IEEE Computer Society Press, c1984.
Related Authors: IEEE Computer Society.
Description: 1 v.: ill.; 28 cm. March 13-15, 1984.
ISSN: 1049-3484
Notes: Issued by: IEEE Computer Society, the Institute of Electrical and Electronics Engineers.
SERBIB/SERLOC merged record Indexed entirely by: Index to IEEE publications 0099-1368
Subjects: Robotics--Congresses.
LC Classification: TJ210.3 I58a
Dewey Class No.: 629.8/92 20

International encyclopedia of robotics: applications and automation / Richard C. Dorf, editor-in-chief; Shimon Y. Nof, consulting editor.
Published/Created: New York: Wiley, c1988.
Related Authors: Dorf, Richard C. Nof, Shimon Y., 1946-
Description: 3 v.: ill.; 29 cm.
ISBN: 0471878685 (set)
Notes: "A Wiley-Interscience publication." Includes bibliographies and index.
Subjects: Robotics--Encyclopedias.
LC Classification: TJ210.4 .I57 1988
Dewey Class No.: 629.8/92/0321 19

International journal of robotics & automation.
Published/Created: Anaheim, Calif.; Calgary, Alta.: ACTA Press, [1986-
Related Authors: International Association of Science and Technology for Development.
Description: v.: ill.; 28 cm. Vol. 1, no. 1-

ISSN: 0826-8185
Notes: "A journal of the International
Association of Science and Technology
for Development-IASTED." Title from
cover. SERBIB/SERLOC merged
record Indx'd selectively by: Computer
& control abstracts 0036-8113 1986-
Electrical & electronics abstracts 0036-
8105 1986- Physics abstracts 0036-8091
1986-
Subjects: Robotics--Periodicals.
Automatic control--Periodicals.
Robotics--Research--Periodicals.
Artificial intelligence--Periodicals.
Robotique--Périodiques.
Automatisation--Périodiques.
LC Classification: TJ210.2 .I6
Dewey Class No.: 629.8/92/05 19

International robotics industry directory.
Published/Created: Conroe, Tex.:
Technical Database Corp., c1984.
Related Authors: Technical Database
Corp.
Description: 1 v.: ill.; 28 cm. 4th ed.
Notes: SERBIB/SERLOC merged
record Updated by: Computerized
manufacturing. Computerized
manufacturing 0746-3405 (DLC) 90-
656036 (OCoLC)9934098
Subjects: Robot industry--Directories.
Robots--Catalogs.
LC Classification: TS191.8 .R6
Dewey Class No.: 629.8/92/0294 19

International Symposium on New Directions
in Computing, August 12-14, 1985,
Norwegian Institute of Technology,
Trondheim, Norway.
Published/Created: Washington, D.C.:
IEEE Computer Society Press; Los
Angeles, CA: Order from IEEE
Computer Society, c1985.
Related Authors: Norges tekniske
høgskole.
Description: ix, 405 p.: ill.; 28 cm.
ISBN: 0818606398 (pbk.) 0818686391
(hard) 081864639X (microfiche)
Notes: Includes bibliographies and
index.

Subjects: Electronic data processing--
Congresses. Artificial intelligence--
Congresses. Robotics--Congresses.
LC Classification: QA75.5 .I635 1985
Dewey Class No.: 004 19

International Workshop on Graphics &
Robotics, April 19-22, 1993, Schloss
Dagstuhl, FRG: abstracts / organizers,
W. Strasser, F. Wahl.
Published/Created: Saarbrücken:
Geschäftsstelle Schloss Dagstuhl,
c1993.
Related Authors: Strasser, Wolfgang,
1941- Wahl, Friedrich M., 1948-
Schloss Dagstuhl International
Conference and Research Center for
Computer Science.
Description: 53 p.; 21 cm.
Notes: At head of title on cover:
Internationales Begegnungs- und
Forschungszentrum für Informatik.
Subjects: Robotics--Congresses.
Computer graphics--Congresses.
Series: Dagstuhl-seminar-report, 0940-
1121; 61
LC Classification: TJ210.3 .I625 1993
Dewey Class No.: 629.8/9266 20

Interoperability of standards for robotics in
CIME / Falk Mikosch, ed.
Published/Created: Berlin; New York:
Springer, c1997.
Related Authors: Mikosch, Falk.
Description: ix, 140 p.: ill.; 24 cm.
ISBN: 3540618848 (Berlin: pbk.: acid-
free ppaer)
Notes: Includes bibliographical
references (p. [139]-140).
Subjects: Robotics--Standards.
Series: Research reports ESPRIT.
Subseries PDT (product data
technology). Project 6457, InterRob; v.
1
LC Classification: TJ211 .I55 1997
Dewey Class No.: 629.8/92 21

Iovine, John.
Robots, androids, and animatrons: 12
incredible projects you can build / John

Iovine.
Edition Information: 2nd ed.
Published/Created: New York:
McGraw-Hill, c2002.
Description: xix, 332 p.: ill.; 23 cm.
ISBN: 0071376836
Notes: Includes index.
Subjects: Robotics--Amateurs' manuals.
LC Classification: TJ211.5 .I58 2002
Dewey Class No.: 629.8/92 21

Iovine, John.
Robots, androids, and animatrons: 12
incredible projects you can build / John
Iovine.
Published/Created: New York:
McGraw-Hill, c1998.
Description: 270 p.: ill.; 24 cm.
ISBN: 0070328048
Notes: Includes index.
Subjects: Robotics--Amateurs' manuals.
LC Classification: TJ211.5 .I58 1998
Dewey Class No.: 629.8/92 21

IROS '93, proceedings of the 1993
IEEE/RSJ International Conference on
Intelligent Robots and Systems:
intelligent robots for flexibility, July 26-
30, 1993, ... Yokohama, Japan / co-
sponsored by the IEEE Industrial
Electronics Society ... [et al.] in
cooperation with Japan Society of
Mechanical Engineers ... [etal.].
Published/Created: [New York]:
Institute of Electrical and Electronics
Engineers, c1993.
Related Authors: IEEE Industrial
Electronics Society.
Description: 3 v. (2317 p.): ill.; 29 cm.
ISBN: 0780308239 (softbound)
0780308247 (casebound) 0780308255
(microfiche)
Notes: "IEEE catalog number
93CH3213-6"--T.p. verso. Includes
bibliographical references and indexes.
Subjects: Robotics--Congresses.
Intelligent control systems--Congresses.
LC Classification: TJ210.3 .I447 1993
Dewey Class No.: 629.8/92 20

IROS '94: proceedings of the IEEE/RSJ/GI
international conference on intelligent
robots and systems: advanced robotic
systems and the real world, September
12-16, 1994, Federal Armed Forces
University, Munich, Germany /
sponsors, Gesellschaft für Informatik ...
[et al.] in cooperation with Institut für
die Technik Intelligenter Systeme,
VDE/VDI Gesellschaft für
Mikroelektronik, VDI/VDE
Gesellschaft für Mess- und
Automatisierungstechnik.
Published/Created: [New York: Institute
of Electrical and Electronics Engineers;
Piscataway, NJ: Available from IEEE
Service Center, c1994.
Related Authors: Gesellschaft für
Informatik.
Description: 3 v. (xxv, 2191 p.): ill.; 28
cm.
ISBN: 0780319338 (softbound)
0780319346 (casebound) 0780319354
(microfiche)
Notes: "IEEE catalog number:
94CH3447-0"--T.p. verso. Includes
bibliographical references and indexes.
Subjects: Robotics--Congresses.
LC Classification: TJ210.3 .I447 1994
Dewey Class No.: 629.8/92 20

ISA/Mid-America: proceedings of the
Instrumentation & Control Systems
Conference and Exhibit, March 17-19,
1987, O'Hare Exposition Center,
Rosemont, Illinois / sponsored by
Instrument Society of America and ISA
sections in District 6.
Published/Created: Research Triangle
Park, N.C.: ISA, 1987.
Related Authors: Instrument Society of
America. Instrument Society of
America. District 6.
Description: 228 p.: ill.; 28 cm.
ISBN: 1556170025 (pbk.)
Notes: "Includes papers from the
following ISA divisions: Automatic
Control Systems, Computer
Technology, Electro-Optics, Metals
Industries, Process Measurement &

Control, Robotics & Expert Systems."
Includes bibliographies and index.
Subjects: Automatic control--
Congresses. Process control--
Congresses.
LC Classification: TJ212.2 .I534 1987
Dewey Class No.: 629.8 19

ISAIRAS '99: Fifth International
Symposium on Artificial Intelligence,
Robotics, and Automation in Space,
ESTEC, Noordwijk, the Netherlands, 1-
3 June 1999 / [editor, Michael Perry].
Published/Created: Noordwijk,
Netherlands: European Space Agency:
Contact ESA Publications Division,
ESTEC, c1999.
Related Authors: Perry, Michael.
European Space Agency.
Description: xi, 728 p.: ill.; 30 cm.
ISBN: 9290927607
Notes: "August 1999." Includes
bibliographical references.
Subjects: Space robotics--Congresses.
Space flight--Data processing--
Congresses. Artificial intelligence--
Congresses.
Series: ESA SP; 440
LC Classification: TL1097 .I54 1999
Dewey Class No.: 629.47 21

ISMCR '94, topical workshop on virtual
reality: proceedings of the Fourth
International Symposium on
Measurement and Control in Robotics,
Houston, Texas, USA, November 30-
December 3, 1994 / organized by the
IMEKO Technical Committee on
Robotics (TC-17).
Published/Created: Linthicum Heights,
Md.: NASA Center for AeroSpace
Information, 1994.
Related Authors: IMEKO Technical
Committee on Robotics (TC-17)
Instrument Society of America. Clear
Lake--Galveston Section.
Description: vii, 154 p.: ill.; 28 cm.
ISBN: 0962711616
Notes: "November 1994." "Sponsored
by ISA Clear Lake Section.

Cosponsored by IEEE Galveston Bay
Section, AIAA Houston Section"--
Cover. Includes bibliographical
references and indexes.
Subjects: Robotics--Congresses.
Robots--Control--Congresses.
Series: NASA conference publication;
10163
LC Classification: TJ210.3 .I59 1994
Dewey Class No.: 006 20

ISMCR '95, proceedings of the Fourth
International Symposium on
Measurement and Control in Robotics,
June 12-16, 1995, Convention Centre of
the Slovak Academy of Sciences,
Smolenice Castle, Slovakia / [sponsored
by] IMEKO Technical Committee on
Robotics (TC-17) ... [et al.; edited by
Vladimír Chudý and Eva Kureková].
Published/Created: Bratislava, Slovakia:
Slovak Technical University, c1995.
Related Authors: Chudý, Vladimír.
Kureková, Eva. IMEKO Technical
Committee on Robotics (TC-17)
Description: 546 p.: ill.; 30 cm.
ISBN: 8022707600
Notes: Includes bibliographical
references and indexes.
Subjects: Robotics--Congresses.
Robots--Control systems--Congresses.
LC Classification: TJ210.3 .I59 1995
Dewey Class No.: 629.8/92 21

Jacak, Witold.
Intelligent robotic systems: design,
planning, and control / Witold Jacak.
Published/Created: New York: Kluwer
Academic: Plenum, c1999.
Description: x, 310 p.: ill.; 24 cm.
ISBN: 0306460629
Notes: Includes bibliographical
references (p. 295-302)
Subjects: Robotics. Intelligent control
systems.
Series: IFSR international series on
systems science and engineering; v. 14.
LC Classification: TJ211 .J33 1999
Dewey Class No.: 629.8/92 21

James, Samuel D. K., 1943-
The impact of cybernation technology on Black automotive workers in the U.S. / by Samuel D.K. James.
Published/Created: Ann Arbor, MI: UMI Research Press, c1985.
Description: xii, 125 p.; 24 cm.
ISBN: 0835717194 (alk. paper)
Notes: Originally presented as the author's thesis (Ph.D.)--University of Delaware, 1984. Includes index.
Bibliography: p. [111]-121.
Subjects: African American automobile industry workers--Effect of technological innovations on. Robotics--United States.
Series: Research for business decisions; no. 84
LC Classification: HD6331.18.A8 J36 1985
Dewey Class No.: 331.6/3/96073 19

Jefferis, David.
Artificial intelligence: robotics and machine evolution / David Jefferis.
Published/Created: New York: Crabtree Pub. Co., 1999.
Description: 32 p.: col. ill.; 29 cm.
ISBN: 0778700461 0778700569 (pbk.)
Summary: An introduction to the past, present, and future of artificial intelligence and robotics, discussing early science fiction predictions, the dawn of AI, and today's use of robots in factories and space exploration.
Notes: Includes index.
Subjects: Robotics--Juvenile literature. Artificial intelligence--Juvenile literature. Robotics. Robots. Artificial intelligence.
Series: Megatech
LC Classification: TJ211.2 .J44 1999
Dewey Class No.: 629.8/9263 21

Jolion, Jean-Michel.
A pyramid framework for early vision: multiresolutional computer vision / by Jean-Michel Jolion and Azriel Rosenfeld.
Published/Created: Dordrecht; Boston: Kluwer Academic Publishers, c1994.
Related Authors: Rosenfeld, Azriel, 1931-
Description: xii, 218 p.: ill.; 25 cm.
ISBN: 079239402X (alk. paper)
Notes: Includes bibliographical references (p. 187-215) and index.
Subjects: Computer vision--Mathematical models.
Series: Kluwer international series in engineering and computer science; SECS 251. Kluwer international series in engineering and computer science. Robotics.
LC Classification: TA1634 .J65 1994
Dewey Class No.: 006.3/7 20

Journal of intelligent & robotic systems.
Published/Created: Dordrecht [Netherlands]; Boston: Kluwer, c1988-
Description: v.: ill.; 24 cm. Vol. 1, no. 1-
ISSN: 0921-0296 CODEN: JIRSES
Notes: SERBIB/SERLOC merged record Indexed entirely by: Computer & control abstracts 0036-8113 1988- Electrical & electronic abstracts 0036-8105 1988- Physics abstracts 0036-8091 1988-
Subjects: Robotics--Periodicals. Manipulators (Mechanism)--Periodicals.
LC Classification: TJ210.2 .J68
Dewey Class No.: 629.8/92 19

Journal of robotic systems.
Published/Created: [New York, NY]: Wiley, [c1984-
Description: v.: ill.; 26 cm. Vol. 1, no. 1 (spring 1984)-
ISSN: 0741-2223 Incorrect
ISSN: 0471-2223
Notes: Title from cover.
SERBIB/SERLOC merged record Indexed entirely by: Applied science & technology index 0003-6986 1991- Indx'd selectively by: International aerospace abstracts 0020-5842 1984-
Subjects: Robotics--Periodicals. Robots--Periodicals.
LC Classification: TJ210.3 .J68

Dewey Class No.: 629.8/92 19

JSME international journal. Series C,
Dynamics, control, robotics, design and
manufacturing.
Published/Created: Tokyo: Japan
Society of Mechanical Engineers,
c1993-c1997.
Related Authors: Nihon Kikai Gakkai.
Description: 5 v.: ill.; 30 cm. Vol. 36,
no. 1 (Mar. 1993)-v. 40, no. 1 (Mar.
1997).
ISSN: 1340-8062 Incorrect
ISSN: 0914-8825
Notes: Title from cover.
SERBIB/SERLOC merged record
Subjects: Mechanical engineering--
Periodicals.
LC Classification: TJ1 .J753
Dewey Class No.: 620.1 20

JSME international journal. Series C,
Mechanical systems, machine elements
and manufacturing.
Published/Created: Tokyo, Japan: Japan
Society of Mechanical Engineers,
c1997-
Related Authors: Nihon Kikai Gakkai.
Description: v.: ill.; 30 cm. Vol. 40, no.
2 (June 1997)-
ISSN: 1344-7653 Incorrect
ISSN: 1340-8062
Notes: Title from cover.
SERBIB/SERLOC merged record
Subjects: Mechanical engineering--
Periodicals.
LC Classification: TJ1 .J753
Dewey Class No.: 620.1 20

JSME international journal. Series III,
Vibration, control engineering,
engineering for industry.
Published/Created: Tokyo, Japan: Japan
Society of Mechanical Engineers,
c1988-c1992.
Related Authors: Nihon Kikai Gakkai.
Description: 5 v.: ill.; 30 cm. Continues
volume number, but not the issue
numbers, of its former title. Vol. 31, no.
1 (Mar. 1988)-v. 35, no. 4 (Dec. 1992).

ISSN: 0914-8825
Notes: Title from cover.
SERBIB/SERLOC merged record
Indx'd selectively by: Computer &
control abstracts 0036-8113 Mar. 1988-
Electrical & electronic abstracts 0036-
8105 Mar. 1988- Physics abstracts
0036-8091 Mar. 1988-
Subjects: Mechanical engineering--
Periodicals.
LC Classification: TJ1 .J753
Dewey Class No.: 620.1 20

Junge, Hans Dieter.
Pocket dictionary of robotics: English
German = Taschenwörterbuch
Robotertechnik: Deutsch Englisch /
Hans-Dieter Junge.
Published/Created: Berlin: Verlag für
Architektur und technische
Wissenschaften, c1986.
Description: 104 p.; 17 cm.
ISBN: 343302801X (pbk.)
Notes: Cover Robotics =
Robotertechnik.
Subjects: Robotics--Dictionaries.
English language--Dictionaries--
German.
LC Classification: TJ210.4 .J86 1986
Dewey Class No.: 629.8/92 19

Kafrissen, Edward.
Industrial robots and robotics / Edward
Kafrissen, Mark Stephans.
Published/Created: Reston, Va.: Reston
Pub. Co., c1984.
Related Authors: Stephans, Mark.
Description: xiv, 396 p.: ill.; 25 cm.
ISBN: 0835930718 :
Notes: Includes index. Bibliography: p.
379-382.
Subjects: Robots, Industrial. Robotics.
LC Classification: TS191.8 .K34 1984
Dewey Class No.: 629.8/92 19

Kanatani, Ken'ichi, 1947-
Statistical optimization for geometric
computation: theory and practice /
Kenichi Kanatani.
Published/Created: Amsterdam; New

York: Elsevier, 1996.
Description: xiv, 509 p.: ill.; 23 cm.
ISBN: 0444824278 (acid-free paper)
Notes: Includes bibliographical
references (p. 485-499) and index.
Subjects: Robotics. Computer vision.
Mathematical statistics.
Series: Machine intelligence and pattern
recognition; v. 18
LC Classification: TJ211 .K354 1996
Dewey Class No.: 006.4/2/0151954 20

Keller, Charles.
Ohm on the range: robot and computer
jokes / compiled by Charles Keller;
illustrated by Art Cumings.
Published/Created: Englewood Cliffs,
N.J.: Prentice-Hall, c1982.
Related Authors: Cumings, Art, ill.
Description: [46] p.: ill.; 24 cm.
ISBN: 0136335527 :
Summary: A collection of robot and
computer jokes and riddles.
Subjects: Riddles, Juvenile. Computers-
-Juvenile humor. Robotics--Juvenile
humor. Riddles. Computers--Wit and
humor. Robots--Wit and humor.
LC Classification: PN6371.5 .K39 1982
Dewey Class No.: 818.5402 19

Kelly, Derek A.
A layman's introduction to robotics /
Derek Kelly.
Published/Created: Princeton, N.J.:
Petrocelli Books, c1986.
Description: xxvii, 209 p.: ill.; 24 cm.
ISBN: 0894332651
Notes: Includes indexes. Bibliography:
p. 191-196.
Subjects: Robotics--Popular works.
LC Classification: TJ211.15 .K45 1986
Dewey Class No.: 629.8/92 19

Kelly, Derek A.
Ollie the robot-- talks about robots /
Derek Kelly; illustrations by Christine
Fedorowicz.
Published/Created: [Princeton, N.J.]:
Petrocelli Books, c1987.
Description: xiii, 121 p., [2] p. of plates:

ill.; 22 cm.
ISBN: 0894332791
Subjects: Robotics. Robots.
LC Classification: TJ211 .K45 1987
Dewey Class No.: 629.8/92 19

Keramas, James G.
Robot technology fundamentals / James
G. Keramas.
Published/Created: Albany, N.Y.:
Delmar Publishers, c1999.
Description: xv, 407 p.: ill.; 25 cm.
ISBN: 0827382367
Notes: Includes bibliographical
references and index.
Subjects: Robotics. Robots, Industrial.
LC Classification: TJ211 .K47 1999
Dewey Class No.: 670.42/72 21

Kerker, Alexander.
Robots / by Alexander Kerker;
illustrated by Tom LaPadula.
Published/Created: New York: Golden
Book; Racine, Wis.: Western Pub. Co.,
c1984.
Related Authors: LaPadula, Tom, ill.
Description: 45 p.: col. ill.; 28 cm.
ISBN: 0307125092 0307625095 (lib.
bdg.)
Summary: A simple introduction to
various types of robots and their
functions.
Notes: Includes index. Bibliography: p.
44.
Subjects: Robots--Juvenile literature.
Robots. Robotics.
LC Classification: TJ211.2 .K46 1984
Dewey Class No.: 629.8/92 19

Key abstracts. Robotics & control.
Published/Created: Hitchin, U.K.:
INSPEC, Institution of Electrical
Engineers; New York, N.Y.: Institute of
Electrical and Electronics Engineers,
1987-
Related Authors: INSPEC (Information
service) Institute of Electrical and
Electronics Engineers.
Description: v.; 30 cm. Jan. 1987-
ISSN: 0950-4842

Notes: Title from cover. Place of U.S. publisher varies: Piscataway, NJ, . SERBIB/SERLOC merged record
Subjects: Automatic control--Abstracts--Periodicals. Robotics--Abstracts--Periodicals. Industrial engineering--Abstracts--Periodicals.
LC Classification: Z5853.A8 K49 TJ213
Dewey Class No.: 629.8 20

Kim, Y. H. (Young Ho)
High-Level feedback control with neural networks / Y.H. Kim, F.L. Lewis.
Published/Created: Singapore; River Edge, NJ: World Scientific, c1998.
Related Authors: Lewis, Frank L.
Description: x, 216 p.: ill.; 23 cm.
ISBN: 9810233760
Notes: Includes bibliographical references and index.
Subjects: Feedback control systems. Neural networks (Computer science)
Series: World Scientific series in robotics and intelligent systems; vol. 21
LC Classification: TJ216 .K47 1998
Dewey Class No.: 629.8/3 21

Klafter, Richard David.
Robotic engineering: an integrated approach / Richard D. Klafter, Thomas A. Chmielewski, Michael Negin.
Published/Created: Englewood Cliffs, N.J.: Prentice Hall, c1989.
Related Authors: Chmielewski, Thomas A. Negin, Michael.
Description: xxiii, 744 p.: ill.; 24 cm.
ISBN: 0134687523
Notes: Includes bibliographies and index.
Subjects: Robotics.
LC Classification: TJ211 .K555 1989
Dewey Class No.: 629.8/92 19

Knight, David C.
Robotics, past, present, & future / by David C. Knight.
Published/Created: New York: W. Morrow, 1983.
Description: 122 p.: ill.; 22 cm.

ISBN: 0688014909 :
Summary: Discusses the history and workings of robots and automata; their many uses in industry, homes, offices, medicine, and space; and their possible future applications.
Notes: Includes index.
Subjects: Robotics--Juvenile literature. Robotics. Robots.
LC Classification: TJ211 .K58 1983
Dewey Class No.: 629.8/92 19

Knight, Timothy Orr.
Probots and people: the age of the personal robot / Timothy O. Knight.
Published/Created: New York: McGraw-Hill, c1984.
Description: xii, 126 p.: ill.; 21 cm.
ISBN: 0070351066 (pbk.)
Notes: Includes index. Bibliography: p. 115.
Subjects: Robotics.
Series: Byte books.
LC Classification: TJ211 .K59 1984
Dewey Class No.: 629.8/92 19

Koivo, Antti J.
Fundamentals for control of robotic manipulators / Antti J. Koivo.
Published/Created: New York: Wiley, c1989.
Description: xi, 468 p.: ill.; 25 cm.
ISBN: 0471857149
Notes: Includes bibliographical references.
Subjects: Robotics. Manipulators (Mechanism)
LC Classification: TJ211 .K65 1989
Dewey Class No.: 629.8/92 19

Korein, James Urey.
A geometric investigation of reach / James Urey Korein.
Published/Created: Cambridge, Mass.: MIT Press, c1985.
Description: 208 p.: ill.; 24 cm.
ISBN: 0262111047
Notes: Includes index. Thesis (Ph. D.)--University of Pennsylvania, 1984.
Bibliography: p. [196]-206.

Subjects: Links and link-motion.
Machinery, Kinematics of. Robotics.
Series: ACM distinguished dissertations
LC Classification: TJ182 .K67 1985
Dewey Class No.: 620.8/2 19

Koren, Yoram.
Robotics for engineers / by Yoram
Koren.
Published/Created: New York:
McGraw-Hill, c1985.
Description: xvii, 347 p.: ill.; 24 cm.
ISBN: 0070353999 :
Notes: Includes bibliographies and
index.
Subjects: Robotics. Robots, Industrial.
LC Classification: TJ211 .K66 1985
Dewey Class No.: 629.8/92 19

Kozlowski, Krzysztof.
Modelling and identification in robotics
/ Krzysztof Kozlowski.
Published/Created: Berlin; New York:
Springer, c1998.
Description: xxvi, 261 p.: ill.; 24 cm.
ISBN: 354076240X (hardcover: alk.
paper)
Notes: Includes bibliographical
references (p. [249]-261) and index.
Subjects: Robotics--Mathematical
models. Robots--Dynamics--
Mathematical models. System
identification.
Series: Advances in industrial control
LC Classification: TJ211 .K69 1998
Dewey Class No.: 629.8/92 21

Krasnoff, Barbara.
Robots, reel to real / Barbara Krasnoff.
Published/Created: New York: Arco
Pub., c1982.
Description: 154 p.: ill.; 25 cm.
ISBN: 0668051396 :
Notes: Includes index. Bibliography: p.
147-150.
Subjects: Robotics. Robots, Industrial.
Series: The Arco how it works series
LC Classification: TJ211 .K7
Dewey Class No.: 629.8/92 19

Laboratory robotics and automation.
Published/Created: New York, N.Y.:
VCH Publishers, c1988-
Description: v.: ill. Vol. 1, no. 1 (Jan.
1989)- Ceased with v. 12, no. 6 (Dec.
2000).
ISSN: 0895-7533
Notes: Title from cover. Publisher:
Wiley, SERBIB/SERLOC merged
record Indx'd selectively by: Chemical
abstracts 0009-2258 1989- Computer &
control abstracts 0036-8113 Jan. 1989-
Electrical & electronics abstracts 0036-
8105 Jan. 1989- Physics abstracts 0036-
8091 Jan. 1989-
Subjects: Laboratories--Automation--
Periodicals. Robotics--Periodicals.
Automation--periodicals. Laboratories-
periodicals. Robotics--periodicals.
LC Classification: WMLC 93/4740
Q183.A1 L3
Dewey Class No.: 629 11

Laird, Delbert.
MAR-13-M: mimetic assisted robotics,
version 13, modular series / [written by
Delbert Laird; illustrations by Jason
Juta, Fleetwood, Michael D.
Kowalczyk].
Published/Created: [S.l.]: Maelstrom
Hobby, c1994.
Description: 32 p.: ill.; 28 cm.
ISBN: 1885755015
Notes: "Marauder 2107 supplement"--
T.p. verso.
Subjects: Marauder 2107.
LC Classification: GV1469.25.M346
L35 1994
Dewey Class No.: 793.93/2 20

Lambert, Mark, 1946-
50 facts about robots / by Mark
Lambert.
Edition Information: A Warwick Press
Library ed.
Published/Created: New York, N.Y.:
Warwick Press, 1983.
Description: 32 p.: col. ill.; 29 cm.
ISBN: 0531092186
Summary: Discusses, in a question and

answer format, the development of various types of robots and their many uses.
Notes: Includes index.
Subjects: Robotics--Juvenile literature. Robotics. Questions and answers.
Series: 50 facts
LC Classification: TJ211 .L26 1983
Dewey Class No.: 629.8/92 19

Lambert, Mark, 1946-
Robotics / Mark Lambert.
Published/Created: Vero Beach, FL: Rourke Enterprises, c1985.
Description: 45 p.: ill. (some col.); 27 cm.
ISBN: 0865929092
Summary: Traces the history of robots, discussing their components, programming, maintenance, applications, and future uses in homes, schools, work places, and space.
Notes: Includes index. Bibliography: p. 42.
Subjects: Robotics--Juvenile literature. Robots--Juvenile literature. Robotics. Robots.
Series: Just look at--
LC Classification: TJ211.2 .L36 1985
Dewey Class No.: 629.8/92 19

Languages for sensor-based control in robotics / edited by Ulrich Rembold, Klaus Hörmann.
Published/Created: Berlin; New York: Springer-Verlag, c1987.
Related Authors: Rembold, Ulrich. Hörmann, Klaus.
Description: ix, 625 p.: ill.; 25 cm.
ISBN: 0387176659 (U.S.)
Notes: Proceedings of the NATO Advanced Research Workshop on Languages for Sensor-Based Control in Robotics held in Il Ciocco, Castelvecchio, Pascoli/Italy, Sept. 1-5, 1986. Includes bibliographies.
Subjects: Robotics--Congresses. Robots--Programming. Programming languages (Electronic computers)--Congresses.

Series: NATO ASI series. Series F, Computer and systems sciences; no. 29.
LC Classification: TJ210.3 .N37 1986
Dewey Class No.: 629.8/92 19

Larsen, Lawrence P., 1944-
Industrial robotics and automation / revised and edited by Lawrence P. Larsen, Karl Wojcikiewicz; experiments by Lawrence P. Larsen.
Published/Created: Benton Harbor, Mich.: Heathkit/Zenith Educational Systems, c1986.
Related Authors: Wojcikiewicz, Karl. Hoekstra, Robert L. Robotics and automated systems.
Description: 2 v.: ill.; 28 cm.
ISBN: 0871191415 (pbk.)
Notes: Adaptation of: Robotics and automated systems / Rpbert L. Hoekstra. c1986. With new experiments and examinations. Includes index.
Subjects: Robots, Industrial. Automation.
LC Classification: TS191.8 .L37 1986
Dewey Class No.: 629.8 19

Latombe, Jean-Claude.
Robot motion planning / Jean-Claude Latombe.
Published/Created: Boston: Kluwer Academic Publishers, c1991.
Description: xviii, 651 p.: ill.; 25 cm.
ISBN: 0792391292
Notes: Includes bibliographical references (p. [613]-642) and index.
Subjects: Robots--Motion.
Series: Kluwer international series in engineering and computer science; SECS 124. Kluwer international series in engineering and computer science. Robotics.
LC Classification: TJ211.4 .L38 1991
Dewey Class No.: 629.8/92 20

Lechner, H. D.
The computer chronicles / H.D. Lechner.
Published/Created: Belmont, Calif.: Wadsworth/Continuing Education,

c1984.
Description: xvi, 391 p.: ill.; 24 cm.
ISBN: 0534033962 (pbk.)
Notes: Includes index.
Subjects: Electronic digital computers.
Electronic data processing. Robotics.
Artificial intelligence.
Series: The Wadsworth continuing
education professional series The
Wadsworth series in continuing
education
LC Classification: QA76.5 .L378 1984b
Dewey Class No.: 001.64 19

Lee, Mark H.
Intelligent robotics / Mark H. Lee.
Published/Created: New York: Halsted
Press; Milton Keynes: Open University
Press, 1989.
Description: xiv, 210 p.: ill.; 25 cm.
ISBN: 0470213930
Notes: Includes bibliographical
references.
Subjects: Robotics. Artificial
intelligence.
Series: Open University Press robotics
series
LC Classification: TJ211 .L44 1989
Dewey Class No.: 629.8/92 19

Lee, Mary Price.
Exploring careers in robotics / by Mary
Price Lee and Richard S. Lee.
Edition Information: 1st ed.
Published/Created: New York: Rosen
Pub. Group, 1984.
Related Authors: Lee, Richard S.
(Richard Sandoval), 1927-
Description: 128 p.: ill.; 22 cm.
ISBN: 0823906205 :
Summary: Introduces careers involving
work with artificial intelligence, robot
technology, and use of robots in
industry, medicine, and space
exploration.
Notes: Includes index. Bibliography: p.
117-118.
Subjects: Robotics--Vocational
guidance. Robotics--Vocational
guidance. Robots. Vocational guidance.

LC Classification: HD9696.R622 L44
1984
Dewey Class No.: 629.8/92 19

Leger, Chris, 1973-
Darwin2K: an evolutionary approach to
automated design for robotics / by Chris
Leger.
Published/Created: Boston, MA:
Kluwer Academic, 2000.
Description: xiii, 271 p.: ill.; 25 cm.
ISBN: 0792379292 (alk. paper)
Notes: Includes bibliographical
references (p. [261]-266) and index.
Subjects: Darwin2K. Robotics.
Computer-aided design.
Series: Kluwer international series in
engineering and computer science;
SECS 574. Kluwer international series
in engineering and computer science.
Robotics.
LC Classification: TJ211 .L45 2000
Dewey Class No.: 629.8/92 21

Leonard, John J.
Directed sonar sensing for mobile robot
navigation / John J. Leonard, Hugh F.
Durrant-Whyte.
Published/Created: Boston: Kluwer
Academic Publishers, c1992.
Related Authors: Durrant-Whyte, Hugh
F., 1961-
Description: xix, 183 p.: ill.; 25 cm.
ISBN: 0792392426 (acid-free paper)
Notes: Revision of Leonard's thesis (D.
Phil.)--University of Oxford, 1990.
Includes bibliographical references (p.
169-179) and index.
Subjects: Robots--Control. Mobile
robots.
Series: Kluwer international series in
engineering and computer science;
SECS 175. Kluwer international series
in engineering and computer science.
Robotics.
LC Classification: TJ211.35 .L46 1992
Dewey Class No.: 629.8/92 20

Lerner, Eric, 1947-
New directions in robots and automated

manufacturing / Eric Lerner.
Published/Created: Stamford, Conn.:
Business Communications Co., 1982.
Description: vii, 132 leaves: ill.; 29 cm.
ISBN: 0893362190 (soft)
Notes: Bibliography: leaves 120-132.
Subjects: Robotics. Automatic control
equipment industry. Market surveys.
Series: Business opportunity report; G-
053
LC Classification: HD9696.R622 L47
1982
Dewey Class No.: 338.4/7629892 19

Lilly, Kathryn W.
Efficient dynamic simulation of robotic
mechanisms / by Kathryn W. Lilly.
Published/Created: Boston: Kluwer
Academic, c1993.
Description: 136 p.: ill.; 25 cm.
ISBN: 0792392868 (acid-free paper)
Notes: Includes bibliographical
references (p. 129-132) and index.
Subjects: Robots--Dynamics--
Mathematical models. Robots--
Computer simulation.
Series: Kluwer international series in
engineering and computer science;
SECS 203. Kluwer international series
in engineering and computer science.
Robotics.
LC Classification: TJ211.4 .L55 1992
Dewey Class No.: 629.8/92 20

Lim, Li Li.
Development of robotics technology in
Singapore, April 1987 [microform].
Published/Created: Singapore: DBS
Bank, Economic Research Dept., [1987]
Description: 1 v. (various foliations):
ill.; 30 cm.
Notes: Microfiche. Jakarta: Library of
Congress Office; Washington, D.C.:
Library of Congress Photoduplication
Service, 1990. 1 microfiche; 11 x 15
cm.
LC Classification: Microfiche 90/51828
(H)

Lindblom, Steven.
How to build a robot / by Steven
Lindblom.
Edition Information: 1st ed.
Published/Created: New York: T.Y.
Crowell, c1985.
Description: 78 p.: ill.; 24 cm.
ISBN: 0690044410: 0690044429 (lib.
bdg.) :
Summary: Discusses the nature and
history of robots and the technological
requirements of making them move,
sense, and "think."
Notes: Includes index.
Subjects: Robotics--Juvenile literature.
Robots--Juvenile literature. Robots.
Robotics.
LC Classification: TJ211.2 .L558 1985
Dewey Class No.: 629.8/92 19

Lindeberg, Tony.
Scale-space theory in computer vision /
Tony Lindeberg.
Published/Created: Boston: Kluwer
Academic, c1994.
Description: xii, 423 p.: ill.; 25 cm.
ISBN: 0792394186 (alk. paper)
Notes: Includes bibliographical
references (p. 399-414) and index.
Subjects: Computer vision.
Series: Kluwer international series in
engineering and computer science;
SECS 256. Kluwer international series
in engineering and computer science.
Robotics.
LC Classification: TA1634 .L56 1994
Dewey Class No.: 006.3/7 20

Liptak, Karen.
Robotics basics / by Karen Liptak;
drawings by Mike Petronella.
Published/Created: Englewood Cliffs,
N.J.: Prentice-Hall, c1984.
Related Authors: Petronella, Michael,
1952- ill.
Description: 48 p.: ill.; 24 cm.
ISBN: 0137820879 :
Summary: Explains what robots are, and
discusses the future role of robots in
industry and the home and the

relationship between robots and humans.
Notes: Includes index.
Subjects: Robotics--Juvenile literature. Robots--Juvenile literature. Robotics. Robots.
LC Classification: TJ211.2 .L56 1984
Dewey Class No.: 629.8/92 19

Litterick, Ian.
Robots and intelligent machines / Ian Litterick; [illustrated and designed by David Anstey].
Published/Created: New York: Bookwright Press, 1984.
Related Authors: Anstey, David, ill.
Description: 47 p.: col. ill.; 26 cm.
ISBN: 0531047733 (lib. bdg.)
Summary: Describes the development of robots, how they work, and their many uses in industry, medicine, transportation, scientific experiments, and the home.
Notes: Includes index. Bibliography: p. 46.
Subjects: Robotics--Juvenile literature. Artificial intelligence--Juvenile literature. Robotics.
Series: The Age of computers
LC Classification: TJ211.2 .L57 1984
Dewey Class No.: 629.8/92 19

Lockman, Darcy, 1972-
Robots / by Darcy Lockman.
Published/Created: New York: Benchmark Books, c2001.
Description: 48 p.: col. ill.; 21 cm.
ISBN: 0761410473
Summary: Provides a brief history of robotics, describes tasks for which robots are useful, and suggests future development.
Notes: Includes bibliographical references and index.
Subjects: Robots--Juvenile literature. Robots.
Series: Kaleidoscope (Tarrytown, N.Y.)
LC Classification: TJ211.2 .L63 2001
Dewey Class No.: 629.8/92 21

Logic, artificial intelligence and robotics: laptec '2001 / [edited by] Jair Minoro Abe.
Published/Created: Burke, VA: IOS Press, 2001.
Description: p.; cm.
ISBN: 1586032062 (ios press)
Series: Frontiers in artificial intelligence and applications

Logsdon, Tom, 1937-
The robot revolution / by Tom Logsdon.
Published/Created: New York: Simon and Schuster, c1984.
Description: 207 p.: ill.; 22 cm.
ISBN: 0671467050 (pbk.): 0671507117
Notes: Includes index. Bibliography: p. 195-196.
Subjects: Robotics. Robots, Industrial.
LC Classification: TJ211 .L63 1984
Dewey Class No.: 629.8/92 19

Lonergan, Tom.
The VOR / Tom Lonergan and Carl Frederick.
Published/Created: Rochelle Park, N.J.: Hayden Book Co., c1983.
Related Authors: Frederick, Carl.
Description: 120 p.: ill.; 23 cm.
ISBN: 0810451867 (pbk.) :
Subjects: Robotics. Artificial intelligence.
LC Classification: TJ211 .L65 1983
Dewey Class No.: 629.8/92 19

Loofbourrow, Tod.
How to build a computer-controlled robot / Tod Loofbourrow.
Published/Created: Rochelle Park, N.J.: Hayden Book Co., c1978.
Description: 132 p.: ill.; 23 cm.
ISBN: 0810456818
Notes: Includes index.
Subjects: Robotics--Amateurs' manuals. Microcomputers--Amateurs' manuals.
LC Classification: TJ211 .L66
Dewey Class No.: 629.8/92

Lowe, David G.
Perceptual organization and visual

recognition / by David G. Lowe.
Published/Created: Boston: Kluwer
Academic Publishers; Hingham, MA,
U.S.A.: Distributors for North America,
Kluwer Academic Publishers, c1985.
Description: xi, 162 p.: ill.; 25 cm.
ISBN: 089838172X
Notes: Includes index. Bibliography: p.
154-162.
Subjects: Computer vision. Visual
perception.
Series: Kluwer international series in
engineering and computer science;
SECS 5. Kluwer international series in
engineering and computer science.
Robotics.
LC Classification: TA1632 .L68 1985
Dewey Class No.: 001.64/4 19

Lukoff, Herman.
From dits to bits: a personal history of
the electronic computer / Herman
Lukoff.
Published/Created: Portland, Or.:
Robotics Press, c1979.
Description: xvi, 219 p.: ill.; 22 cm.
ISBN: 0896610020
Notes: Bibliography: p. [210]-211.
Subjects: Lukoff, Herman. Moore
School of Electrical Engineering--
Biography. Computers--History.
Computer engineers--United States--
Biography. Computer industry--United
States--History.
LC Classification: TK7885.22.L84 A33
Dewey Class No.: 621.3819/58/0924 B

Machine intelligence and autonomy for
aerospace systems / edited by Ewald
Heer, Henry Lum.
Published/Created: Washington, DC:
American Institute of Aeronautics and
Astronautics, c1988.
Related Authors: Heer, Ewald. Lum,
Henry, 1936-
Description: xv, 355 p.: ill.; 24 cm.
ISBN: 0930403487
Notes: Includes bibliographical
references.
Subjects: Space stations--Automation.

Artificial intelligence. Space robotics.
Series: Progress in astronautics and
aeronautics; v. 115
LC Classification: TL507 .P75 vol. 115
TL797
Dewey Class No.: 629.1 s
629.47/028/563 19

Machine intelligence and knowledge
engineering for robotic applications /
edited by Andrew K. C. Wong, Alan
Pugh.
Published/Created: Berlin; New York:
Springer-Verlag, c1987.
Related Authors: Wong, Andrew K. C.
Pugh, A. (Alan) North Atlantic Treaty
Organization. Scientific Affairs
Division.
Description: xiv, 486 p.: ill.; 25 cm.
ISBN: 0387178449 (U.S.)
Notes: "Published in cooperation with
NATO Scientific Affairs Division."
"Proceedings of the NATO Advanced
Research Workshop on Machine
Intelligence and Knowledge
Engineering for Robotic Applications
held at Maratea, Italy, May 12-16,
1986."--T.p. verso. Includes
bibliographies.
Subjects: Robotics--Congresses.
Artificial intelligence--Congresses.
Series: NATO ASI series. Series F,
Computer and systems sciences; no. 33.
LC Classification: TJ210.3 .N375 1986
Dewey Class No.: 629.8/92 19

Machine intelligence and robotics: report of
the NASA Study Group: final report.
Published/Created: [Washington, D.C.?:
National Aeronautics and Space
Administration], 1980.
Related Authors: United States.
National Aeronautics and Space
Administration. NASA Study Group on
Machine Intelligence and Robotics.
Description: xiii, 400 p.: ill.; 28 cm.
Notes: Mar. 1980. Item 830-C Includes
bibliographies.
Subjects: Robotics. Artificial
intelligence.

LC Classification: TJ211 .M32 1980
Dewey Class No.: 629.8/92 19
Govt. Doc. No.: NAS 1.2:M 18

Machine learning: proceedings of the
seventh international conference (1990),
University of Texas, Austin, Texas,
June 21-23, 1990 / editor/workshop
chairs, Bruce Porter and Raymond
Mooney; sponsors, Office of Naval
Research, Artificial Intelligence and
Robotics Program ... [et al.].
Published/Created: San Mateo, Calif.:
Morgan Kaufmann Publishers, c1990.
Related Authors: Porter, Bruce, 1956-
Mooney, Raymond J. (Raymond
Joseph) Artificial Intelligence and
Robotics Program (U.S.)
Description: v, 427 p.: ill.; 28 cm.
ISBN: 1558601414
Notes: "Proceedings of the Seventh
International Conference on Machine
Learning"--Cover. Includes
bibliographical references and index.
Subjects: Machine learning--
Congresses.
LC Classification: Q325.5 .M34 1990
Dewey Class No.: 006.3/1 20

Mair, Gordon M., 1949-
Industrial robotics / Gordon M. Mair.
Published/Created: New York: Prentice
Hall, 1988.
Description: xiv, 354 p.: ill.; 25 cm.
ISBN: 0134632176 0134632095 (pbk.)
Notes: Includes index. Bibliography: p.
344-345.
Subjects: Robots, Industrial.
LC Classification: TS191.8 .M32 1988
Dewey Class No.: 670.42/7 19

Malcolm, Douglas R., 1948-
Robotics: an introduction / Douglas R.
Malcolm, Jr.
Edition Information: 2nd ed.
Published/Created: Boston, Mass.:
PWS-KENT Pub. Co., c1988.
Description: xvi, 395 p.: ill.; 25 cm.
ISBN: 0534914748
Notes: Includes index.

Subjects: Robotics.
Series: The PWS-Kent series in
technology
LC Classification: TJ211 .M337 1988
Dewey Class No.: 629.8/92 19

Malcolm, Douglas R., 1948-
Robotics: an introduction / Douglas R.
Malcolm, Jr.
Published/Created: Boston, Mass.:
Breton Publishers, c1985.
Description: xiv, 368 p.: ill.; 25 cm.
ISBN: 0534047521
Notes: Includes index.
Subjects: Robotics.
LC Classification: TJ211 .M337 1985
Dewey Class No.: 629.8/92 19

Manko, David J., 1957-
A general model of legged locomotion
on natural terrain / by David J. Manko;
foreword by William L. Whittaker.
Published/Created: Boston: Kluwer
Academic Publishers, c1992.
Description: xii, 116 p.: ill.; 24 cm.
ISBN: 0792392477 (alk. paper)
Notes: Includes bibliographical
references (p. [109]-113) and index.
Subjects: Robotics. Artificial legs.
Mobile robots.
Series: The Kluwer international series
in engineering and computer science.
Robotics
LC Classification: TJ211 .M355 1992
Dewey Class No.: 629.8/92 20

Man-machine interface in the nuclear
industry: proceedings of an International
Conference on Man-Machine Interface
in the Nuclear Industry (Control and
Instrumentation, Robotics, and Artificial
Intelligence) / organized by the
International Atomic Energy Agency in
co-operation with the Commission of
the European Communities and the
Nuclear Energy Agency of the OECD
and held in Tokyo, 15-19 February
1988.
Published/Created: Vienna:
International Atomic Energy Agency,

1988.
Related Authors: International Atomic
Energy Agency. Commission of the
European Communities. OECD Nuclear
Energy Agency.
Description: 825 p.: ill.; 24 cm.
ISBN: 9200205887
Notes: English, French, and Russian.
Includes bibliographical references and
indexes.
Subjects: Nuclear power plants--Control
rooms--Congresses. Human-machine
systems--Congresses. Nuclear power
plants--Human factors--Congresses.
Human engineering--Congresses.
Series: Proceedings series (International
Atomic Energy Agency)
LC Classification: TK9220 .I58 1988
Dewey Class No.: 621.48/3 20

Mapping the course between automation
technologies: proceedings of the
DIGICOM '87 Symposium, October 21-
22, 1987, Montreal, Quebec, Canada /
DIGICOM '87 Symposium;
programmed by ISA divisions, Robotics
and Expert Systems ... [et al.].
Published/Created: Research Triangle
Park, NC: Instrument Society of
America, c1988.
Related Authors: Instrument Society of
America. Robotics & Expert Systems
Division.
Description: 208 p.: ill.; 28 cm.
ISBN: 155617103X
Notes: Includes bibliographies and
index.
Subjects: Automation--Congresses.
LC Classification: T59.5 .D52 1987
Dewey Class No.: 670.42/7 19

Marker, A.
Industrial robots in Australia: an
introduction / A. Marker.
Published/Created: Canberra:
Legislative Research Service, Dept. of
Parliamentary Library, 1985.
Related Authors: Australia. Parliament.
Description: 10 p.; 30 cm.
Notes: At head of The Parliament of the

Commonwealth of Australia.
Bibliography: p. 10.
Subjects: Robots, Industrial. Robotics--
Australia.
Series: Current issues brief; 1985/3.
LC Classification: TS191.8 .M33 1985
Dewey Class No.: 670.42/7 19

Marrs, Texe W.
Careers with robots / Texe W. Marrs.
Published/Created: New York, N.Y.:
Facts on File, c1988.
Description: viii, 213 p.: ill.; 24 cm.
ISBN: 0816012229 :
Notes: Includes index. Bibliography: p.
206-207.
Subjects: Robotics--Vocational
guidance.
LC Classification: TJ211.25 .M37 1988
Dewey Class No.: 629.8/92 19

Marsh, Peter, 1952-
Robots / Peter Marsh.
Edition Information: Warwick Press
library ed.
Published/Created: New York, N.Y.:
Warwick Press, 1983.
Description: 37 p.: ill. (some col.); 28
cm.
ISBN: 0531092232
Summary: Explains the nature of robots,
describes the kinds of tasks they can
perform, and predicts their future uses.
Notes: Includes index.
Subjects: Robotics--Juvenile literature.
Robots--Juvenile literature. Robotics.
Robots.
Series: Science in action
LC Classification: TJ211.2 .M37 1983
Dewey Class No.: 629.8/92 19

Marshall, A. D. (A. Dave)
Computer vision, models, and
inspection / A.D. Marshall and R.R.
Martin.
Published/Created: Singapore; River
Edge, NJ: World Scientific, 1992.
Related Authors: Martin, R. R. (Ralph
R.)
Description: xviii, 437 p.: ill.; 23 cm.

ISBN: 9810207727
Notes: Includes bibliographical references (p. 399-422) and index.
Subjects: Quality control--Optical methods--Automation. Computer vision-- Industrial applications. Engineering inspection--Automation. Three-dimensional display systems.
Series: World Scientific series in robotics and automated systems; vol. 4
LC Classification: TS156.2 .M37 1992
Dewey Class No.: 670.42/5 20

Martin, Fred G.
Robotic explorations: a hands-on introduction to engineering / Fred G. Martin.
Published/Created: Upper Saddle River, N.J.: Prentice Hall, c2001.
Description: x, 462 p.: ill. 27 cm.
ISBN: 0130895687
Notes: Includes bibliographical references (p. 449) and index.
Subjects: Robotics. Engineering design.
LC Classification: TJ211 .M36645 2001
Dewey Class No.: 629.8/92 21

Mason, Matthew T.
Mechanics of robotic manipulation / Matthew T. Mason.
Published/Created: Cambridge, Mass.: MIT Press, c2001.
Description: xi, 253 p.: ill.; 24 cm.
ISBN: 0262133962 (hc.: alk. paper)
Notes: "A Bradford book." Includes bibliographical references (p. [241]-245) and index.
Subjects: Manipulators (Mechanism) Robotics.
Series: Intelligent robots and autonomous agents
LC Classification: TJ211 .M345 2001
Dewey Class No.: 629.8/92 21

Masterson, James W.
Robotics / James W. Masterson, Elmer C. Poe, Stephen W. Fardo.
Published/Created: Reston, Va.: Reston Pub. Co., c1985.
Related Authors: Poe, Elmer. Fardo,

Stephen W.
Description: viii, 263 p.: ill.; 25 cm.
ISBN: 0835966925 :
Notes: Includes index.
Subjects: Robotics.
LC Classification: TJ211 .M367 1985
Dewey Class No.: 629.8/92 19

Masterson, James W.
Robotics technology / by James W. Masterson, Robert L. Towers, Stephen W. Fardo.
Published/Created: Tinley Park, IL: Goodheart-Willcox Co., 2001.
Related Authors: Towers, Robert L. Fardo, Stephen W.
Description: p. cm.
ISBN: 1566378656
Subjects: Robotics.
LC Classification: TJ211 .M3672 2001
Dewey Class No.: 629.8/92 21

Masterson, James W.
Robotics technology / by James W. Masterson, Robert L. Towers, Stephen W. Fardo.
Published/Created: South Holland, Ill.: Goodheart-Willcox Co., c1996.
Related Authors: Towers, Robert L. Fardo, Stephen W.
Description: 320 p.: ill. (some col.); 26 cm.
Notes: Includes index.
Subjects: Robotics.
LC Classification: TJ211 .M3672 1996
Dewey Class No.: 670.42/72 20

Maus, Rex.
Robotics: a manager's guide / Rex Maus, Randall Allsup.
Published/Created: New York: Wiley, c1986.
Related Authors: Allsup, Randall.
Description: x, 238 p.: ill.; 24 cm.
ISBN: 0471842648: 0471842656 (pbk.) :
Notes: Includes index. Bibliography: p. 209.
Subjects: Robotics.
LC Classification: TJ211 .M38 1986

Dewey Class No.: 629.8/92 19

McCloy, D.
Robotics: an introduction / D. McCloy
and D.M.J. Harris.
Published/Created: New York: Halsted
Press, 1986.
Related Authors: Harris, D. M. J. (D.
Michael J.)
Description: xiv, 304 p.: ill.; 25 cm.
ISBN: 0470203250
Notes: "A Halsted Press book." Includes
index. Bibliography: p. [291]-300.
Subjects: Robotics.
Series: Open University Press robotics
series
LC Classification: TJ211 .M39 1986b
Dewey Class No.: 629.8/92 19

McComb, Gordon.
The robot builder's bonanza / Gordon
McComb.
Edition Information: 2nd ed.
Published/Created: New York:
McGraw-Hill, 2001.
Description: xiii, 753 p.: ill.; 24 cm.
ISBN: 0071362967
Notes: Includes bibliographical
references (p. 709-717) and index.
Subjects: Robotics.
LC Classification: TJ211 .M418 2000
Dewey Class No.: 629.8/92 21

McComb, Gordon.
The robot builder's bonanza: 99
inexpensive robotics projects / Gordon
McComb.
Edition Information: 1st ed.
Published/Created: Blue Ridge Summit,
PA: Tab Books, c1987.
Description: ix, 326 p.: ill.; 25 cm.
ISBN: 0830608001: 0830628002 (pbk.)
:
Notes: Includes index. Bibliography: p.
313-315.
Subjects: Robotics.
LC Classification: TJ211 .M418 1987
Dewey Class No.: 629.8/92 19

McCullough, L. E.
Plays of people at work: grades K-3 /
L.E. McCullough.
Published/Created: Lyme, NH: Smith
and Kraus, 1998.
Description: p. cm.
ISBN: 157525140X
Summary: A collection of twelve plays
in which Janice, Steve, Paula, and
Burton investigate a variety of
occupations, including newspaper
reporter, robotics engineer, and crime
lab technician.
Subjects: Labor--Juvenile drama.
Children's plays, American.
Occupations--Drama. Plays.
Series: Young actor series
LC Classification: PS3563.C35297
P586 1998
Dewey Class No.: 812/.54 21

McFadden, Thomas, 1946-
A selected bibliography on robotics /
compiled by T.G. McFadden; abstracted
by C. David Davis; edited by Patricia
Knittel.
Published/Created: Rochester, N.Y.:
Technical and Education Center of the
Graphic Arts, College of Graphic Arts
and Photography, Rochester Institute of
Technology, [1983]-
Related Authors: Davis, C. David.
Knittel, Patricia, 1952- Rochester
Institute of Technology. Technical and
Education Center of the Graphic Arts.
Description: v. <1,; 28 cm.
ISBN: 0899380123 (pbk.: v. 1) :
Notes: Includes index.
Subjects: Robotics--Bibliography.
Robots--Bibliography.
LC Classification: Z5167 .M38 1983
TJ211
Dewey Class No.: 016.6298/92 19

McInroy, John E.
Reliable plan selection by intelligent
machines / John E. McInroy, Joseph C.
Musto, George N. Saridis.
Published/Created: Singapore; River
Edge, NJ: World Scientific,c 1995.

Related Authors: Musto, Joseph C.
Saridis, George N., 1931-
Description: ix, 154 p.: ill.; 23 cm.
ISBN: 9810223366
Notes: Includes bibliographical
references (p. 147-152) and index.
Subjects: Intelligent control systems.
Robotics. Reliability (Engineering)
Series: Series in intelligent control and
intelligent automation; vol. 1
LC Classification: TJ217.5 .M34 1995
Dewey Class No.: 629.8/92 20

McKerrow, Phillip, 1949-
Introduction to robotics / Phillip John
McKerrow.
Published/Created: Sydney; Reading,
Mass.: Addison-Wesley Pub. Co.,
c1991.
Description: xvi, 811 p.: ill.; 24 cm.
ISBN: 0201182408
Notes: Includes bibliographical
references (p. 775-794) and index.
Subjects: Robotics.
Series: Electronic systems engineering
series
LC Classification: TJ211 .M4184 1991
Dewey Class No.: 629.8/92 20

Mechanical engineering handbook
[electronic resource].
Published/Created: Boca Raton, FL:
CRC Press, c1999-
Related Authors: Kreith, Frank.
Description: computer optical discs; 4
3/4 in. 1999-
ISSN: 1097-9417
Computer File Info.: Text in PDF file
format; Adobe Acrobat Reader included
on disc.
Summary: Reference tool for
mechanical engineers. Covers
traditional areas of engineering as well
as modern manufacturing and design,
robotics, computer engineering,
environmental engineering, economics
and project management, patent law,
bioengineering, and communication and
information systems. Includes chapter
and appendix on physical properties as

well as mathematical and computational
methods. Features proximity search,
zoom, and hyperlinking functions.
Notes: Editor: 1999- Frank Kreith. Title
from disc label. System requirements:
IBM PC or compatible; 486 or higher;
Windows 3.1 or higher; 8MB RAM;
CD-ROM drive.
Subjects: Mechanical engineering--
Handbooks, manuals, etc.
LC Classification: TJ151
Dewey Class No.: 621 13

Mechatronics & robotics, I / P.A
MacConaill, P. Drews, K.-H. Robrock,
editors.
Published/Created: Amsterdam;
Washington: IOS Press, 1991.
Related Authors: MacConaill, P. Drews,
P. Robrock, Karl-Heinz, 1943-
Description: ix, 346 p.: ill.; 25 cm.
ISBN: 905199057X
Notes: Includes bibliographical
references and index.
Subjects: Mechatronics. Robotics.
Series: Advances in design and
manufacturing
LC Classification: TJ163.12 .M43 1991
Dewey Class No.: 670.42/7 20

Mechatronics and machine vision / edited by
John Billingsley.
Published/Created: Baldock,
Hertfordshire, England; Philadephia,
PA: Research Studies Press, c2000.
Related Authors: Billingsley, J. (John)
Description: xiii, 370 p.: ill.; cm.
ISBN: 0863802613
Notes: Includes bibliographical
references and indexes.
Subjects: Mechatronics. Computer
vision.
Series: Robotics and mechatronics
series; 3
LC Classification: TJ163.12. M433
2000
Dewey Class No.: 621 21

Mechatronics and robotics II / P. Drews,
editor.

Published/Created: Burke, VA: IOS
Press, 1994.
Description: p. cm.
Series: Advance in design and
manufacturing; v. 3
LC Classification: 9404 BOOK NOT
YET IN LC

Mechatronics: designing intelligent
machines: international conference, 12-
13 September 1990, Robinson College,
University of Cambridge / sponsored by
Solid Mechanics/Machines Systems
Group of the Institution of Mechanical
Engineers in association with Institution
of Electrical Engineers.
Published/Created: Bury St. Edmunds,
Suffolk: Published for the Institution of
Mechanical Engineers by Mechanical
Engineering Publications, 1990.
Related Authors: Institution of
Mechanical Engineers (Great Britain).
Solid Mechanics/Machines Systems
Group. Institution of Electrical
Engineers.
Description: 280 p.: ill.; 30 cm.
ISBN: 0852987226
Notes: Includes bibliographical
references.
Subjects: Mechatronics--Congresses.
Robotics--Congresses.
Series: Proceedings of the Institution of
Mechanical Engineers (Unnumbered)
LC Classification: TJ230 .M444 1990
Dewey Class No.: 621.8/15 20

Medicine meets virtual reality: art, science,
technology: healthcare (r)evolution /
edited by James D. Westwood ... [et al.].
Published/Created: Amsterdam;
Washington, D.C.: IOS Press, c1998.
Related Authors: Westwood, James D.
Description: xv, 409 p.: ill.; 25 cm.
ISBN: 9051993862 (IOS Press)
4274902102 (Ohmsha)
Notes: "Proceedings of Medicine Meets
Virtual Reality 6, San Diego, California,
January 28-31, 1998." Includes
bibliographical references and index.
Subjects: Computer vision in medicine--

Congresses. Virtual reality in medicine-
-Congresses. Robotics in medicine--
Congresses. Surgery--Data processing--
Congresses.
Series: Studies in health technology and
informatics; v. 50
LC Classification: R859.7.C67 M43
1998
Dewey Class No.: 610/.28 21

Medicine meets virtual reality: health care in
the information age / edited by Suzanne
J. Weghorst, Hans B. Sieburg, Karen S.
Morgan.
Published/Created: Amsterdam;
Washington, DC: IOS Press, 1996.
Related Authors: Weghorst, Suzanne J.
Sieburg, Hans B. Morgan, Karen S.
Description: xvi, 734 p.: ill.; 25 cm.
ISBN: 9051992505
Notes: "Proceedings of Medicine Meets
Virtual Reality 4, San Diego, California,
January 17-20, 1996." Includes
bibliographical references and index.
Subjects: Virtual reality in medicine--
Congresses. Computer vision in
medicine--Congresses. Robotics in
medicine--Congresses. Surgery--Data
processing--Congresses.
Series: Studies in health technology and
informatics; v. 29
LC Classification: R859.7.C67 M43
1996
Dewey Class No.: 610/.285/6 21

Megahed, Saïd M.
Principles of robot modelling and
simulation / Saïd M. Megahed.
Published/Created: Chichester; New
York: J. Wiley, c1993.
Description: xviii, 312 p.: ill.; 26 cm.
ISBN: 0471933481
Notes: Includes bibliographical
references and index.
Subjects: Robotics--Mathematical
models. Robotics--Computer
simulation.
LC Classification: TJ211.47 .M44 1993
Dewey Class No.: 629.8/92/015118 20

MEMS '99: Twelfth IEEE International Conference on Micro Electro Mechanical Systems: technical digest: Orlando, Florida, USA, January 17-21, 1999 / sponsored by the IEEE Robotics and Automation Society.
Published/Created: Piscataway, NJ: IEEE, c1999.
Related Authors: IEEE Robotics and Automation Society.
Description: xxxvi, 660 p.: ill.; 28 cm.
ISBN: 0780351940 (softbound edition) 0780351959 (casebound edition) 0780351967 (microfiche edition)
Notes: "IEEE Catalog Number 99CH36291"--verso of T.p. Includes index and bibliographic references.
Subjects: Microelectromechanical systems--Congresses. Detectors--Congresses. Actuators--Congresses.
LC Classification: TK7875 .I35 1999
Dewey Class No.: 621.381 21

Menzel, Peter, 1948-
Robo sapiens: evolution of a new species / Peter Menzel and Faith D'Aluisio.
Published/Created: Cambridge, Mass.: MIT Press, 2000.
Related Authors: D'Aluisio, Faith, 1957-
Description: 239 p.: ill.; 28 cm.
ISBN: 0262133822 (hc.: alk. paper)
Notes: "A Material world book." Includes bibliographical references and index.
Subjects: Robotics. Artificial intelligence. Intelligent control systems.
LC Classification: TJ211 . M45 2000
Dewey Class No.: 629.8/92 21

Metos, Thomas H.
Robots A2Z / by Thomas H. Metos.
Published/Created: New York: J. Messner, c1980.
Description: 80 p.: ill.; 22 cm.
ISBN: 0671340271 (lib. bdg.) :
Summary: Discusses robots, automatons, and other mechanical devices, the earliest of which was developed as part of an Egyptian water clock around 2000 B.C.
Notes: Includes index.
Subjects: Robots--Juvenile literature. Robots, Industrial--Juvenile literature. Androids--Juvenile literature. Robotics. Robots.
LC Classification: TJ211 .M47 1980
Dewey Class No.: 629.8/92 19

MFI '96, 1996 IEEE/SICE/RSJ International Conference on Multisensor Fusion and Integration for Intelligent Systems, December 8-11, 1996, Washington, D.C., U.S.A.
Published/Created: [New York]: Institute of Electrical and Electronics Engineers, c1996.
Related Authors: Institute of Electrical and Electronics Engineers. Keisoku Jid⁻o Seigyo Gakkai (Japan) Robotics Society of Japan.
Description: xv, 848 p.: ill.; 28 cm.
ISBN: 078033700X (softbound) 0780337018 (microfiche)
Notes: "96TH8242." Includes bibliographical references and index.
Subjects: Intelligent control systems--Congresses. Multisensor data fusion--Congresses.
LC Classification: TJ217.5 .I342 1996
Dewey Class No.: 629.8/9 21

MHS'99: proceedings of 1999 International Symposium on Micromechatronics and Human Science: Nagoya Congress Center & Nagoya Municicpal Industrial Research Institute, November 23-26, 1999 / cosponsored by IEEE Industrial Electronics Society, IEEE Robotics and Automation Society, Nagoya University; technically sponsored by Japan Society of Mechanical Engineers ... [et al.].
Published/Created: Piscataway, NJ, USA: Obtained from IEEE Operations Center, c1999.
Related Authors: IEEE Industrial Electronics Society. IEEE Robotics and Automation Society. Nagoya Daigaku.

Description: vi, 287 p.: ill.; 28 cm.
ISBN: 0780357906
Notes: Includes bibliographical
references and index.
Subjects: Mechatronics--Congresses.
Micromechanics--Congresses.
Microelectromechanical systems--
Congresses.
LC Classification: TJ163.12 .I574 1999
Dewey Class No.: 621.381 21

Micro electro mechanical systems: an
investigation of micro structures,
sensors, actuators, machines and robots:
proceedings / IEEE, Salt Lake City,
Utah, February 20-22, 1989; sponsored
by the IEEE Robotics and Automation
Council.
Published/Created: Piscataway, NJ:
Institute of Electrical and Electronics
Engineers, 1989.
Related Authors: IEEE Robotics and
Automation Council. IEEE Worskshop
on Micro Electro Mechanical Systems
(2nd: 1989: Salt Lake City, Utah)
Description: xii, 156 p.: ill.; 28 cm.
Notes: Papers presented at the second
IEEE Workshop on Micro Electro
Mechanical Systems. "IEEE catalog
number 89TH0249-3." Includes
bibliographies.
Subjects: Electromechanical devices--
Congresses. Miniature objects--
Congresses. Microelectronics--
Congresses. Robotics--Congresses.
LC Classification: TK153 .M468 1989
Dewey Class No.: 621.3 20

Micro electro mechanical systems: an
investigation of micro structures,
sensors, actuators, machines, and
robots: proceedings / IEEE, Napa
Valley, California, 11-14, February
1990; sponsored by the IEEE Robotics
and Automation Society, and in
cooperation with the ASME Dynamic
Systems and Control Division.
Published/Created: New York, NY:
IEEE, c1990.
Related Authors: IEEE Robotics and

Automation Society. American Society
of Mechanical Engineers. Dynamic
Systems and Control Division. IEEE
Workshop on Micro Electro Mechanical
Systems (3rd: 1990: Napa Valley,
Calif.)
Description: xiii, 226 p.: ill.; 28 cm.
Notes: Papers presented at the third
IEEE Workshop on Micro Electro
Mechanical Systems. "IEEE catalog
number 90CH2832-4." Title on spine:
IEEE micro electro mechanical systems
workshop. Includes bibliographical
references and indexes.
Subjects: Electromechanical devices--
Congresses. Miniature objects--
Congresses. Microelectronics--
Congresses. Robotics--Congresses.
LC Classification: TK153 .M468 1990
Dewey Class No.: 621.3 20

Micro electro mechanical systems: an
investigation of micro structures,
sensors, actuators, machines, and
systems: proceedings / IEEE, Fort
Lauderdale, Florida, February 7-10,
1993; sponsored by the IEEE Robotics
and Automation Society in cooperation
with the ASME Dynamic Systems and
Control Division.
Published/Created: [New York]:
Institute of Electrical and Electronics
Engineers; Piscataway, NJ: IEEE
Service Center, c1993.
Related Authors: IEEE Robotics and
Automation Society. American Society
of Mechanical Engineers. Dynamic
Systems and Control Division. IEEE
Workshop on Micro Electro Mechanical
Systems (1993: Fort Lauderdale, Fla.)
Description: xvii, 294 p.: ill.; 28 cm.
ISBN: 078030957X (softbound)
0780309588 (casebound) 0780309596
(microfiche)
Notes: "1993 IEEE Workshop on Micro
Electro Mechanical Systems"--P. iii.
"IEEE catalog number 93CH3265-6."
Title on spine: IEEE Micro Electro
Mechanical Systems Workshop.
Includes bibliographical references and

index.
Subjects: Electromechanical devices--
Congresses. Microelectronics--
Congresses. Detectors--Congresses.
Actuators--Congresses.
LC Classification: TK153 .M4683 1993
Dewey Class No.: 620/.4 20

Micro-electronics, robotics, and jobs.
Published/Created: Paris: Organisation
for Economic Co-operation and
Development; [Washington, D.C.:
OECD Publications and Information
Center, distributor], 1982.
Related Authors: Organisation for
Economic Co-operation and
Development. Working Party on
Information, Computer, and
Communications Policy.
Description: 265 p.: ill.; 24 cm.
ISBN: 9264123849 (pbk.)
Notes: Selected papers presented at the
second special session of OECD's
Committee for Scientific and
Technological Policy, Working Party on
Information, Computer, and
Communications Policy, held at its
headquarters in Oct. 1981. Includes
bibliographies.
Subjects: Microelectronics--Social
aspects--Congresses. Labor supply--
Effect of technological innovations on
Congresses. Robots, Industrial--
Congresses.
Series: Information, computer,
communications policy; 7.
LC Classification: HD6331 .M48 1982
Dewey Class No.: 331.12 19

Microrobotics and microassembly III: 29-30
October, 2001, Newton,
[Massachusetts] USA / Bradley J.
Nelson, Jean-Marc Breguet,
chairs/editors; sponsored and published
by SPIE--the International Society for
Optical Engineering.
Published/Created: Bellingham,
Washington: SPIE, c2001.
Related Authors: Nelson, Bradley J.
Breguet, Jean-Marc. Society of Photo-

optical Instrumentation Engineers.
Description: ix, 326 p.: ill.; 28 cm.
ISBN: 0819442968
Notes: Includes bibliographic references
and author index.
Subjects: Robotics--Congresses.
Microelectromechanical systems--
Congresses. Microfabrication--
Congresses.
Series: Proceedings of SPIE--the
International Society for Optical
Engineering; v. 4568.

Microrobotics and microassembly: 21-22
September, 1999, Boston,
Massachusetts / Bradley J. Nelson,
Jean-Marc Breguet, chairs/editors;
sponsored and published by SPIE--the
International Society for Optical
Engineering.
Published/Created: Bellingham,
Washington: SPIE, c1999.
Related Authors: Nelson, Bradley J.
Breguet, Jean-Marc. Society of Photo-
optical Instrumentation Engineers.
Description: vii, 222 p.: ill.; 28 cm.
ISBN: 0819434272 0819434272
Notes: Includes bibliographic references
and author index.
Subjects: Robotics--Congresses.
Microelectromechanical systems--
Congresses. Microfabrication--
Congresses.
Series: Proceedings of SPIE--the
International Society for Optical
Engineering; v. 3834.
LC Classification: TJ210.3 .M515 1999
Dewey Class No.: 629.8/92 21

Microrobotics and micromanipulation: 4-5
November 1998, Boston, Massachusetts
/ Armin Sulzmann, Bradley J. Nelson,
chairs/editors; sponsored by SPIE--the
International Society for Optical
Engineering; endorsed by SME--the
Society of Manufacturing Engineers.
Published/Created: Bellingham, Wash.,
USA: SPIE, c1998.
Related Authors: Sulzmann, Armin.
Nelson, Bradley J. Society of Photo-

optical Instrumentation Engineers.
Society of Manufacturing Engineers.
Description: x, 224 p.: ill.; 28 cm.
ISBN: 0819429805
Notes: Includes bibliographical
references and index.
Subjects: Robotics--Congresses.
Manipulators (Mechanism)--
Congresses. Microelectromechanical
systems--Congresses.
Series: Proceedings of SPIE--the
International Society for Optical
Engineering; v. 3519.
LC Classification: TJ210.3 .M52 1998
Dewey Class No.: 629.8/92 21

Microrobotics and micromechanical
systems: 25 October 1995, Philadelphia,
Pennsylvania / Lynne E. Parker,
chair/editor; sponsored and published by
SPIE--the International Society for
Optical Engineering.
Published/Created: Bellingham, Wash.,
USA: SPIE, c1995.
Related Authors: Parker, Lynne E.
Society of Photo-optical
Instrumentation Engineers.
Microrobotics and Micromechanical
Systems Conference (1995:
Philadelphia, Pa.)
Description: vii, 132 p.: ill.; 28 cm.
ISBN: 0819419575
Notes: Proceedings of the Microrobotics
and Micromechanical Systems
Conference. Includes bibliographical
references and index.
Subjects: Robotics--Congresses.
Microelectromechanical systems--
Congresses.
Series: Proceedings of SPIE--the
International Society for Optical
Engineering; v. 2593.
LC Classification: TJ210.3 .M53 1995
Dewey Class No.: 629.8/92 21

Microrobotics and microsystem fabrication:
16-17 October 1997, Pittsburgh,
Pennsylvania / Armin Sulzmann,
chair/editor; sponsored ... by SPIE--the
International Society for Optical

Engineering; cooperating organizations,
NIST--National Institute of Standards
and Technology, CIMS--Coalition for
Intelligent Manufacturing Systems,
[and] A-CIMS--Academic Coalition for
Intelligent Manufacturing Systems.
Published/Created: Bellingham, Wash.,
USA: SPIE, c1998.
Related Authors: Sulzmann, Armin.
Society of Photo-optical
Instrumentation Engineers. National
Institute of Standards and Technology
(U.S.) Coalition for Intelligent
Manufacturing Systems. Academic
Coalition for Intelligent Manufacturing
Systems.
Description: x, 238 p.: ill.; 28 cm.
Notes: Includes bibliographical
references and index.
Subjects: Robotics--Congresses.
Microelectromechanical systems--
Congresses. Microfabrication--
Congresses.
Series: Proceedings of SPIE--the
International Society for Optical
Engineering; v. 3202.
LC Classification: TJ210.3 .M534 1998
Dewey Class No.: 629.8/92 21

Microrobotics: components and
applications: 21-22 November, 1996,
Boston, Massachusetts / Armin
Sulzmann, chair/editor; sponsored and
published by SPIE--the International
Society for Optical Engineering.
Published/Created: Bellingham, Wash.:
SPIE, c1996.
Related Authors: Sulzmann, Armin.
Society of Photo-optical
Instrumentation Engineers.
Description: viii, 212 p.: ill.; 28 cm.
ISBN: 0819423084
Notes: Includes bibliographical
references and index.
Subjects: Robotics--Congresses.
Microelectromechanical systems--
Congresses.
Series: Proceedings of SPIE--the
International Society for Optical
Engineering; v. 2906.

LC Classification: TJ210.3 .M54 1996
Dewey Class No.: 629.8/92 21

Miller, Rex, 1929-
　Fundamentals of industrial robots and
　robotics / Rex Miller.
　Published/Created: Boston, Mass.:
　PWS-KENT Pub. Co., c1988.
　Description: xv, 288 p.: ill.; 24 cm.
　ISBN: 0534914705
　Notes: Includes index. Bibliography: p.
　283-284.
　Subjects: Robotics.
　Series: PWS-Kent series in technology
　LC Classification: TJ211 .M535 1988
　Dewey Class No.: 629.8/92 19

Miller, Richard Kendall, 1946-
　Machine vision for robotics and
　automated inspection / by Richard K.
　Miller.
　Published/Created: Fort Lee, N.J., USA:
　Technical Insights, [c1983]
　Description: 3 v.: ill.; 28 cm.
　ISBN: 0896710467 (pbk.: set)
　Contents: v. 1. Fundamentals -- v. 2.
　Applications -- v. 3.
　Manufacturers/systems.
　Notes: "20 January 1983." Includes
　bibliographical references.
　Subjects: Robots, Industrial. Image
　processing.
　LC Classification: TS191.8 .M5527
　1983
　Dewey Class No.: 629.8/92 19

Miller, Richard Kendall, 1946-
　Manufacturing simulation: a new tool
　for robotics, FMS, and industrial
　process design / Richard K. Miller.
　Published/Created: Lilburn, GA:
　Fairmont Press: Englewood Cliffs, NJ:
　Distributed by Prentice-Hall, c1990.
　Description: 1 v. (various pagings): ill.;
　28 cm.
　ISBN: 0881731048 :
　Notes: Includes bibliographical
　references.
　Subjects: Manufacturing processes--
　Computer simulation.

LC Classification: TS183 .M56 1990
Dewey Class No.: 670.42/01/13 20

Miller, Richard Kendall, 1946-
　Manufacturing simulation: a new tool
　for robotics, FMS, and industrial
　process design / written by Richard K.
　Miller.
　Published/Created: Madison, GA: SEAI
　Technical Publications; Fort Lee, NJ:
　Technical Insights, [c1985]
　Description: 148 leaves: ill.; 29 cm.
　ISBN: 0896710661 (pbk.) 0914993267
　(Technical Insights: pbk.)
　Notes: Bibliography: leaves 143-148.
　Subjects: Manufacturing processes--
　Mathematical models. Manufacturing
　processes--Data processing.
　LC Classification: TS183 .M56 1985
　Dewey Class No.: 670.42 19

Miller, Richard Kendall, 1946-
　Survey on clean room robotics / by
　Richard K. Miller and Terri C. Walker.
　Published/Created: Madison, GA:
　Future Technology Surveys, [c1988]
　Related Authors: Walker, Terri C.,
　1964-
　Description: 35 leaves; 29 cm.
　ISBN: 1558650458
　Notes: List of manufacturers of
　commercial clean room robots: leaves
　13-23. Includes bibliographical
　references (leaves 18-19).
　Subjects: Robot industry--United States.
　Market surveys--United States. Clean
　rooms.
　Series: Survey report (Future
　Technology Surveys, Inc.); #46.
　LC Classification: HD9696.R623
　U6355 1988

Miller, Steven M.
　Impacts of industrial robotics: potential
　effects on labor and costs within the
　metalworking industries / Steven M.
　Miller.
　Published/Created: Madison, Wis.:
　University of Wisconsin Press, c1989.
　Description: xxv, 255 p.: ill.; 24 cm.

ISBN: 0299105008: 0299105040 (pbk.)
:
Notes: Includes index. Bibliography: p.
243-249.
Subjects: Metalworking industries--
Automation--Economic aspects. Metal-
workers--Effect of technological
innovations on. Robots, Industrial.
Series: The Economics of technological
change
LC Classification: HD9506.A2 M527
1989
Dewey Class No.: 338.4/5671 19

Milton, Joyce.
Here come the robots / by Joyce Milton;
illustrated with photographs, prints, and
cartoons by Peter Stern.
Published/Created: New York: Hastings
House, c1981.
Related Authors: Stern, Peter, ill.
Description: 118 p.: ill.; 24 cm.
ISBN: 0803863632
Summary: Discusses the history of
robots, real robots that have existed, and
some fictional ones.
Notes: Includes index. Bibliography: p.
113-114.
Subjects: Robots--Juvenile literature.
Robots, Industrial--Juvenile literature.
Robots. Robotics.
LC Classification: TJ211 .M54
Dewey Class No.: 629.8/92 19

Miyazaki, Masahiro.
Gunji robotto sens̄o: nerawareru Nihon
no saisentan gijutsu / Miyazaki
Masahiro.
Published/Created: T̄okȳo:
Daiyamondosha, Sh̄owa 57 [1982]
Description: v, 236 p.: ill.; 19 cm.
Subjects: Electronics in military
engineering. Robotics--Japan.
LC Classification: UG485 .M56 1982

Mobile robotics in healthcare / edited by
Nikos Katevas.
Published/Created: Amsterdam;
Washington, DC: IOS Press, c2001.
Related Authors: Katevas, Nikos.

Description: ix, 380 p.: ill.; 25 cm.
ISBN: 1586030795 (IOS Press)
427490413X (Ohmsha)
Notes: Includes bibliographical
references and index.
Series: Assistive technology research
series, 1381-813X; v. 7
LC Classification: IN PROCESS

Modeling and control of robotic
manipulators and manufacturing
processes: presented at the Winter
Annual Meeting of the American
Society of Mechanical Engineers,
Boston, Massachusetts, December 13-
18, 1987 / sponsored by the Dynamic
Systems and Controls Division, ASME;
edited by R. Shoureshi, K. Youcef-
Toumi, H. Kazeroonni.
Published/Created: New York, N.Y.
(345 E. 47th St., New York 10017):
ASME, c1987.
Related Authors: Shoureshi, R.
(Rahmatallah) Youcef-Toumi, Kamal.
Kazerooni, H. (Homayoon) American
Society of Mechanical Engineers.
Dynamic Systems and Control Division.
Description: v, 422 p.: ill.; 28 cm.
Notes: Includes bibliographies.
Subjects: Robotics--Congresses.
Manipulators (Mechanism)--
Congresses. Robots, Industrial--
Congresses.
Series: DSC (Series); vol. 6.
LC Classification: TJ210.3 .A54 1987
Dewey Class No.: 670.42/7 19

Modelling and control of compliant and
rigid motion systems: presented at the
Winter Annual Meeting of the
American Society of Mechanical
Engineers, Atlanta, Georgia, December
1-6, 1991 / sponsored by the Robotics
Technical Panel of the Dynamic
Systems and Control Division, ASME;
edited by Wayne J. Book, Frank Paul.
Published/Created: New York, N.Y.:
ASME, c1991.
Related Authors: Book, Wayne J. Paul,
Frank W. American Socviety of

Mechanical Engineers. Winter Meeting
(1991: Atlanta, Ga.) American Society
of Mechanical Engineers. Dynamic
Systems and Control Division. Robotics
Technical Panel.
Description: vi, 187 p.: ill.; 28 cm.
ISBN: 0791808629
Notes: Spine Modeling and control of
compliant and rigid motion systems.
Includes bibliographical references and
index.
Subjects: Manipulators (Mechanism)--
Congresses. Robotics--Congresses.
Series: DSC (Series); vol. 31.
LC Classification: TJ210.3 .M64 1991
Dewey Class No.: 671.42/72 20

Modelling the innovation: communications,
automation, and information systems:
proceedings of the IFIP TC7
Conference on Modelling the
Innovation--Communications,
Automation, and Information Systems,
Rome, Italy, 21-23 March 1990 / edited
by M. Carnevale, M. Lucertini, S.
Nicosia.
Published/Created: Amsterdam,
Netherlands; New York: North-Holland;
New York, N.Y., U.S.A.: Distributors
for the U.S. and Canada, Elsevier
Science Pub. Co., 1990.
Related Authors: Carnevale, M.
Lucertini, M. (Mario) Nicosia, S.
Description: xv, 593 p.: ill.; 23 cm.
ISBN: 044488565X
Notes: Includes bibliographical
references and index.
Subjects: Telecommunication--
Congresses. Robotics--Congresses.
Information technology--Congresses.
LC Classification: TK5101.A1 I36 1990
Dewey Class No.: 621.382 20

Modern robot engineering / edited by E.P.
Popov; translated from the Russian by
Grigory Pasechnik.
Published/Created: Moscow: MIR;
Chicago, Ill.: Imported Publications
[distributor], 1982.
Related Authors: Popov, E. P. (Evgenii

Pavlovich)
Description: 198 p.: ill.; 22 cm.
Notes: Distributor from label on p. 4 of
cover. Includes index. Bibliography: p.
[192]-195.
Subjects: Robotics.
Series: Advances in science and
technology in the USSR. Technology
series.
LC Classification: TJ211 .M63 1982
Dewey Class No.: 629.8/92 19

Moravec, Hans P.
Mind children: the future of robot and
human intelligence / Hans Moravec.
Published/Created: Cambridge, Mass.:
Harvard University Press, 1988.
Description: 214 p.: ill.; 25 cm.
ISBN: 0674576160 (alk. paper)
Notes: Includes index. Bibliography: p.
197-201.
Subjects: Artificial intelligence.
Robotics.
LC Classification: Q335 .M65 1988
Dewey Class No.: 006.3 19

Moravec, Hans P.
Robot: mere machine to transcendent
mind / by Hans Moravec.
Published/Created: New York: Oxford
University Press, 1999.
Description: ix, 227 p.: ill.; 24 cm.
ISBN: 0195116305
Notes: Includes index.
Subjects: Robotics.
LC Classification: TJ211 .M655 1999
Dewey Class No.: 303.48/34/0112 21

Morgan, Chris.
Robots: planning and implementation /
Chris Morgan.
Published/Created: Kempston, Bedford,
U.K.: IFS Publications; Berlin; New
York: Springer-Verlag, 1984.
Description: 195 p.: ill.; 24 cm.
ISBN: 0387125841 (Springer: U.S.)
Notes: Includes index.
Subjects: Robotics. Robots, Industrial.
LC Classification: TJ211 .M664 1984

Dewey Class No.: 629.8/92 19

Morriss, S. Brian.
Automated manufacturing systems: actuators, controls, sensors, and robotics / S. Brian Morriss.
Published/Created: New York: Glencoe, c1995.
Description: xiii, 301 p.: ill.; 25 cm.
ISBN: 0028023315
Notes: Includes index.
Subjects: Automatic control. Process control.
LC Classification: TJ213 .M553 1995
Dewey Class No.: 670.42/7 20

Motors in robotics / Motor Tech Trends.
Published/Created: [United States]: Motor Tech Trends, c1985.
Related Authors: Motor Tech Trends (Firm)
Description: 1 v. (unpaged): ill.; 28 cm.
Notes: Cover Motion systems in robotics. "December 1985." "A U.S. market and technology forecast for the exclusive use of the Library of Congress"--Cover.
Subjects: Robot industry--United States. Motor industry--United States. Market surveys--United States.
LC Classification: HD9696.R623 U64 1985
Dewey Class No.: 381/.45629892/0973 20

Mulhall, Douglas.
Our molecular future: how nanotechnology, robotics, genetics, and artificial intelligence will transform our world / Douglas Mulhall.
Published/Created: Amherst, NY: Prometheus Books, 2002.
Description: p. cm.
ISBN: 1573929921 (alk. paper)
Notes: Includes bibliographical references and index.
Subjects: Nanotechnology. Robotics. Genetics. Artificial intelligence.
LC Classification: T174.7 .M85 2002

Dewey Class No.: 303.48/3 21

Multirobot systems / edited by Rajiv Mehrotra, Murali R. Varanasi.
Published/Created: Los Alamitos, Calif.: IEEE Computer Society Press, c1990.
Related Authors: Mehrotra, Rajiv. Varanasi, Murali R.
Description: x, 123 p.: ill.; 28 cm.
ISBN: 0818619775 (pbk.) 0818659777 (microfiche)
Notes: "SAN 264-620-X"--T.p. verso. Includes bibliographical references.
Subjects: Robotics. Robots, Industrial.
Series: IEEE Computer Society robot technology series
LC Classification: TJ211 .M85 1990
Dewey Class No.: 670.42/72 20

Multisensor integration and fusion for intelligent machines and systems / edited by Ren C. Luo and Michael G. Kay.
Published/Created: Norwood, N.J.: Ablex Pub., c1995.
Related Authors: Luo, Ren C. Kay, Michael G.
Description: xvi, 688 p.: ill.; 24 cm.
ISBN: 0893918636
Notes: Includes bibliographical references and indexes.
Subjects: Robotics. Artificial intelligence. Multisensor data fusion.
LC Classification: TJ211 .M86 1995
Dewey Class No.: 629.8/92 20

Murphy, Robin, 1957-
Introduction to AI robotics / Robin R. Murphy.
Published/Created: Cambridge, Mass.: MIT Press, c2000.
Description: xix, 466 p.: ill.; 24 cm.
ISBN: 0262133830 (alk. paper)
Notes: "A Bradford book" Includes bibliographical references (p. [449]-458) and index.
Subjects: Robotics. Artificial intelligence.
Series: Intelligent robotics and

autonomous agents
LC Classification: TJ211 .M865 2000
Dewey Class No.: 629.8/6263 21

Murray, Richard M.
A mathematical introduction to robotic
manipulation / Richard M. Murray,
Zexiang Li, S. Shankar Sastry.
Published/Created: Boca Raton: CRC
Press, c1994.
Related Authors: Li, Zexiang, 1961-
Sastry, Shankar.
Description: xix, 456 p.: ill.; 25 cm.
ISBN: 0849379814 (acid-free paper)
Notes: Includes bibliographical
references (p. 441-448) and index.
Subjects: Robotics.
LC Classification: TJ211 .M87 1994
Dewey Class No.: 629.8/92 20

Museum madness [computer file].
Edition Information: Version 1.0.
Published/Created: Minneapolis, Minn.:
MECC, c1994.
Related Authors: Minnesota
Educational Computing Corporation.
Description: 6 computer disks: col.; 3
1/2 in. + 1 manual.
ISBN: 0792907884
Computer File Info.: Computer data and
program. Macintosh System 7.0
Summary: An educational program
designed to enhance students' higher-
order thinking skills and reading
comprehension as they learn about
astronomy, American history, robotics,
computers, prehistoric people, natural
history, and other topics in the course of
organizing museum exhibits.
Notes: Title from disk label. Issued also
for IBM-compatible PCs. System
requirements: color-capable Macintosh;
4MB RAM; System 7.0 or later; 256-
color monitor; high-density floppy disk
drive; hard disk with 10MB free space;
external speakers recommended.
Subjects: Science--Study and teaching
(Secondary)--Juvenile software.
Engineering--Study and teaching
(Secondary)--Juvenile software.

Thought and thinking--Juvenile
software. Reading comprehension--
Juvenile software. Museums--
Educational aspects--Juvenile software.
Thought and thinking--Software.
Classification--Software. Museums--
Software.
LC Classification: Q181
Dewey Class No.: 500 12

Nakamura, Yoshihiko.
Advanced robotics: redundancy and
optimization / Yoshihiko Nakamura.
Published/Created: Reading, Mass.:
Addison-Wesley Pub. Co., c1991.
Description: xi, 337 p.: ill.; 25 cm.
ISBN: 0201151987
Notes: Includes bibliographical
references and index.
Subjects: Robotics.
Series: Addison-Wesley series in
electrical and computer engineering.
Control engineering
LC Classification: TJ211 .N34 1991
Dewey Class No.: 629.8/92 20

Nazaretov, V. M.
Engineering artificial intelligence /
[authors, V.M. Nazaretov and D.P. Kim;
translated by P.N. Budzilovich]; editor,
I.M. Makarov, English ed. editor, E.I.
Rivin.
Published/Created: New York:
Hemisphere Pub. Corp., c1990.
Related Authors: Kim, D. P. (Dmitrii
Petrovich) Makarov, Igor Mikhailovich.
Rivin, Eugene I.
Description: vii, 118 p.: ill.; 24 cm.
Notes: Translation of: Tekhnicheskaia
imitatsiia intellekta. Title on verso t.p.:
Tekhnicheskaya imitatsiya intellekta.
Includes bibliographical references
(p.113-114).
Subjects: Robotics. Automation.
Artificial intelligence.
LC Classification: TJ211 .N3913 1990
Dewey Class No.: 629.8/92 20

NBS/RIA Robotics Research Workshop:
proceedings of the NBS/RIA Workshop

on Robotic Research held at
Williamsburg, Virginia, July 12-13,
1977 / sponsored by the National
Bureau of Standards and the Robot
Institute of America; edited by John M.
Evans, Jr., James S. Albus, and Anthony
J. Barbera.
Published/Created: [Washington]: U.S.
Dept. of Commerce, National Bureau of
Standards: for sale by the Supt. of
Docs., U.S. Govt. Print. Off., 1978.
Related Authors: Evans, John M. (John
Martin), 1942- Albus, James Sacra.
Barbera, Anthony J. United States.
National Bureau of Standards. Robot
Institute of America.
Description: iv, 34 p.; 26 cm.
Subjects: Robotics--Congresses.
Series: NBS special publication; 500-
29. NBS special publication. Computer
science & technology.
LC Classification: QC100 .U57 no. 500-
29 TJ211
Dewey Class No.: 602/.1 s 629.8/92

NBS/RIA Robotics Research Workshop:
proceedings of the NBS/RIA Workshop
on Robotic Research, held at the
National Bureau of Standards in
Gaithersburg, MD, on November 13-15,
1979 / edited by James S. Albus ... [et
al.]; sponsored by the Robot Institute of
America.
Published/Created: Washington, D.C.:
U.S. Dept. of Commerce, National
Bureau of Standards: For sale by the
Supt. of Docs., U.S. G.P.O., 1981.
Related Authors: Albus, James Sacra.
Center for Mechanical Engineering and
Process Technology (U.S.). Industrial
Systems Division. Robot Institute of
America. United States. National
Bureau of Standards.
Description: iv, 49 p.: ill.; 26 cm.
Notes: "Industrial Systems Division,
Center for Mechanical Engineering and
Process Technology, National
Engineering Laboratory, National
Bureau of Standards." "Issued April
1981." S/N 003-003-02307-8 Item 247

Subjects: Robotics--Congresses.
Series: NBS special publication; 602
LC Classification: QC100 .U57 no. 602
TS191.8
Dewey Class No.: 602/.18 s 629.8/92 19
Govt. Doc. No.: C 13.10:602

Nehmzow, Ulrich, 1961-
Mobile robotics: a practical introduction
/ Ulrich Nehmzow.
Published/Created: London; New York:
Springer, c2000.
Description: xii, 243 p.: ill.; 23 cm.
ISBN: 1852331739 (alk. paper)
Notes: Includes bibliographical
references and index.
Subjects: Mobile robots.
Series: Applied computing
LC Classification: TJ211.415 .N44 2000
Dewey Class No.: 629.8/92 21

Neural adaptive control technology / editors,
Rafal Zbikowski, Kenneth J. Hunt.
Published/Created: Singapore; River
Edge, N.J.: World Scientific, c1996.
Related Authors: Zbikowski, R. (Rafal)
Hunt, K. J. (Kenneth J.), 1963-
Description: viii, 347 p.: ill.; 23 cm.
ISBN: 9810225571
Notes: Includes bibliographical
references and index.
Subjects: Adaptive control systems.
Neural networks (Computer science)
Nonlinear control theory.
Series: World Scientific series in
robotics and intelligent systems; vol. 15
LC Classification: TJ217 .N47 1996
Dewey Class No.: 629.8/9 20

Neural network perspectives on cognition
and adaptive robotics / edited by Antony
Browne.
Published/Created: Bristol;
Philadelphia: Institute of Physics Pub.,
c1997.
Related Authors: Browne, Antony.
Description: xiv, 270 p.: ill.; 24 cm.
ISBN: 0750304553
Notes: Includes bibliographical
references (p. 251-267) and index.

Subjects: Neural networks (Computer science) Robotics.
LC Classification: QA76.87 .N4767 1997
Dewey Class No.: 006.3/2 21

Neural networks for control / edited by W. Thomas Miller, III, Richard S. Sutton, and Paul J. Werbos.
Published/Created: Cambridge, Mass.: MIT Press, c1990.
Related Authors: Miller, W. Thomas. Sutton, Richard S. Werbos, Paul J. (Paul John), 1947- National Science Foundation (U.S.)
Description: xviii, 524 p.: ill.; 24 cm.
ISBN: 0262132613
Notes: "A Bradford book." "Based on a workshop held at the University of New Hampshire in October, 1988, sponsored by the National Science Foundation, and entitled 'The application of neural networks to robotics and control'"--Pref. Includes bibliographical references and index.
Subjects: Automatic control--Congresses. Neural computers--Congresses.
Series: Neural network modeling and connectionism
LC Classification: TJ223.M53 N48 1990
Dewey Class No.: 629.8/9 20

Neural networks in manufacturing and robotics / presented at the Winter Annual Meeting of the American Society of Mechanical Engineers, Anaheim, California, November 8-13, 1992; sponsored by the Production Engineering Division, ASME; edited by Yung C. Shin, Ahmed H. Abodelmonem, Soundar Kumara.
Published/Created: New York: American Society of Mechanical Engineers, 1992.
Related Authors: Shin, Yung C. Abodelmonem, Ahmed H. Kumara, Soundar T., 1952- American Society of Mechanical Engineers. Production

Engineering Division.
Description: v, 165 p.: ill.; 26 cm.
ISBN: 0791810623 (pbk.)
Notes: Includes bibliographical references and index.
Subjects: Neural networks (Computer science)--Congresses. Robotics--Congresses. Manufacturing processes--Congresses.
Series: PED (Series); vol. 57.
LC Classification: QA76.87 .A44 1992
Dewey Class No.: 670/.285/63 20

Neural networks in robotics / edited by George A. Bekey, Kenneth Y. Goldberg.
Published/Created: Boston: Kluwer Academic, c1993.
Related Authors: Bekey, George A., 1928- Goldberg, Ken. University of Southern California. Center for Neural Engineering. Workshop on Neural Networks in Robotics (1st: 1991: Los Angeles, Calif.)
Description: xi, 563 p.: ill.; 24 cm.
ISBN: 079239268X
Notes: "Most of the papers contained in this book were presented at the First Workshop on Neural Networks in Robotics, sponsored by University of Southern California's Center for Neural Engineering, October 1991." Includes bibliographical references and index.
Subjects: Robots--Control systems--Congresses. Neural networks (Computer science)--Congresses.
LC Classification: TJ211.35 .N48 1992
Dewey Class No.: 629.8/92 20

Neural systems for robotics / edited by Omid Omidvar, Patrick van der Smagt.
Published/Created: San Diego: Academic Press, c1997.
Related Authors: Omidvar, Omid. Smagt, Patrick van der.
Description: xvii, 346 p.: ill.; 24 cm.
ISBN: 0125262809 (acid-free paper)
Notes: Includes bibliographical references and index.
Subjects: Robots--Control systems.

Neural networks (Computer science)
LC Classification: TJ211.35 .N473 1997
Dewey Class No.: 629.8/92632 21

Neurotechnology for biomimetic robots /
edited by Joseph Ayers, Joel Davis, and
Alan Rudolph.
Published/Created: Cambridge, Mass.:
MIT Press, 2002.
Related Authors: Ayers, Joseph. Davis,
Joel L., 1942- Rudolph, Alan.
Description: p. cm.
ISBN: 026201193X (hc.: alk. paper)
Notes: "A Bradford book." Includes
bibliographical references and index.
Subjects: Robotics. Neural networks
(Computer science) Biosensors.
LC Classification: TJ211 .N478 2002
Dewey Class No.: 629.8/92 21

Neutral interfaces in design, simulation, and
programming for robotics / I. Bey ... [et
al.].
Published/Created: Berlin; New York:
Springer-Verlag, c1994.
Related Authors: Bey, I. (Ingward)
Description: xv, 334 p.: ill.; 24 cm.
ISBN: 3540575316 0387575316 (U.S.)
Notes: Includes bibliographical
references and index.
Subjects: Robotics. Neural networks
(Computer science)
Series: Research reports ESPRIT.
Subseries PDT. Project 5109, NIRO
LC Classification: TJ211 .N48 1994
Dewey Class No.: 629.8 20

Niku, Saeed B.
An introduction to robotics analysis,
systems, applications / Saeed B. Niku.
Published/Created: Upper Saddle River,
N.J.: Prentice Hall, c2001.
Description: xiv, 349 p.: ill.; 24 cm.
ISBN: 0130613096
Notes: Includes bibliographical
references and index.
Subjects: Robotics.

N-Nagy, Francis.
Engineering foundations of robotics /

Francis N-Nagy, Andras Siegler.
Published/Created: Englewood Cliffs,
NJ: Prentice-Hall International, c1987.
Related Authors: Siegler, Andras, 1952-
Description: xiii, 263 p.: ill.; 25 cm.
ISBN: 0132788055 (pbk.)
Notes: Includes bibliographies and
index.
Subjects: Robotics.
LC Classification: TJ211 .N16 1987
Dewey Class No.: 629.8/92 19

Noble, P. J. W. (Peter J. W.)
Printed circuit board assembly: the
complete works / P.J.W. Noble.
Published/Created: New York: Halsted
Press; Milton Keynes: Open University
Press, 1989.
Description: x, 203 p.: ill.; 25 cm.
ISBN: 0470212667
Subjects: Printed circuits--Design and
construction.
Series: Open University Press robotics
series
LC Classification: TK7868.P7 N63
1989
Dewey Class No.: 621.381/74 19

Nolfi, Stefano.
Evolutionary robotics: the biology,
intelligence, and technology of self-
organizing machines / Stefano Nolfi and
Dario Floreano.
Published/Created: Cambridge, Mass.:
MIT Press, c2000.
Related Authors: Floreano, Dario.
Description: 320 p.: ill.; 24 cm.
ISBN: 0262140705 (hc)
Notes: "A Bradford book." Includes
bibliographical references (p. [295]-
315) and index.
Subjects: Evolutionary robotics.
Series: Intelligent robots and
autonomous agents
LC Classification: TJ211.37 .N65 2000
Dewey Class No.: 629.8/92 21

Nonholonomic motion planning / edited by
Zexiang Li, J.F. Canny.
Published/Created: Boston: Kluwer

Academic, c1993.
Related Authors: Li, Zexiang, 1961-
Canny, John.
Description: xv, 448 p.: ill.; 24 cm.
ISBN: 0792392752 (alk. paper)
Notes: Includes bibliographical
references and index.
Subjects: Robots--Motion.
Series: The Kluwer international series
in engineering and computer science.
Robotics
LC Classification: TJ211.4 .N66 1993
Dewey Class No.: 629.8/92 20

Nourbakhsh, Illah Reza, 1970-
Interleaving planning and execution for
autonomous robots / by Illah Reza
Nourbakhsh.
Published/Created: Boston: Kluwer
Academic Publishers, 1997.
Description: xvi, 145 p.: ill.; 24 cm.
ISBN: 0792398289 (alk. paper)
Notes: Includes bibliographical
references and index.
Subjects: Autonomous robots. Robots--
Control systems. Artificial intelligence.
Uncertainty--Mathematical models.
Series: Kluwer international series in
engineering and computer science;
SECS 385. Kluwer international series
in engineering and computer science.
Robotics.
LC Classification: TJ211.415 .N68 1997
Dewey Class No.: 629.8/92 21

Occupational health and safety in
automation and robotics: the
proceedings of the 5th UOEH
International Symposium, Kitakyushu,
Japan, 20-21 September 1985 / edited
by Kageyu Noro.
Published/Created: London; New York:
Taylor & Francis, 1987.
Related Authors: Noro, Kageyu.
Description: xviii, 437 p.: ill.; 26 cm.
ISBN: 0850663512 :
Subjects: Robotics--Health aspects--
Congresses. Robots, Industrial--Health
aspects--Congresses. Automation--
Health aspects--Congresses. Robots,

Industrial--Safety measures--
Congresses. Automation--Safety
measures--Congresses. Accidents,
Occupational--prevention & control--
congresses. Automation--congresses.
Human Engineering--congresses.
LC Classification: RC965.R64 U54
1985
Dewey Class No.: 670.42/7 19

Optical 3D measurement techniques II:
applications in inspection, quality
control, and robotics: 4-7 October 1993,
Zürich, Switzerland / Armin Gruen,
Heribert Kahmen, chairs/editors;
organized by Institute of Geodesy and
Photogrammetry, ETH Zürich, Institute
of National Surveying and Engineering
Geodesy, University of Technology,
Vienna, International Society for
Photogrammetry and Remote Sensing
(ISPRS); cooperating organizations,
ISPRS Commission V: Close-Range
Photogrammetry and Machine Vision ...
[et al.].
Published/Created: Bellingham, Wash.:
SPIE--the International Society for
Optical Engineering, c1994.
Related Authors: Gruen, A. (Armin)
Kahmen, Heribert, 1940-
Eidgenössische Technische Hochschule
Zürich. Institut für Geodäsie und
Photogrammetrie. Technische
Universität Wien. Institut für
Landesvermessung und
Ingieurgeodäsie. International Society
for Photogrammetry and Remote
Sensing.
Description: x, 624 p.: ill.; 29 cm.
ISBN: 0819415618
Notes: Includes bibliographical
references and index.
Subjects: Optical detectors--Congresses.
Photogrammetry--Congresses. Three-
dimensional display systems--
Congresses. Optical measurements--
Congresses.
Series: Proceedings of SPIE--the
International Society of Optical
Engineering; v. 2252.

LC Classification: TA1750 .O62 1994
Dewey Class No.: 681/.2 20

Optical 3-D measurement techniques III:
applications in inspection, quality
control, and robotics: papers presented
to the conference organized at Vienna,
Oktober 2-4, 1995 / Armin Gruen,
Heribert Kahmen, eds.
Published/Created: Heidelberg:
Wichmann, 1995.
Related Authors: Gruen, A. (Armin)
Kahmen, Heribert, 1940- Technische
Universität Wien. Institut für
Landesvermessung und
Ingieurgeodäsie.
Description: xii, 533 p.: ill.; 22 cm.
ISBN: 3879072752
Notes: "Organized by the Institute of
National Surveying and Engineering
Geodesy, Dept. of Engin. Geodesy,
University of Technology, Vienna... [et
al.]"--T.p. verso. Text in English and
German. Includes bibliographical
references.
Subjects: Optical detectors--Congresses.
Three dimensional display systems--
Congresses. Optical measurements--
Congresses. Optical detectors--
Industrial applications--Congresses.
LC Classification: TA1630 .O58 1995
Dewey Class No.: 681/.25 21

Optical 3-D measurement techniques:
applications in inspection, quality
control, and robotics: papers presented
to the conference organized at Vienna,
Austria, September 18-20, 1989 /
[edited by] A. Gruen, H. Kahmen.
Published/Created: Karlsruhe:
Wichmann, c1989.
Related Authors: Gruen, A. (Armin)
Kahmen, Heribert, 1940- Technische
Universität Wien. Institut für
Landesvermessung und
Ingieurgeodäsie. Eidgenössische
Technische Hochschule Zürich. Institut
für Geodäsie und Photogrammetrie.
Description: xii, 495 p.: ill.; 21 cm.
ISBN: 3879072000

Notes: English and German. Conference
organized by the Institute of National
Surveying and Engineering Geodesy,
Dept. of Engin. Geodesy, University of
Technology, Vienna and the Institute of
Geodesy and Photogrammetry, ETH,
Swiss Federal Institute of Technology,
Zürich. Includes bibliographical
references.
Subjects: Optical detectors--Congresses.
Three dimensional display systems--
Congresses. Optical measurements--
Congresses. Optical detectors--
Industrial applications--Congresses.
LC Classification: TA1632 .O63 1989
Dewey Class No.: 681/.2 20

Palei, S. M. (Sergei Markovich)
Illustrated dictionary of robotics:
English, German, French, Russian /
S.M. Paley.
Published/Created: Moskva: MNTK
"Robot", 1993.
Description: viii, 347 p.: ill.; 25 cm.
ISBN: 5900442012
Notes: Includes indexes.
Subjects: Robotics--Dictionaries--
Polyglot. Robots, Industrial--
Dictionaries--Polyglot. Dictionaries,
Polyglot.
LC Classification: TJ210.4 .P35 1993
Dewey Class No.: 629.8/92/03 20

Paltrowitz, Stuart.
Robotics / Stuart and Donna Paltrowitz.
Edition Information: Messner certified
ed.
Published/Created: New York: J.
Messner, c1983.
Related Authors: Paltrowitz, Donna.
Description: 63 p.: ill.; 22 cm.
ISBN: 0671440772 0671497200 (pbk.)
Summary: Describes various types of
robots, their purpose, and how they
function. Also discusses briefly the
future of robots and some of the tasks
they will perform.
Notes: "A Jem book." Includes index.
Bibliography: p. 61.
Subjects: Robots--Juvenile literature.

Robots.
LC Classification: TJ211 .P33 1983
Dewey Class No.: 629.8/92 19

Pantelidis, Veronica S.
Robotics in education: an information
guide / by Veronica Sexauer Pantelidis.
Published/Created: Metuchen, N.J.:
Scarecrow Press, 1991.
Description: xiv, 435 p.; 23 cm.
ISBN: 0810824663 (alk. paper)
Notes: Includes bibliographical
references and index.
Subjects: Robotics--Study and teaching.
LC Classification: TJ211.26 .P35 1991
Dewey Class No.: 629.8/92/071 20

Parallel computation systems for robotics:
algorithms and architectures / edited by
A. Fijany and A. Bejczy.
Published/Created: Singapore; River
Edge, NJ: World Scientific, c1992.
Related Authors: Fijany, A. Bejczy, A.
Description: vii, 247 p.: ill.; 23 cm.
ISBN: 9810206631 (hc) 981020664X
(sc)
Notes: Includes bibliographical
references.
Subjects: Robots--Control systems.
Robots--Programming. Parallel
processing (Electronic computers)
Series: World Scientific series in
robotics and automated systems; vol. 2
LC Classification: TJ211.35 .P37 1992
Dewey Class No.: 629.8/92 20

Parsons, Thomas W.
Voice and speech processing / Thomas
W. Parsons.
Published/Created: New York:
McGraw-Hill, c1986.
Description: xiii, 402 p.: ill.; 24 cm.
ISBN: 0070485410 :
Notes: Includes bibliographies and
index.
Subjects: Speech processing systems.
Series: McGraw-Hill series in electrical
engineering. CAD/CAM, robotics, and
computer vision McGraw-Hill series in
electrical engineering. Communications

and signal processing
LC Classification: TK7882.S65 P37
1986
Dewey Class No.: 006.4/54 19

Paul, Richard P.
Robot manipulators: mathematics,
programming, and control: the computer
control of robot manipulators / Richard
P. Paul.
Published/Created: Cambridge, Mass.:
MIT Press, c1981.
Description: 279 p.: ill.; 24 cm.
ISBN: 026216082X
Notes: Includes bibliographies and
index.
Subjects: Robotics. Manipulators
(Mechanism)
Series: The MIT Press series in artificial
intelligence
LC Classification: TS191.8 .P38
Dewey Class No.: 629.8/92 19

Perry, Robert L. (Robert Louis), 1950-
Artificial intelligence / Robert L. Perry.
Published/Created: New York: Franklin
Watts, 2000.
Description: 63 p.: col. ill.; 25 cm.
ISBN: 053111757X
Summary: Describes different types of
artificial intelligence from heuristic to
case-based reasoning and beyond,
explores robotics, softbots, and games,
and defines how AI relates to computers
and humans.
Notes: Includes bibliographical
references (p. 25-26) and index.
Subjects: Artificial intelligence--
Juvenile literature. Artificial
intelligence.
Series: Watts library
LC Classification: Q335.4 .P47 2000
Dewey Class No.: 006.3 21

Petit, Jean-Pierre, 1937-
Run, robot, run / Jean-Pierre Petit;
translated by Ian Stewart; edited by
Wendy Campbell.
Published/Created: Los Altos, Calif.: W.
Kaufmann, c1985.

Related Authors: Campbell, Wendy.
Description: 72 p.: ill.; 28 cm.
ISBN: 0865760837 (pbk.)
Summary: Cartoons stories in which a young inventor's experiments with robots and other devices introduce the concepts of robotics and artificial intelligence.
Notes: Translation of: A quoi rêvent les robots?
Subjects: Robotics--Juvenile literature. Artificial intelligence--Juvenile literature. Robots--Cartoons and comics. Robotics--Cartoons and comics. Artificial intelligence--Cartoons and comics. Cartoons and comics.
Series: Petit, Jean-Pierre, 1937-
Aventures d'Anselme Lanturlu. English.
LC Classification: TJ211.2 .P4813 1985
Dewey Class No.: 629.8/92 19

Poole, Harry H.
Fundamentals of robotics engineering / Harry H. Poole.
Published/Created: New York: Van Nostrand Reinhold, c1989.
Description: x, 436 p.: ill.; 24 cm.
ISBN: 0442272987
Notes: Includes bibliographical references.
Subjects: Robotics.
LC Classification: TJ211 .P65 1989
Dewey Class No.: 629.8/92 19

Popchev, Ivan Petkov.
Decentralized systems / Ivan Popchev.
Published/Created: Sofia: Pub. House of the Bulgarian Academy of Sciences, 1989.
Description: 278 p.: ill.; 21 cm.
Notes: Text in English and transliterated Russian; summaries in Bulgarian and Russian. At head of Bulgarian Academy of Sciences, Institute of Industrial Cybernetics and Robotics.
LC Classification: MLCS 93/02457 (T)

Practical motion planning in robotics: current approaches and future directions / edited by Kamal Gupta, Angel P. del Pobil.
Published/Created: Chichester, West Sussex; New York: Wiley, c1998.
Related Authors: Gupta, Kamal. Pobil, Angel Pasqual del.
Description: x, 356 p.: ill.; 25 cm.
ISBN: 047198163X
Notes: Includes bibliographical references and index.
Subjects: Robots--Motion--Planning.
LC Classification: TJ211.4 .G87 1998
Dewey Class No.: 629.8/92 21

Presence: teleoperators and virtual environments.
Published/Created: Cambridge, MA: MIT Press, c1992-
Description: v.: ill.; 28 cm. Winter 1992 issue called also premier issue. Vol. 1, no. 1 (winter 1992)-
ISSN: 1054-7460
Notes: Title from cover.
SERBIB/SERLOC merged record
Subjects: Human-machine systems--Periodicals. Virtual reality--Periodicals. Human-computer interaction--Periodicals. Robotics--Periodicals. Manipulators (Mechanism)--Periodicals.
LC Classification: TA167 .P69
Dewey Class No.: 006 20

Problem solver in automatic control systems/robotics / staff of Research and Education Association.
Published/Created: New York, N.Y.: The Association, c1982.
Related Authors: Research and Education Association.
Description: xi, 1013 p.: ill.; 26 cm.
ISBN: 0878915427 (pbk.)
Notes: Includes index.
Subjects: Automatic control--Problems, exercises, etc. Robotics--Problems, exercises, etc.
LC Classification: TJ213.8 .P76 1982
Dewey Class No.: 629.8/076 19

Proceedings / ... IEEE International Workshop on Robot and Human Communication, RO-MAN; co-

sponsored by IEEE Industrial
Electronics Society ... [et al.].
Distinctive Robot and human
communication
Published/Created: Piscataway, N.J.:
Institute of Electrical and Electronics
Engineers, 1992-
Related Authors: IEEE Industrial
Electronics Society.
Description: v.: ill.; 28 cm. [1st] (1992)-
Ceased with: 7th, held in 1998.
Notes: SERBIB/SERLOC merged
record
Subjects: Robotics--Congresses.
Human-machine systems--Congresses.
LC Classification: TJ210.3 .I4438a
Dewey Class No.: 629.8/92 B 21

Proceedings / 1987 Symposium on
   Advanced Manufacturing, September
   28-30, 1987; sponsored by University of
   Kentucky Center for Robotics and
   Manufacturing Systems and Office of
   Engineering Continuing Education and
   Extension.
   Published/Created: Lexington, Ky.:
   Office of Engineering Services, College
   of Engineering, University of Kentucky,
   c1987.
   Related Authors: University of
   Kentucky. Center for Robotics and
   Manufacturing Systems. University of
   Kentucky. Office of Engineering
   Continuing Education and Extension.
   Description: 142 p.: ill.; 28 cm.
   ISBN: 0897790693 (pbk.)
   Notes: "1987 theme, implementing
   automation." "September 1987."
   Includes bibliographies.
   Subjects: Production engineering--
   Congresses. Computer integrated
   manufacturing systems--Congresses.
   Flexible manufacturing systems--
   Congresses. Automation--Congresses.
   Series: UKY bulletin; 144.
   LC Classification: TS176 .S886 1987
   Dewey Class No.: 670.42 19

Proceedings / 1993 IEEE/Tsukuba
   International Workshop on Advanced

Robotics: can robots contribute to
preventing environmental deterioration?
November 8-9, 1993, AIST Tsukuba
Research Center, Tsukuba, Japan;
cosponsored by Mechanical
Engineering Laboratory ... [et al.];
technically cosponsored by IEEE
Robotics and Automation Society ... [et
al.].
Published/Created: Piscataway, NJ:
IEEE Service Center, [c1993]
Related Authors: Kikai Gijutsu
Kenky¯ujo.
Description: 119 p.: ill.; 29 cm.
ISBN: 0780314417 (softbound ed.)
0780314425 (Microfiche ed.)
Subjects: Environmental protection--
Congresses. Robotics--Congresses.
LC Classification: TD170.2 .I34 1993
Dewey Class No.: 628 20

Proceedings / 1996 IEEE Conference on
   Emerging Technologies and Factory
   Automation, ETFA '96, Kauai, Hawaii,
   November 18-21, 1996 / sponsored by
   the IEEE Industrial Electronics Society;
   in technical cooperation with Society of
   Instrument and Control Engineers of
   Japan.
   Published/Created: [New York]:
   Institute of Electrical and Electronics
   Engineers, c1996.
   Related Authors: IEEE Industrial
   Electronics Society. Keisoku Jid¯o
   Seigyo Gakkai (Japan)
   Description: 2 v. (xiii, 787 p.): ill.; 30
   cm.
   ISBN: 0780336852 (paper) 0780336860
   (microfiche)
   Notes: "IEEE catalog number
   96TH8238"--T.p. verso. Includes
   bibliographical references and indexes.
   Subjects: Automation--Congresses.
   Artificial intelligence--Congresses.
   Robotics--Congresses.
   LC Classification: T59.5 .I27 1996
   Dewey Class No.: 670.42/7 21

Proceedings / 23rd International Symposium
   on Industrial Robots, 6th-9th October,

1992, Barcelona, Spain.
Published/Created: Barcelona, Spain:
Asociación Espanola de Rob´otica,
1992.
Related Authors: International
Federation of Robotics.
Description: xxxi, 788 p.: ill.; 31 cm.
ISBN: 8460436527
Notes: Includes indexes.
Subjects: Robots, Industrial--
Congresses. Robotics--Congresses.
Robot vision--Congresses.
LC Classification: TS191.8 .I57 1992
Dewey Class No.: 670.42/72 21

Proceedings / 2nd symposium, foundations
of kinematics, dynamics, and control of
manipulation-robots, October 20-24,
1986, Schwerin, GDR.
Published/Created: Karl-Marx-Stadt:
Akademie der Wissenschaften der
DDR, Institut für Mechanik, 1987.
Related Authors: Institut für Mechanik
(Akademie der Wissenschaften der
DDR) Institut po mekhanika i
biomekhanika (Bulgarska academiia na
naukite) Kombinat Zentraler
Industrieanlagenbau der Metallurgie.
Description: 3 v.: ill.; 21 cm.
Notes: English, German, and Russian.
Cover 2nd symposium, foundations of
kinematics, dynamics, and control of
manipulation-robots. "The symposium
was organized by the Institute of
Mechanics of the Academy of Sciences
of the GDR, Institute of Mechanics and
Biomechanics of the Bulgarian
Academy of Sciences, VEB Kombinat
Zentraler Industrieanlagenbau der
Metallurgie"--P. iii. Includes
bibliographies.
Subjects: Robotics--Congresses.
Manipulators (Mechanism)--
Congresses.
Series: Report (Institut für Mechanik
(Akademie der Wissenschaften der
DDR)); No. 3-5.
LC Classification: TJ210.3 .P76 1987
Dewey Class No.: 629.8/92 19

Proceedings / 3rd Canadian CAD/CAM &
Robotics Conference.
Edition Information: 1st ed.
Published/Created: Ancaster, Ont.,
Canada: Canadian Institute of
Metalworking; Dearborn, Mich.:
Society of Manufacturing Engineers,
c1984.
Related Authors: Canadian Institute of
Metalworking. Computer and
Automated Systems Association of
SME.
Description: 1 v. (various pagings): ill.;
28 cm.
ISBN: 0872631508 (pbk.) :
Notes: Sponsored by the Canadian
Institute of Metalworking and the
Computer and Automated Systems
Association of SME. Includes
bibliographical references and index.
Subjects: CAD/CAM systems--
Congresses. Robots, Industrial--
Congresses.
LC Classification: TS155.6 .C375 1984
Dewey Class No.: 670.28/5 19

Proceedings / 3rd Symposium on Theory
and Practice of Robots and
Manipulators, Third International
CISM-IFToMM Symposium, Udine
Italy, September 12-15, 1978; sponsored
by CISM, Centre international des
sciences mécaniques, IFToMM,
International Federation for the Theory
of Machines and Mechanisms, in
association with the Technical Division
of the Polish Academy of Sciences;
edited by A. Morecki, G. Bianchi, K.
Kedzior.
Published/Created: Amsterdam; New
York: Elsevier Scientific Pub. Co.;
Warszawa: PWN-Polish Scientific
Publishers; New York: distribution for
the U.S.A. and Canada, Elsevier/North-
Holland, 1980.
Related Authors: Morecki, Adam.
Bianchi, G. (Giovanni), 1924- Kedzior,
K. International Centre for Mechanical
Sciences. International Federation for
the Theory of Machines and

Mechanisms. Polska Akademia Nauk. Wydzial IV-Nauk Technicznych. Description: xvii, 596 p.: ill.; 24 cm. ISBN: 0444997725 : Notes: Includes bibliographical references. Subjects: Robotics--Congresses. Manipulators (Mechanism)--Congresses. LC Classification: TJ211 .R64 1978 Dewey Class No.: 629.8/92

Proceedings / Conference on Artificial Intelligence Applications. Published/Created: Washington, D.C.: IEEE Computer Society Press, 1987- Related Authors: IEEE Computer Society. IEEE Computer Society. Technical Committee on Machine Intelligence and Pattern Analysis. American Association for Artificial Intelligence. Canadian Society for Computational Studies of Intelligence. Description: v.: ill.; 28 cm. Vol. for 1991 published in 2 v. 3rd (1987)- ISSN: 1043-0989 Notes: Published: Los Alamitos, Calif., <1991-1993. Sponsored by: Computer Society of the IEEE, 1987-1991; by: IEEE Computer Society Technical Committee on Pattern Analysis and Machine Intelligence, 1992-<1993; in cooperation with: American Association of Artificial Intelligence, 1988 and <1993-; Canadian Society for Computational Studies of Intelligence <1993- SERBIB/SERLOC merged record Subjects: Artificial intelligence--Congresses. Robotics--Congresses. Expert systems (Computer science)--Congresses. LC Classification: Q334 .C66a Dewey Class No.: 006.3 20

Proceedings / IECON '91, 1991 International Conference on Industrial Electronics, Control and Instrumentation, International Conference Center Kobe, Kobe, Japan,

October 28-November 1, 1991; co-sponsored by the Industrial Electronics Society of the IEEE (IEEE/IES), the Society of Instrument and Control Engineers of Japan (SICE); in cooperation with the Institute of Electrical Engineers of Japan ... [et al.]. Published/Created: New York: IEEE; Tokyo: SICE, c1991. Related Authors: IEEE Industrial Electronics Society. Keisoku Jid¯o Seigyo Gakkai. Description: 3 v. (2591 p.): ill.; 29 cm. ISBN: 0879426888 (softbound) Contents: v. 1. Invited session; Special session; Power electronics and motion control -- v. 2. Signal processing and system control; Intelligent sensors and instrumentation -- v. 3. Robotics, CIM, and automation; Emerging technologies. Notes: "IEEE 91CH2976-9." "SICE 91PR002-1." Includes bibliographical references. Subjects: Industrial electronics--Congresses. Power electronics--Congresses. Electronic control--Congresses. Computer integrated manufacturing systems--Congresses. Robotics--Congresses. Electronic instruments--Congresses. LC Classification: TK7881 .I55 1991 Dewey Class No.: 670.42 20

Proceedings / IEEE International Conference on Robotics and Automation. Published/Created: Washington, D.C.: IEEE Computer Society Press, c1986- Related Authors: IEEE Computer Society. IEEE Robotics and Automation Council. IEEE Robotics and Automation Society. Description: v.: ill.; 28 cm. Issued in 3 or more v. 1986- ISSN: 1050-4729 Notes: Sponsored by: IEEE Council on Robotics and Automation, 1986-1988; by: IEEE Robotics and Automation Society, 1989-<1997. SERBIB/SERLOC merged record

Indexed entirely by: Index to IEEE
publications 0099-1368
Subjects: Robotics--Congresses. Robots,
Industrial--Congresses. Automatic
control--Congresses.
LC Classification: TJ210.3 .I58a
Dewey Class No.: 629.8/92/05 20

Proceedings / IEEE International
    Symposium on Intelligent Control.
    Published/Created: Washington, D.C.:
    Computer Society Press of the IEEE,
    c1987-c1998.
    Related Authors: IEEE Control Systems
    Society. National Institute of Standards
    and Technology (U.S.) National Science
    Foundation (U.S.) United States. Army
    Research Office. International
    Symposium on Computational
    Intelligence in Robotics and
    Automation. Intelligent Systems and
    Semiotics (Conference)
    Description: 12 v.: ill.; 28 cm. 1987-
    1998.
    Notes: Vols. for 1991-1998 have
    Proceedings of the ... IEEE International
    Symposium on Intelligent Control.
    Sponsored by: IEEE Control Systems
    Society, <1995-; co-sponsored by the
    society and the National Institute of
    Standards and Technology, the National
    Science Foundation, and the U.S. Army
    Research Office, 1998. Conference for
    1998 held jointly with the IEEE
    International Symposium on
    Computational Intelligence in Robotics
    and Automation, and Intelligent
    Systems and Semiotics.
    SERBIB/SERLOC merged record
    Indexed entirely by: Index to IEEE
    publications 0099-1368
    Subjects: Intelligent control systems--
    Congresses.
    LC Classification: TJ212.2 .I325a
    Dewey Class No.: 629.8 20

Proceedings / IEEE International Workshop
    on Robot and Human Interactive
    Communication; co-organized by IEEE
    Industrial Electronics Society ... [et al.].

Published/Created: Piscataway, NJ:
Institute of Electrical and Electronics
Engineers,
Related Authors: IEEE Industrial
Electronics Society.
Description: v.: ill.; 28 cm.
Notes: Description based on: 9th (2000).
Subjects: Robotics--Congresses.
Human-machine systems--Congresses.
LC Classification: TJ210.3 .I4438a

Proceedings / International Workshop on
    Industrial Automation Systems--Seiken
    Symposium, Tokyo, Japan, February 4-
    6, 1987; co-sponsored by IEEE-
    Industrial Electronics Society, Institute
    of Industrial Science, University of
    Tokyo, in cooperation with IEEE
    Robotics and Automation Council, the
    Society of Instrument and Control
    Engineers of Japan (SICE), Foundation
    for Promotion of Industrial Science.
    Published/Created: New York, N.Y.:
    Institute of Electrical and Electronics
    Engineers; Piscataway, NJ: May be
    ordered from Order Dept., IEEE, c1987.
    Related Authors: IEEE Industrial
    Electronics Society. T⁻oky⁻o Daigaku.
    Seisan Gijutsu Kenky⁻ujo.
    Description: x, 230 p.: ill.; 28 cm.
    Notes: "IEEE catalog number
    87TH0165-1." Includes bibliographies.
    Subjects: Automation--Congresses.
    LC Classification: T59.5 .I575 1987
    Dewey Class No.: 670.42/7 19

Proceedings / the Third NASA/DOD
    Workshop on Evolvable Hardware, EH-
    2001, 12-14 July 2001, Long Beach,
    California, USA; sponsored by National
    Aeronautics and Space Administration
    (NASA), Defense Advanced Research
    Projects Agency (DARPA); co-hosted
    by JPL Center for Integrated Space
    Microsystems (CISM), JPL Center for
    Space Microelectronics Technology
    (CSMT), NASA Ames Information
    Sciences and Technology Directorate;
    edited by Didier Keymeulen ... [et al.].
    Published/Created: Los Alamitos, CA:

IEEE Computer Society, 2001.
Related Authors: Keymeulen, Didier.
United States. National Aeronautics and
Space Administration. United States.
Defense Advanced Research Projects
Agency.
Description: x, 287 p.: ill.; 28 cm.
ISBN: 0769511805 0769511813
(bookbroker) 0769511821 (microfiche)
Notes: Includes bibliographical
references and author index.
Subjects: Electronic apparatus and
appliances--Automatic control
Congresses. Evolutionary robotics--
Congresses. Adaptive control systems--
Congresses. Biological systems--
Computer simulation--Congresses.
LC Classification: TK7870 .N34 2001
Dewey Class No.: 629.8 21

Proceedings / Third Annual Conference on
Intelligent Robotic Systems for Space
Exploration, Rensselaer Polytechnic
Institute, Troy, New York, November
18-19, 1991; sponsored by NASA
Center for Intelligent Robotic Systems
for Space Exploration.
Published/Created: Los Alamitos,
Calif.: IEEE Computer Society Press,
c1991.
Related Authors: NASA Center for
Intelligent Robotic Systems for Space
Exploration.
Description: viii, 131 p.: ill.; 28 cm.
ISBN: 0818625953
Notes: Includes bibliographical
references and index.
Subjects: Space robotics--Congresses.
Outer space--Exploration--Congresses.
LC Classification: TL787 .C634 1991
Dewey Class No.: 629.4 20

Proceedings 7th European Space
Mechanisms & Tribology Symposium:
ESTEC, Noordwijk, The Netherlands 1-
3 October 1997 / [compiled by B.H.
Kaldeich-Schürmann]; sponsored by
ESA, NIVR & CNES.
Published/Created: Noordwijk,
Netherlands: ESA Publications

Division, c1997.
Related Authors: Kaldeich-Schürmann,
Brigitte. European Space Agency.
Nederlands Instituut voor
Vliegtuigontwikkeling en Ruimtevaart.
Centre national d'études spatiales
(France)
Description: vii, 304 p.: ill.; 30 cm.
ISBN: 9290926406
Notes: Includes bibliographical
references.
Subjects: Space vehicles--Lubrication--
Congresses. Tribology--Congresses.
Space robotics--Congresses.
Series: ESA SP; 410.
LC Classification: TL917 .E93 1997
Dewey Class No.: 629.47 21

Proceedings IEEE International Workshop
on Intelligent Robots and Systems:
toward the next generation robot and
system: Tokyo, Japan Oct. 31-Nov. 2,
1988 / cosponsored by IEEE Industrial
Electronics Society ... [et al.].
Published/Created: New York, N.Y.:
IEEE, c1989.
Related Authors: IEEE Industrial
Electronics Society.
Description: xx, 826 p.: ill.; 28 cm.
Notes: Cover Proceedings 1988 IEEE
International Workshop on Intelligent
Robots and Systems. Page 1-2, two
reports (10 p.), and a separate copy of
the program (10 p.) inserted. "IEEE
catalog number 88TH0234-5."
Subjects: Robotics--Congresses.
Artificial intelligence--Congresses.
LC Classification: TJ210.3 .I443 1988
Dewey Class No.: 629.8/92 20

Proceedings International Workshop on
Industrial Applications of Machine
Vision and Machine Intelligence:
Seiken Symposium, Tokyo, Japan
February 2-5, 1987 / co-sponsored by
IEEE-Industrial Electronics Society
[and the] Institute of Industrial Science,
University of Tokyo; in cooperation
with IEEE Robotics and Automation
Council, International Association for

Pattern Recognition (TC8), Foundation for Promotion of Industrial Science. Published/Created: New York, N.Y.: Publishing Service, Institute of Electrical and Electronics Engineers; Piscataway, N.J., U.S.A.: Additional copies may be order from Order Dept., IEEE, c1987.
Related Authors: IEEE Industrial Electronics Society. T⁻oky⁻o Daigaku. Seisan Gijutsu Kenky⁻ujo.
Description: xii, 386 p.: ill.; 28 cm.
Notes: Includes bibliographical references.
Subjects: Computer vision--Congresses. Robots, Industrial--Congresses.
LC Classification: TA1632 .I595 1987
Dewey Class No.: 006.3/7 19

Proceedings of 1996 IEEE International Conference on Evolutionary Computation (ICEC '96): May 20-22, 1996, Symposion & Toyoda Auditorium, Nagoya University, Japan / co-sponsored by IEEE Neural Network Council (NNC), Society of Instrument and Control Engineers (SICE); technically co-sponsored by Robotics Society of Japan (RSJ), Japan Society for Fuzzy Theory and Systems (SOFT), Japan Society of Mechanical Engineers (JSME).
Published/Created: Piscataway, N.J.: Institute of Electrical and Electronics Engineers, c1996.
Related Authors: IEEE Neural Networks Council. Keisoku Jid⁻o Seigyo Gakkai (Japan)
Description: xxii, 891, 4 p.: ill.; 28 cm.
ISBN: 0780329023 (softbound) 0780329031(microfiche)
Notes: "IEEE catalog number: 96TH8114"--T.p. verso. Includes bibliographical references and index.
Subjects: Artificial intelligence--Congresses. Evolutionary computation--Congresses. Genetic algorithms--Congresses. Neural networks (Computer science)--Congresses.
LC Classification: Q334 .I428 1996

Dewey Class No.: 005.1 21

Proceedings of International Symposium on Automation and Robotics in Production Engineering, May 24-27, 1988, Xi`an, China.
Published/Created: Xi`an, China: Xi`an Jiaotong University, [1988]
Description: v, 385 p.: ill.; 26 cm.
LC Classification: MLCM 92/07713 (T)

Proceedings of the ... Conference on Remote Systems Technology / sponsored by the Remote Systems Technology Division.
Published/Created: Hinsdale, Ill.: American Nuclear Society, c1964- c1993.
Related Authors: American Nuclear Society. Remote Systems Technology Division.
Description: 29 v.: ill.; 28 cm. 12th (Nov. 30-Dec. 3, 1964)-40th (1992).
ISSN: 0069-8644
Notes: SERBIB/SERLOC merged record Indx'd selectively by: Engineering index annual (1968) 0360-8557 Engineering index monthly (1984) 0742-1974 Engineering index bioengineering abstracts 0736-6213 Engineering index energy abstracts 0093-8408 Computer & control abstracts 0036-8113 Electrical & electronics abstracts 0036-8105 Physics abstracts. Science abstracts. Series A 0036-8091 Chemical abstracts 0009-2258
Subjects: Remote handling (Radioactive substances)--Congresses. Hot laboratories (Radioactive substances)--Equipment and supplies--Congresses. Robotics--Congresses.
LC Classification: TK9151.6 .C66a
Dewey Class No.: 621.48 19

Proceedings of the ... Conference on Robotics and Remote Systems.
Published/Created: La Grange Park, Ill.: American Nuclear Society, c1994-
Related Authors: American Nuclear Society. Robotics and Remote Systems

Division.
Description: v.: ill.; 28 cm. 41st (1993)-
ISSN: 0069-8644
Notes: Sponsored by: the Robotics and
Remote Systems Division.
SERBIB/SERLOC merged record
Indx'd selectively by: Chemical
abstracts 0009-2258
Subjects: Remote handling (Radioactive
substances)--Congresses. Hot
laboratories (Radioactive substances)--
Equipment and supplies--Congresses.
Robotics--Congresses.
LC Classification: TK9151.6 .C66a
Dewey Class No.: 621.48 19

Proceedings of the 13th World Congress:
International Federation of Automatic
Control, San Francisco, USA, 30th
June-5th July 1996 / edited by Janos J.
Gertler, Jose B. Cruz, Jr., Michael
Peshkin.
Edition Information: 1st ed.
Published/Created: Oxford; New York:
Published for the International
Federation of Automatic Control by
Pergamon, 1997-
Related Authors: Gertler, Janos. Cruz,
Jose B. Peshkin, M. A. International
Federation of Automatic Control.
Description: v. <1-17: ill.; 31 cm.
ISBN: 0080429092 (pbk.: v. A)
0080429106 (pbk.: v. B) 0080428114
(pbk.: v. C) 0080429122 (pbk.: v. D)
0080429130 (pbk.: v. E) 0080429149
(pbk.: v. F) 0080429157 (pbk.: v. G)
0080429165 (pbk.: v. H) 0080429173
(pbk.: v. I) 0080429181 (pbk.: v. J)
008042919X (pbk.: v. K) 0080429203
(pbk.: v. L) 0080429211 (pbk.: v. M)
008042922X (pbk.: v. N) 0080429238
(pbk.: v. O) 0080429246 (pbk.: v. P)
0080429254 (pbk.: v. Q)
Contents: v. A. Robotics, components
and instruments -- v. B. Manufacturing,
social effects, bio-production,
biomedical, environment -- v. C.
Control design I -- v. D. Control design
II, optimization -- v. E. Nonlinear
systems I -- v. F. Nonlinear systems II --

v. G. Education, robust control I -- v. H.
Robust control II, stochastic systems --
v. I. Identification I -- v. J. Identification
II, discrete event systems -- v. K.
Adaptive control -- v. L. Systems
engineering and management -- v. M.
Chemical process control, mineral,
mining, metals -- v. N. Fault detection,
pulp and paper, biotechnology -- v. O.
Power plants and systems, computer
control -- v. P. Aerospace,
transportation systems -- v. Q.
Automotive, marine, autonomous
vehicles.
Notes: Includes bibliographical
references and indexes.
Subjects: Automatic control--
Congresses. Control theory--
Congresses. Robotics--Congresses.
LC Classification: TJ212.2 .I58 1996
Dewey Class No.: 629.8 21

Proceedings of the 14th World Congress,
International Federation of Automatic
Control: Beijing, P.R. China, 5-9 July
1999: (in 18 volumes) / edited by Han-
Fu Chen, Dai-Zhan Cheng, and Ji-Feng
Zhang.
Edition Information: 1st ed.
Published/Created: New York:
Published for the International
Federation of Automatic Control by
Pergamon, c1999.
Related Authors: Ch`en, Han-fu. Cheng,
Dai-Zhan. Zhang, Ji-Feng.
Description: v. A-[R]: ill.; 30 cm.
Contents: v. A. Manufacturing, social
effects, scheduling -- v. B. Robotics
automation -- v. C. Control design -- v.
D. Linear system, robust control I -- v.
E. Robust control II -- v. F. Nonlinear
system I -- v. G. Nonlinear system II,
optimal control -- v. H. Modeling,
identification, signal processing I -- v. I.
Modeling, identification, signal
processing II, adaptive control -- v. J.
Discrete event systems, stochastic
systems, fuzzy and neural systems I -- v.
K. Fuzzy and neural systems II, control
in agricultural processes -- v. L.

Biomedical and environmental systems, systems engineering -- v. M. Management, global, and educational issues -- v. N. Chemical process control, mineral and metal processing -- v. O. Power systems, biotechnological processes, fault detection I -- v. P. Fault detection II, aerospace, marine systems -- v. Q. Transportation systems, computer control -- v. [R]. Plenary volume.
Notes: Includes bibliographical references and indexes.
Subjects: Automatic control--Congresses.
Series: IFAC conference proceedings
LC Classification: TJ212.2 I58 1999
Dewey Class No.: 629.8 21

Proceedings of the 1988 Symposium on Advanced Manufacturing / [sponsored by] Center for Robotics and Manufacturing Systems, University of Kentucky, September 26, 27, 28, 1988, Radisson Plaza Hotel, Lexington, Kentucky.
Published/Created: Lexington, Ky.: OES Publications, College of Engineering, University of Kentucky, c1988.
Related Authors: University of Kentucky. Center for Robotics and Manufacturing Systems.
Description: 148 p.: ill.; 28 cm.
ISBN: 0897790723
Notes: Cover Kentucky industry on the rise. Includes bibliographical references.
Subjects: Production engineering--Congresses. Manufacturing processes--Automation--Congresses. Flexible manufacturing systems--Congresses.
Series: UKY bulletin; 147.
LC Classification: TS176 .S886 1988
Dewey Class No.: 670.42 20

Proceedings of the 1992 International Conference on Industrial Electronics, Control, Instrumentation, and Automation: November 9-13, 1992, Marriot Mission Valley, San Diego,

USA / co-sponsored by the Industrial Electronics Society of the IEEE (IEEE/IES), the Society of Instrument and Control Engineers of Japan (SICE).
Published/Created: Piscataway, NJ: IEEE; Tokyo: SICE, c1992.
Related Authors: IEEE Industrial Electronics Society. Keisoku Jid̄o Seigyo Gakkai.
Description: 3 v. (1649 p.): ill.; 28 cm.
ISBN: 0780305825 (softbound) 0780305833 (casebound) 0780305841 (microfiche)
Contents: v. 1. Power electronics and motion control -- v. 2. Robotics, CIM, and automation emerging technologies -- v. 3. Signal porocessing [sic] and systems control, intelligent sensors and instrumentation.
Notes: "IECON '92"--Cover. Includes bibliographical references and indexes.
Subjects: Industrial electronics--Congresses. Automation--Congresses. Robotics--Congresses. Manufacturing processes--Data processing--Congresses.
LC Classification: TK7881 .I55 1992
Dewey Class No.: 670.42/7 20

Proceedings of the 1992 Japan-U.S.A. Symposium on Flexible Automation: presented at the 1992 Japan-U.S.A. Symposium on Flexible Automation, a Pacific Rim conference, San Francisco, California, July 13-15, 1992 / sponsoring societies, the American Society of Mechanical Engineers [and] Institute of Systems, Control, and Information Engineers of Japan; participating societies, IEEE Robotics and Automation Society ... [et al.]; editor, Ming Leu.
Published/Created: New York: American Society of Mechanical Engineers; Kyoto: Institute of Systems, Control, and Information Engineers, c1992.
Related Authors: Leu, M. C. American Society of Mechanical Engineers. Shisutemu Seigyo J̄oh̄o Gakkai. IEEE

Robotics and Automation Society.
Description: 2 v. (xix, 1760 p.): ill.; 28 cm.
ISBN: 0791806758 (set)
Notes: Spine Japan-U.S.A. flexible automation--1992. Includes bibliographical references and index.
Subjects: Flexible manufacturing systems--Congresses. Robotics--Congresses.
LC Classification: TS155.6 .J36 1992
Dewey Class No.: 670.42/7 20

Proceedings of the 1994 / Second Australian and New Zealand Conference on Intelligent Information Systems, Brisbane, Australia, 29 November-2 December 1994.
Published/Created: [New York]: Institute of Electrical and Electronics Engineers, c1994.
Related Authors: Institute of Electrical and Electronics Engineers.
Description: xiv, 496 p.: ill.; 30 cm.
ISBN: 0780324048 (softbound) 0780324056 (microfiche)
Contents: Neural networks 1 -- Neural networks 2 -- Neural networks for control-- Neural networks for pattern recognition -- Neural networks 3 -- Robotics -- Parallel and signal processing algorithms -- Fuzzy logic - general -- Fuzzy control -- Fuzzy logic - applications -- Evolutionary programming -- Other methods -- Computer vision 1 -- Computer vision 2 -- Computer vision 3 -- Artificial intelligence - learning -- Artificial intelligence - planning -- Artificial intelligence - reasoning -- Expert systems -- Intelligent information systems 1 -- intelligent information systems 2.
Notes: Conference was called by the IEEE Australia and New Zealand Councils. "IEEE Catalog Number 94TH8019"--P. 2 of cover. Includes bibliographical references and author index.
Subjects: Expert systems (Computer

science)--Congresses. Intelligent control systems--Congresses. Neural networks (Computer science)--Congresses.
LC Classification: QA76.76.E95 A96 1994

Proceedings of the 1995 IEEE IECON: 21st International Conference on Industrial Electronics, Control, and Instrumentation.
Published/Created: [New York]: Institute of Electrical and Electronics Engineers; Piscataway, N.J.: Can be ordered from IEEE Service Center, c1995.
Related Authors: Institute of Electrical and Electronics Engineers.
Description: 2 v. (xlv, 1651 p.): ill.; 28 cm.
ISBN: 0780330269 (softbound) 0780330277(casebound) 0780330285 (michofiche)
Contents: v. 1. Plenary sessions. Invited sessions. Power electronics -- v. 2. Signal processing & control. Robotics vision & sensors. Emerging technologies. Factory automation.
Notes: "November 6-10, 1995, Hyatt Regency, Orlando International Airport Hotel, Orlando, Florida, USA"--Cover. "IEEE catalog number: 95CH35868"--T.p. verso. Includes bibliographical references and index.
Subjects: Industrial electronics--Congresses. Automation--Congresses. Robotics--Congresses. Electronic control--Congresses.
LC Classification: TK7881 .I55 1995
Dewey Class No.: 670.42/7 21

Proceedings of the 1996 IEEE IECON: 22nd International Conference on Industrial Electronics, Control, and Instrumentation.
Published/Created: [New York]: Institute of Electrical and Electronics Engineers; Piscataway, N.J.: Can be ordered from IEEE Service Center, c1996.
Related Authors: Institute of Electrical

and Electronics Engineers.
Description: 3 v. (lxxiv, 1995 p.): ill.;
28 cm.
ISBN: 0780327756 (softbound)
0780327764 (casebound) 0780327772
(microfiche)
Notes: "Held at the Lai Lai Sheraton
Hotel in Taipai, Taiwan from August 5-
10, 1996"--P. iv. "IEEE catalog number
96CH35830"--T.p. verso. Includes
bibliographical references and indexes.
Subjects: Industrial electronics--
Congresses. Automation--Congresses.
Robotics--Congresses. Electronic
control--Congresses. Power electronics-
-Congresses.
LC Classification: TK7881 .I55 1996
Dewey Class No.: 670.42/7 21

Proceedings of the 1996 IEEE/RSJ
International Conference on Intelligent
Robots and Systems: IROS 96: robotic
intelligence interacting with dynamic
worlds, November 4-8, 1996, Senri Life
Science Center, Osaka, Japan /
sponsored by IEEE Industrial
Electronics Society ... [et al.].
Published/Created: [New York]: IEEE;
Piscataway, N.J.: IEEE Service Center,
[1996], c1990.
Related Authors: IEEE Industrial
Electronics Society.
Description: 3 v. (xxxv, 1746 p.): ill.; 28
cm.
ISBN: 078033213X (softbound)
0780332148 (softbound) 0780332156
(microfiche)
Notes: "IEEE catalog number:
95CH35908"--T.p. verso. Includes
bibliographical references and indexes.
Subjects: Robotics--Congresses.
Artificial intelligence--Congresses.
Intelligent control systems--Congresses.
LC Classification: TJ210.3 .I447 1996
Dewey Class No.: 629.8/92 21

Proceedings of the 1997 IEEE/RSJ
International Conference on Intelligent
Robots and System, IROS '97s:
innovative robotics for real-world

applications, September 7-11, 1997,
World Trade Center Atria, Grenoble,
France / organized by INRIA Rhône-
Alpes; sponsored by IEEE Industrial
Electronics Society ... [et al.].
Published/Created: [New York]:
Institute of Electrical and Electronics
Engineers, c1996.
Related Authors: INRIA Rhône-Alpes.
Description: 3 v. (xxiii, 1850 p.): ill.; 28
cm.
ISBN: 0780341198 (softbound)
0780341201 (casebound) 078034121X
(microfiche)
Notes: "IEEE catalog number:
97CH36108"--Label on t.p. verso.
Includes bibliographical references and
indexes.
Subjects: Robotics--Congresses.
Artificial intelligence--Congresses.
Intelligent control systems--Congresses.
LC Classification: TJ210.3 .I447 1997
Dewey Class No.: 629.8/92 21

Proceedings of the 1999 IEEE International
Symposium on Assembly and Task
Planning (ISATP '99): towards flexible
and agile assembly and manufacturing,
July 21-24, 1999, Porto, Portugal /
sponsored by IEEE Robotics &
Automation Society.
Published/Created: Piscataway, NJ:
IEEE, 1999.
Related Authors: IEEE Robotics and
Automation Society.
Description: xiii, 466 p.: ill.; 28 cm.
ISBN: 0780357043 (softbound edition)
Notes: "99TH8470." Includes
bibliographical references and index.
Subjects: Assembly-line methods--
Congresses. Production planning--
Congresses. Robots, Industrial--
Congresses. Flexible manufacturing
systems--Congresses.
LC Classification: TS178.4 .I34 1999
Dewey Class No.: 670.42 21

Proceedings of the 2000 Japan-USA
Symposium on Flexible Automation:
International Conference on New

Technological Innovation for the 21st Century, Ann Arbor, Michigan, July 23-26, 2000 / sponsored by the Dynamic Systems and Control Division, ASME [and] the Institute of Systems, Control, and Instrumentation Engineers, ISCIE; edited by Steven Y. Liang [and] Tatsuo Arai.
Published/Created: New York: American Society of Mechanical Engineers, c2000.
Related Authors: Liang, Steven Y. Arai, Tatsuo. American Society of Mechanical Engineers Dynamic Systems and Control Division. Institute of Systems Control and Instrumentation Engineers.
Description: 2 v., xxii, (1405 p.): ill.; 28 cm.
ISBN: 079183509X
Notes: Paged continuously. Includes bibliographical references and index.
Subjects: Flexible manufacturing systems--Congresses. Robotics--Congresses. Robots, Industrial--Congresses.

Proceedings of the 2nd International Conference on Robot Vision and Sensory Controls, November 2-4, 1982, Stuttgart, Germany / organised jointly by Institute of Production Automation (IPA) and IFS (Conferences) Ltd.
Published/Created: Kempston, Bedford, England: IFS Publications, 1982.
Related Authors: Fraunhofer-Institut für Produktionstechnik und Automatisierung. International Fluidics Services Ltd.
Description: iv, 388 p.: ill.; 30 cm.
ISBN: 0903608294 (pbk.)
Notes: Includes bibliographical references.
Subjects: Robotics--Congresses. Robots, Industrial--Congresses. Robot vision--Congresses.
Series: Proceedings of SPIE--the International Society for Optical Engineering; v. 392.
LC Classification: TJ211 .I483 1982

Dewey Class No.: 629.8/92 19

Proceedings of the 6th International Conference on Robot Vision and Sensory Controls, 3-5 June 1986, Paris, France / edited by M. Briot.
Published/Created: Kempston, Bedford, UK: IFS; Berlin; New York: Springer-Verlag, c1986.
Related Authors: Briot, M. IFS (Conferences) Ltd. SEPIC (Firm: Paris, France)
Description: viii, 278 p.: ill.; 30 cm.
ISBN: 0387163271 (Springer-Verlag)
Notes: "RoViSeC 6." "An international event organised jointly by IFS (Conferences) Ltd., Kempston, Bedford, UK, and SEPIC, Paris, France." Includes bibliographies.
Subjects: Robotics--Congresses. Robots, Industrial--Congresses. Robot vision--Congresses.
LC Classification: TJ210.3 .I577 1986
Dewey Class No.: 629.8/92 19

Proceedings of the 8th European Space Mechanisms and Tribology Symposium: Toulouse, France, 29 September - 1 October 1999 / organised by the European Space Agency (ESA) and the Centre National d'Etudes Spatiales (CNES) ... [et al.]; [Editor: Dorothea Danesy].
Published/Created: Noordwijk, Netherlands: ESA Publications Division, c1999.
Related Authors: Danesy, D. European Space Agency. Centre national d'études spatiales (France)
Description: vii, 345 p.: ill.; 30 cm.
ISBN: 9290927526
Notes: Includes bibliographical references.
Subjects: Space vehicles--Lubrication--Congresses. Space vehicles--Equipment and supplies--Congresses. Space vehicles--Design and construction--Congresses. Tribology--Congresses. Space robotics--Congresses.
Series: ESA SP; 438.

LC Classification: TL875 .E87 1999
Dewey Class No.: 629.47 21

Proceedings of the ANS Seventh Topical
    Meeting on Robotics and Remote
    Systems, April 27 to May 1, 1997,
    Radisson Riverfront Hotel and
    Conference Center, Augusta, Georgia /
    sponsored by American Nuclear
    Society, Savannah River Section,
    American Nuclear Society, Robotics
    and Remote Systems Division in
    cooperation with Association for
    Robotics in Hazardous Environments ...
    [et al.].
    Published/Created: La Grange Park, Ill.:
    American Nuclear Society, c1997.
    Related Authors: American Nuclear
    Society. Savannah River Section.
    American Nuclear Society. Robotics
    and Remote Systems Division.
    Association for Robotics in Hazardous
    Environments.
    Description: 2 v.: ill.; 28 cm.
    ISBN: 0894486179
    Notes: Includes bibliographical
    references and indexes.
    Subjects: Remote handling (Radioactive
    substances)--Congresses. Hot
    laboratories (Radioactive substances)--
    Equipment and supplies--Congresses.
    Robotics--Congresses.
    LC Classification: TK9151.6 .A57 1997
    Dewey Class No.: 620/.46 21

Proceedings of the ASME Dynamic Systems
    and Control Division / sponsored by the
    1996 International Mechanical
    Engineering Congress and Exposition,
    November 17-22, 1996, Atlanta,
    Georgia; edited by Kourosh Danai;
    organizers, A. Alleyne ... [et al.].
    Published/Created: New York, N.Y.:
    American Society of Mechanical
    Engineers, c1996.
    Related Authors: Danai, Kouresh.
    American Society of Mechanical
    Engineers. Dynamic Systems and
    Control Division. International
    Mechanical Engineering Congress and

Exposition (1996: Atlanta, Ga.)
    Description: xi, 937 p.: ill.; 28 cm.
    ISBN: 0791815285
    Contents: Advanced transportations
    systems -- Modeling, sensing, and
    control of manufacturing processes --
    Open-architecture systems for
    intelligent manufacturing -- Precision
    machine control and metrology --
    Haptic interfaces for virtual
    environment and teleoperator systems --
    Impedance planning and
    implementation of biomechanical
    systems -- Robotics -- Automated
    modeling -- Advances in
    instrumentation and components for
    mechanical systems -- Neural and fuzzy
    control -- Robust control of mechanical
    systems -- Dynamics, control, and
    measurement of biomechanical systems.
    Notes: Includes bibliographical
    references and index.
    Subjects: Automatic control--
    Congresses. Control theory--
    Congresses. Intelligent control systems-
    -Congresses.
    Series: DSC (Series); vol. 58.
    LC Classification: TJ212.2 .P75 1996
    Dewey Class No.: 629.8 21

Proceedings of the ASME Dynamic Systems
    and Control Division: presented at the
    1997 ASME International Mechanical
    Engineering Congress and Exposition,
    November 16-21, 1997, Dallas, Texas /
    sponsored by the Dynamic Systems and
    Control Division, ASME; edited by
    Giorgio Rizzoni.
    Published/Created: New York, N.Y.:
    American Society of Mechanical
    Engineers, c1997.
    Related Authors: Rizzoni, Giorgio.
    American Society of Mechanical
    Engineers. Dynamic Systems and
    Control Division. International
    Mechanical Engineering Congress and
    Exposition (1997: Dallas, Tex.)
    Description: x, 766 p.: ill.; 28 cm.
    ISBN: 0791818241
    Contents: Haptic interfaces for virtual

environment and teleoperator systems --
Expert and intelligent systems -- Design
and control of smart machines --
Intelligent sensors and instrumentation -
- Biomechanics -- Robotics -- Robust
nonlinear control -- Precision machine
control and metrology -- Advanced
transportation system.
Notes: Includes bibliographical
references and index.
Subjects: Automatic control--
Congresses. Control theory--
Congresses. Intelligent control systems-
-Congresses.
Series: DSC (Series); vol. 61.
LC Classification: TJ212.2 .P762 1997
Dewey Class No.: 629.8 21

Proceedings of the ASME Dynamic Systems
and Control Division: presented at the
1995 ASME International Mechanical
Engineering Congress and Exposition,
November 12-17, 1995, San Francisco,
California / sponsored by the Dynamic
Systems and Control Division, ASME;
edited by Thomas E. Alberts.
Published/Created: New York:
American Society of Mechanical
Engineers, c1995.
Related Authors: Alberts, T. E. (Thomas
E.) American Society of Mechanical
Engineers. Dynamic Systems and
Control Division. International
Mechanical Engineering Congress and
Exposition (1995: San Francisco, Calif.)
Description: 2 v. (xii, 1091 p.): ill.; 28
cm.
ISBN: 0791817466
Contents: V. 1. Vibration control.
Dynamic systems. Robotics.Sliding
mode control. Robust and nonlinear
control. Automated modeling. Control
of manufacturing processes. Precision
control -- v. 2. Haptic interfaces for
virtual environment and teleoperator
systems. Applications of intelligent
techniques. Sensors for identification
and control. Automotive and
agricultural applications of fluid power.
Adaptive control experiments in

emerging technologies.
Micromechanical systems. Automatic
control lab experiments for graduate
engineering education..Neural and fuzzy
control systems.
Notes: Includes bibliographical
references and index.
Subjects: Automatic control--
Congresses. Control theory--
Congresses. Intelligent control systems-
-Congresses.
Series: DSC (Series); vol. 57.
LC Classification: TJ212.2 .P76 1995
Dewey Class No.: 629.8 20

Proceedings of the Embedded Topical
Meeting on DOE Spent Nuclear Fuel
and Fissile Material Management, San
Diego, California, June 4-8, 2000 /
sponsored by the American Nuclear
Society, Fuel Cycle and Waste
Management Division; cosponsored by
the American Nuclear Society, Nuclear
Criticality Safety Division, Robotics
and Remote Systems Division in
cooperation with the American Institute
of Chemical Engineers.
Published/Created: LaGrange Park, Ill.:
American Nuclear Society, c2000.
Related Authors: American Nuclear
Society. Fuel Cycle and Waste
Management Division.
Description: ix, 462 p.: ill.; 28 cm.
ISBN: 0894486489
Notes: "ANS order no. 700273"--T.p.
verso. Includes bibliographical
references and index.
Subjects: United States. Dept. of
Energy--Congresses. Radioactive
wastes--United States--Management--
Congresses. Spent reactor fuels--United
States--Management Congresses.
LC Classification: TD898.118 .E53
2000
Dewey Class No.: 621.48/3/ 21

Proceedings of the Fifth International
Symposium on "Methods and models in
automation and robotics": 25-29 August
1998, Miedzyzdroje, Poland / editors S.

Domek, R. Kaszy'nski, L. Tarasiejski.
Published/Created: Szczecin: Wydawn.
Uczelniane Politechniki Szczeci'nskiej,
1998.
Related Authors: Domek, S. (Stefan)
Kaszy'nski, R. Tarasiejski, L.
Description: v. <1, 3: ill.; 30 cm.
ISBN: 8387423815
Notes: Includes bibliographical
references and index.
Subjects: Automation--Congresses.
Robotics--Congresses.
LC Classification: TJ212.2 .I62 1998
Dewey Class No.: 670.42/7 21

Proceedings of the Fifth International
Symposium on Methods and Models in
Automation and Robotics: 25-29 August
1998, Miedzyzdroje, Poland / editors S.
Domek, R. Kaszy'nski, L. Tarasiejski.
Published/Created: Szczecin: Wydawn.
Uczelniane Politechniki Szczeci'nskiej,
1998.
Description: v. <2
ISBN: 8387423815
LC Classification: IN PROCESS

Proceedings of the First International
Symposium on Measurement and
Control in Robotics / ISMCR '90,
NASA/Lyndon B. Johnson Space
Center, Houston, Texas, USA, June 20-
22, 1990; organized by the IMEKO
Technical Committee on Robotics (TC-
17); host sponsor, Automation and
Robotics Division, NASA/JSC ... [et al.]
Published/Created: Houston, Tex.: Clear
Lake Council of Technical Societies,
[1990]
Related Authors: Havrilla, K. IMEKO
Technical Committee on Robotics (TC-
17) Lyndon B. Johnson Space Center.
automation and Robotics Division.
Description: 3 v.: ill.; 28 cm.
ISBN: 0962711608
Notes: Includes bibliographical
references and indexes.
Subjects: Robotics--Congresses.
Series: IMEKO TC series; no. 27
LC Classification: TJ210.3 .I59 1990

Dewey Class No.: 629.8/92 20

Proceedings of the Fourth International
Conference on Computer Integrated
Manufacturing and Automation
Technology, Troy, New York, October
10-12, 1994 / sponsored by New York
State Center for Advanced Technology
in Robotics and Automation at
Rensselaer Polytechnic Institute in
cooperation with IEEE Robotics and
Automation Society, ASME Material
Handling Engineering Division,
Connecticut State Advanced
Technology Center for Precision
Manufacturing.
Published/Created: Los Alamitos,
Calif.: IEEE Computer Society Press,
c1994.
Related Authors: New York State
Center for Advanced Technology in
Robotics and Automation.
Description: xiii, 463 p.: ill.; 28 cm.
ISBN: 0818665106 (paper) 0818665122
(microfiche)
Notes: Includes bibliographical
references and index.
Subjects: Computer integrated
manufacturing systems--Congresses.
LC Classification: TS155.63 .I57 1994
Dewey Class No.: 670/.285 20

Proceedings of the Fourth International
Symposium on Methods and Models in
Automation and Robotics: 26-29 August
1997, Miedzyzdroje, Poland / editors S.
Domek, Z. Emirsajlow, R. Kaszy'nski.
Published/Created: Szczecin: Wydawn.
Uczelniane Politechniki Szczeci'nskiej,
1997.
Related Authors: Domek, S. (Stefan)
Emirsajlow, Zbigniew. Kaszy'nski, R.
Description: 3 v.(1228 p.): ill.; 30 cm.
ISBN: 8387423300
Notes: Includes bibliographical
references and indexes.
Subjects: Automation--Congresses.
Robotics--Congresses.
LC Classification: TJ212.2 .I62 1997

Dewey Class No.: 670.42/7 21

Proceedings of the IECON '93, International
Conference on Industrial Electronics,
Control, and Instrumentation.
Published/Created: [New York]: IEEE;
Piscataway, NJ: Available from IEEE
Service Center, c1993.
Related Authors: IEEE Industrial
Electronics Society. Keisoku Jid⁻o
Seigyo Gakkai (Japan)
Description: 3 v. (xxxiii, 2393 p.): ill.;
29 cm.
ISBN: 0780308913 (softbound)
0780308921 (casebound)
Contents: v. 1. Plenary session, energing
technologies, and factory automation --
v. 2. Power electronics -- v. 3. Robotics,
vision and sensors; and signal
processing and control.
Notes: "November 15-18, 1993, Hyatt
Regency Hotel, Lahaina, Maui, Hawaii.
Sponsored by the IEEE Industrial
Electronics Society and the Society of
Instrument and Control Engineers of
Japan"--Cover title. "IEEE catalog
number: 93CH3234-2"--T.p. verso.
Includes bibliographical references and
indexes.
Subjects: Industrial electronics--
Congresses. Automation--Congresses.
Robotics--Congresses. Manufacturing
processes--Data processing--
Congresses.
LC Classification: TK7881 .I55 1993

Proceedings of the IECON '97: 23rd
International Conference on Industrial
Electronics, Control, and
Instrumentation / sponsored by IEEE
Industrial Electronics Society; technical
co-sponsor Society of Instrument and
Control Engineers of Japan (SICE).
Published/Created: Piscataway, NJ:
IEEE, c1997.
Related Authors: IEEE Industrial
Electronics Society. Keisoku Jid⁻o
Seigyo Gakkai (Japan)
Description: 4 v. (xxvii, 1555 p.): ill.;
28 cm.

ISBN: 0780339320 (softbound)
0780339339 (casebound) 0780339347
(microfiche) 0780339355 (CD-Rom)
Contents: V. 1. Plenary session, signal
processing and control -- v. 2. Power
electronics -- v. 3. Emerging
technologies, factory automation,
robotics, vision, and sensors -- v. 4.
Tutorials.
Notes: "Held in New Orleans,
Louisiana"--p. iii "IEEE catalog number
97CH36066"--v. 1, t.p. verso. Includes
bibliographical references and indexes.
Subjects: Industrial electronics--
Congresses. Automation--Congresses.
Power electronics--Congresses.
Robotics--Congresses. Electronic
control--Congresses. Electronic
instruments--Congresses.
LC Classification: TK7881 .I55 1997
Dewey Class No.: 621.381 21

Proceedings of the International Conference
on Robotics and Remote Handling in
the Nuclear Industry, 1984 September
23-27, Toronto, Canada.
Published/Created: Toronto: Canadian
Nuclear Society, 1984.
Related Authors: Canadian Nuclear
Society. Canadian Society for
Mechanical Engineering. American
Nuclear Society.
Description: v, 275 p.: ill.; 28 cm.
ISBN: 0919784062 (pbk.) :
Notes: "Organized by the Canaian
Nuclear Society; co-sponsors: Canadian
Society for Mechanical Engineering and
the American Nuclear Society"--Cover.
Includes bibliographies.
Subjects: Remote handling (Radioactive
substances)--Congresses.
LC Classification: TK9151.6 .I58 1984
Dewey Class No.: 621.48 19

Proceedings of the International Topical
Meeting on Remote Systems and
Robotics in Hostile Environments:
March 29-April 2, 1987, Red Lion
Motor Inn, Pasco, Washington /
sponsored by the Remote Systems

Technology Division of the American Nuclear Society, Eastern Washington Section of the American Nuclear Society.
Published/Created: La Grange Park, Ill., USA: American Nuclear Society, c1987.
Related Authors: American Nuclear Society. Remote Systems Technology Division. American Nuclear Society. Eastern Washington Section.
Description: xiii, 669 p.: ill.; 28 cm.
ISBN: 0894481312 (pbk.)
Notes: Includes bibliographies and index.
Subjects: Robotics--Congresses. Manipulators (Mechanism)--Congresses. Remote control--Congresses.
LC Classification: TJ210.3 .I62 1987
Dewey Class No.: 629.8/92 19

Proceedings of the Japan-USA Symposium on Flexible Automation, 1996: presented at the 1996 Japan/USA Symposium on Flexible Automation, July 7-10, 1996, Boston, Massachusetts / sponsors, the American Society of Mechanical Engineers, Institute of Systems, Control, and Information Engineers of Japan; edited by Kim Stelson, Fuminori Oba.
Published/Created: New York: The Society, c1996.
Related Authors: Stelson, K. A. (Kim A.) ¯Oba, Fuminori, 1943- American Society of Mechanical Engineers. Shisutemu Seigyo J¯oh¯o Gakkai.
Description: 2 v. (xvii, 1565 p.): ill.; 28 cm.
ISBN: 0791812316
Notes: Includes bibliographical references and index.
Subjects: Flexible manufacturing systems--Congresses. Robotics--Congresses. Robots, Industrial--Congresses.
LC Classification: TS155.6 .J36 1996
Dewey Class No.: 670.42/7 21

Proceedings of the Military, Government and Aerospace Simulation Symposium: 2000 Advanced Simulation Technologies Conference (ASTC '00), Washington, D.C., April 16-20, 2000, Wyndham City Center Hotel / edited by Michael J. Chinni. Distinctive ASTC '00 2000 Advanced Simulation Technologies Conference
Published/Created: San Diego, Calif.: Society for Computer Simulation International, c2000.
Related Authors: Chinni, Michael J. Society for Computer Simulation. Advanced Simulation Technologies Conference (2000: Washington, D.C.)
Description: viii, 269 p.: ill.; 29 cm.
ISBN: 1565551958
Notes: Held as part of the 2000 Advanced Simulation Technologies Conference (ASTC '00) "Sponsored by the Society for Computer Simulation International"--T.p. Includes bibliographical references and index.
Subjects: Military art and science--Simulation methods--Congresses. Airplanes--Design and construction--Simulation methods Congresses. Computer simulation--Congresses. Robotics--Congresses.
Series: Simulation series; v. 32, no. 3
LC Classification: U104 .M54 2000
Dewey Class No.: 355/.001/1 21

Proceedings of the Second International Symposium on Methods and Models in Automation and Robotics, 30 August-2 September 1995, Miedzyzdroje, Poland / editors, S. Ba´nka, S. Domek, Z. Emirsajlow.
Published/Created: Szczecin, Poland: Institute of Control Engineering, Technical University of Szczecin, 1995.
Related Authors: Ba´nka, Stanislaw. Domek, S. (Stefan) Emirsajlow, Zbigniew. Politechnika Szczeci´nska. Instytut Automatyki Przemyslowej.
Description: 2 v. (x, 863 p.): ill.; 30 cm.
ISBN: 838635917X
Notes: Includes bibliographical

references and index.
Subjects: Automatic control--
Congresses. Robotics--Congresses.
LC Classification: TJ212.2 .I62 1995

Proceedings of the Second International
Workshop on Robot Motion and
Control: RoMoCo'01: October 18-20,
2001, Bukowy Dworek, Poland /
University of Zielona Góra, Pozna'n
University of Technology; [sponsored
by Polish Chapter of the IEEE Robotics
and Automation Society].
Published/Created: Pozna'n, Poland:
Wydawn. Politechniki Pozna'nskiej;
Uniwersytet Zielonogórski, Instytut
Organizacji i Zarzadzania, c2001.
Related Authors: Politechnika
Pozna'nska. Uniwersytet Zielonogórski.
IEEE Robotics and Automation Society.
Description: 316 p.: ill.; 30 cm.
ISBN: 8371435150
Notes: At head of IEEE Robotics &
Automation Society.
Subjects: Robots--Motion--Congresses.
Robots--Control systems--Congresses.

Proceedings of the Third Conference on
Mechatronics and Robotics: "from
design methods to industrial
applications", October 4-6, 1995
Paderborn / edited by Joachim Lükel.
Published/Created: Stuttgart: B.G.
Tuebner, 1995.
Related Authors: Lückel, Joachim.
Description: 551 p.: ill.; 24 cm.
ISBN: 3519026252
Notes: Includes bibliographical
references.
Subjects: Mechatronics--Congresses.
Robotics--Congresses.
LC Classification: TJ163.12 .C66 1995

Proceedings of the Third International
Conference on Computer Integrated
Manufacturing, Rensselaer Polytechnic
Institute, Troy, New York, May 20-22,
1992 / sponsored by New York State
Center for Advanced Technology in
Robotics and Automation at Rensselaer

Polytechnic Institute in cooperation with
IEEE Robotics and Automation Society
... [et al.].
Published/Created: Los Alamitos,
Calif.: IEEE Computer Society Press,
c1992.
Related Authors: New York State
Center for Advanced Technology in
Robotics and Automation.
Description: xii, 510 p.: ill.; 28 cm.
ISBN: 0818626151
Notes: Includes bibliographical
references and index.
Subjects: Computer integrated
manufacturing systems--Congresses.
LC Classification: TS155.6 .I582 1992
Dewey Class No.: 670/.285 20

Proceedings of the Third International
Symposium on Methods and Models in
Automation and Robotics, 10-13
September 1996, Miedzyzdroje, Poland
/ editors, S. Ba'nka, S. Domek, Z.
Emirsajlow.
Published/Created: Szczecin: Institute
of Control Engineering, Technical
University of Szczecin, [1996]
Related Authors: Ba'nka, Stanislaw.
Domek, S. (Stefan) Emirsajlow,
Zbigniew. Politechnika Szczeci'nska.
Instytut Automatyki Przemyslowej.
Description: 3 v.: ill.; 30 cm.
ISBN: 8386359986
Notes: Includes bibliographical
references and indexes.
Subjects: Automatic control--
Congresses. Robotics--Congresses.
LC Classification: TJ212.2 .I62 1996
Dewey Class No.: 670.42/7

Proceedings of the USA-Japan Symposium
on Flexible Automation: crossing
bridges--advances in flexible
automation and robotics / sponsored by
the American Society of Mechanical
Engineers, USA and the Institute of
Systems, Control, and Information
Engineers, Japan: held in Minneapolis,
Minnesota, July 18-20, 1988.
Published/Created: New York, N.Y.

(United Engineering Center, 345 E. 47th St., New York 10017): The Society, c1988.
Related Authors: American Society of Mechanical Engineers. Shisutemu Seigyo J⁻oh⁻o Gakkai.
Description: 2 v. (xvi, 1157 p.): ill.; 28 cm.
Notes: Includes bibliographies and index.
Subjects: Flexible manufacturing systems--Congresses. Robotics--Congresses.
LC Classification: TS155.6 .U83 1988
Dewey Class No.: 670.42 19

Proceedings of the vision for robotics workshop.
Published/Created: Los Alamitos, CA: IEEE Computer Society Press, 1995.
Description: p. cm.
ISBN: 0818671149 (pbk.)
LC Classification: 9508 BOOK NOT YET IN LC

Proceedings, 1999 IEEE International Symposium on Computational Intelligence in Robotics and Automation: CIRA'99: November 8-9, 1999, Monterey, California / sponsored by IEEE Robotics and Automation Society.
Published/Created: Piscataway, NJ: IEEE, c1999.
Related Authors: IEEE Robotics and Automation Society. Institute of Electrical and Electronics Engineers.
Description: xv, 364 p.: ill.; 27 cm.
ISBN: 0780358066 (softbound)
Notes: "IEEE order plan catalog number 99EX375"--t.p. verso. Includes bibliographical references and author index.
Subjects: Robotics--Congresses. Computational intelligence--Congresses.

Proceedings, IEEE, the thirteenth annual International Conference on Micro Electro Mechanical Systems: MEMS

2000: Miyazaki, Japan, January 23-27, 2000 / sponsored by the IEEE Robotics and Automation Society in cooperation with the Micromachine Center.
Published/Created: Piscataway N.J.: Institute of Electrical and Electronics Engineers, c2000.
Related Authors: IEEE Robotics and Automation Society. Micromachine Center (Japan).
Description: xiv, 810 p.: ill.; 30 cm.
ISBN: 0780352734 (softbound) 0780352742 (casebound) 0780352750 (microfiche)
Notes: "IEEE Catalog Number: 00CH36308". Includes bibliographical references and index.
Subjects: Microelectromechanical systems--Congresses.

Proceedings, Rensselaer's Second International Conference on Computer Integrated Manufacturing: May 21-23, 1990, Rensselaer Polytechnic Institute, Troy, New York / co-sponsored by Center for Manufacturing, Productivity, and Technology Transfer's NIST Northeast Manufacturing Technology Center at Rensselaer, Rensselaer Polytechnic Institute; in cooperation with IEEE Robotics and Automation Society, IEEE Computer Society Press.
Published/Created: Los Alamitos, Calif.: IEEE Computer Society Press, c1990.
Related Authors: NIST Northeast Manufacturing Technology Center at Rensselaer. IEEE Robotics and Automation Society. IEEE Computer Society Press.
Description: xiv, 599 p.: ill.; 28 cm.
ISBN: 081861966X (paper) 0818659661 (microfiche)
Notes: "IEEE Computer Society order number 1966"--T.p. verso. "SAN 264-620X"--T.p. verso. Includes bibliographical references and index.
Subjects: Computer integrated manufacturing systems--Congresses.
LC Classification: TS155.6 .I582 1990

Dewey Class No.: 670/.285 20

Proceedings: From Perception to Action
  Conference, September 7-9, 1994,
  Lausanne, Switzerland / edited by P.
  Gaussier, J-D. Nicoud; sponsored by
  International Latsis Foundation, Geneva
  ... [et al.].
  Published/Created: Los Alamitos,
  Calif.: IEEE Computer Society Press,
  c1994.
  Related Authors: Gaussier, P. (Philippe)
  Nicoud, Jean-Daniel. International
  Latsis Foundation.
  Description: xiv, 449 p.: ill.; 24 cm.
  ISBN: 0818664827 (case) 0818664819
  (microfiche)
  Notes: Includes bibliographical
  references and index.
  Subjects: Robotics--Congresses.
  Intelligent control systems--Congresses.
  LC Classification: TJ210.3 .F76 1994
  Dewey Class No.: 629.8/92 20

Proceedings: International Workshop on
  Neural Networks for Identification,
  Control, Robotics, and Signal/Image
  Processing, Venice, Italy, August 21-23,
  1996 / sponsored by the IEEE Computer
  Society, the IEEE Computer Society
  Technical Committee on Pattern
  Analysis and Machine Intelligence
  [PAMI]; in cooperation with ACM
  SIGART, IEEE Circuits and Systems
  Society, IEEE Control Systems Society
  ... [et al.]
  Published/Created: Los Alamitos, CA:
  IEEE Computer Society Press, c1996.
  Related Authors: IEEE Computer
  Society. IEEE Circuits and Systems
  Society. IEEE Computer Society.
  Technical Committee on Pattern
  Analysis and Machine Intelligence.
  IEEE Control Systems Society.
  SIGART.
  Description: xii, 486 p.: ill.; 24 cm.
  ISBN: 0818674563 081867458X
  (microfiche)
  Notes: "IEEE order plan catalog number
  96TB100029"--T.p. verso. Includes

bibliographical references and author
  index.
  Subjects: Neural networks (Computer
  science)--Congresses. Artificial
  intelligence--Congresses. Adaptive
  control systems--Congresses.
  LC Classification: QA76.87 .I5893
  1996
  Dewey Class No.: 629.8/9 21

Proceedings: IROS '90, IEEE International
  Workshop on Intelligent Robots and
  Systems '90: towards a new frontier of
  applications, July 3-6, 1990, Mechanical
  Engineering Research Laboratory,
  Hitachi Ltd., Tsuchiura, Ibaraki, Japan /
  co-sponsored by IEEE Industrial
  Electronics Society ... [et al] in
  cooperation with IEEE Robotics and
  Automation Society ... [et al.].
  Published/Created: New York, NY:
  Institute of Electrical and Electronics
  Engineers, c1990.
  Related Authors: IEEE Industrial
  Electronics Society.
  Description: 2 v. (xxxx 1020 p.): ill.; 28
  cm.
  Notes: Accompanied by: "IROS '90 late
  papers" (5 articles) in envelope. "IEEE
  catalog number: 90TH0332-7"--T.p.
  verso. Includes bibliographical
  references and indexes.
  Subjects: Robotics--Congresses.
  LC Classification: TJ210.3 .I443 1990
  Dewey Class No.: 629.8/92 20

Proceedings: IROS '91, IEEE/RSJ
  International Workshop on Intelligent
  Robots and Systems '91: intelligence for
  mechanical systems: November 3-5,
  1991, International House Osaka,
  Osaka, Japan.
  Published/Created: New York: Institute
  of Electrical and Electronics Engineers,
  c1991.
  Related Authors: IEEE Industrial
  Electronics Society.
  Description: 3 v. (xxxiv, 1674 p.): ill.;
  29 cm.
  ISBN: 078030067X (softbound)

0780300688 (microfiche)
Notes: "Co-sponsored by IEEE
Industrial Electronics Society ... [et al.]
in cooperation with IEEE Robotics and
Automation Society, Japan Society of
Mechanical Engineers (JSME), the
Institue of Systems, Control, and
Information Engineers"--P.[i]. "IEEE
catalog number: 91TH0375-6"--T.p.
verso. Includes bibliographical
refrences.
Subjects: Robotics--Congresses.
Intelligent control systems--Congresses.
LC Classification: TJ210.3 .I443 1991
Dewey Class No.: 629.8/92 20

Proceedings: technology for the intelligent
factory / IEEE International Workshop
on Emerging Technologies and Factory
Automation, August 11-14, 1992,
World Congress Centre Melbourne
Australia; editors, R. Zurawski, T.S.
Dillon.
Published/Created: East Caulfield, Vic.:
CRL Publishing; Piscataway, NJ, USA:
IEEE Service Center, 1992.
Related Authors: Zurawski, Richard.
Dillon, Tharam S., 1943- IEEE
Industrial Electronics Society.
Swinburne Institute of Technology.
Description: xvi, 711 p.: ill.; 30 cm.
ISBN: 0780308867 (softbound: label on
t.p.) 0780308875 (hardbound: label on
t.p.) 0646103237 (Aus CIP on t.p.
verso) :
Notes: "Sponsored by the IEEE
Industrial Electronics Society,
Swinburne Institute of Technology in
cooperation with Applied Computing
Research Institute (Latrobe University),
Laboratory for Concurrent Computing
Systems (Swinburne Institute), IEEE
Victorian Section"--Cover. Includes
bibliographical references and index.
Subjects: Automation--Congresses.
Artificial intelligence--Congresses.
Robotics--Congresses.
LC Classification: T59.5 .I28 1992
Dewey Class No.: 670.42/7 20

Progress in robotics and intelligent systems /
edited by George W. Zobrist and C.Y.
(Pete) Ho.
Published/Created: Norwood, N.J.:
Ablex Pub. Corp., c1994-
Related Authors: Zobrist, George W.
(George Winston), 1934- Ho, C. Y.
(Chung You), 1933-
Description: v. <1-3: ill.; 24 cm.
ISBN: 0893915807 (v. 1)
Notes: Includes bibliographical
references and indexes.
Subjects: Robotics. Artificial
intelligence.
LC Classification: TJ211 .P76 1994
Dewey Class No.: 670.42/72 20

Progress in system and robot analysis and
control design / S.G. Tzafestas and G.
Schmidt (eds.).
Published/Created: London; New York:
Springer, c1999.
Related Authors: Tzafestas, S. G., 1939-
Schmidt, Günther, 1935- European
Robotics, Intelligent Systems, and
Control Conference (1998: Athens,
Greece)
Description: xxvii, 588 p.: ill.; 24 cm.
ISBN: 1852331232 (pbk.: alk. paper)
Notes: Includes bibliographical
references and index.
Subjects: Robotics--Congresses.
Intelligent control systems--Congresses.
Series: Lecture notes in control and
information sciences; 243
LC Classification: TJ210.3 .P78 1999
Dewey Class No.: 629.8/92 21

Prototype crawling robotics system for
remote visual inspection of high-mast
light poles / Margarit G. Lozev ... [et
al.]
Published/Created: Charlottesville, Va.:
Virginia Transportation Research
Council; [Springfield, VA: Available
through the National Technical
Information Service, 1997]
Related Authors: Lozev, Margarit G.
Virginia. Dept. of Transportation.
Virginia Transportation Research

Council.
Description: iii, 42 p.: ill.; 28 cm.
Std. Tech. Rept. No.: FHWA/VTRC 98-R2
Notes: "April 1997." Includes bibliographical references (p. 50-55). Final report. Prepared in cooperation with the Federal Highway Administration and the Virginia Dept. of Transportation under contract no. 0566-010.
Subjects: Roads--Lighting--Virginia--Maintenance and repair. Roads--Lighting--Inspection--Virginia--Automation. Robotics. Robot vision.
Series: VTRC (Series); 98-R2.
LC Classification: IN PROCESS (COPIED) (lccopycat)

Quality use of the computer: computational mechanics, artificial intelligence, robotics, and acoustic sensing / presented at the 1989 ASME Pressure Vessels and Piping Conference--JSME Co-Sponsorship, Honolulu, Hawaii, July 23-27, 1989; edited by J.F. Cory, Jr. ... [et al.].
Published/Created: New York, N.Y.: American Society of Mechanical Engineers, c1989.
Related Authors: Cory, James F. American Society of Mechanical Engineers. Nihon Kikai Gakkai.
Description: vi, 238 p.: ill.; 28 cm.
ISBN: 0791803333
Notes: Includes bibliographical references.
Subjects: Electronic digital computers--Congresses. Mechanics, Applied--Data processing--Congresses. Artificial intelligence--Congresses. Robotics--Congresses. Acoustooptical devices--Congresses.
Series: PVP (Series); vol. 177.
LC Classification: QA75.5 .A74 1989
Dewey Class No.: 004 20

Raibert, Marc H.
Legged robots that balance / Marc H. Raibert.
Published/Created: Cambridge, Mass.: MIT Press, 1986.
Description: xiii, 233 p.: ill.; 24 cm.
ISBN: 0262181177
Notes: Includes index. Bibliography: p. [203]-227.
Subjects: Robotics. Artificial intelligence.
Series: The MIT Press series in artificial intelligence
LC Classification: TJ211 .R35 1986
Dewey Class No.: 629.8/92 19

RAMSETE: articulated and mobile robotics for services and technologies / Salvatore Nicosia ... eds. [et al.].
Published/Created: Berlin; New York: Springer, c2001.
Related Authors: Nicosia, Salvatore, Professor.
Description: xx, 273 p.: ill.; 24 cm.
ISBN: 3540420908 (pbk.: alk. paper)
Notes: Includes bibliographical references.
Subjects: Robots--Control systems. Mobile robots.
Series: Lecture notes in control and information sciences; 270
LC Classification: TJ211.35 b.R37 2001
Dewey Class No.: 629.8/92 21

Rath, Alan, 1959-
Alan Rath: robotics.
Published/Created: Santa Fe, N.M.: SITE Santa Fe; Santa Monica, Calif.: Smart Art Press, c1999.
Related Authors: Site Santa Fe (Gallery) Austin Museum of Art. Scottsdale Museum of Contemporary Art.
Description: 64 p.: col. ill.; 23 x 30 cm.
ISBN: 0965058352
Notes: Published on the occasion of an exhibition organized by SITE Santa Fe and held Oct. 31, 1998-Jan. 24, 1999 and at the Austin Museum of Art, Apr. 17-June 13, 1999 and the Scottsdale Museum of Contemporary Art, Sept. 10, 1999-Jan. 2, 2000.
Subjects: Rath, Alan, 1959- --Exhibitions. Kinetic sculpture--United

States--Exhibitions. Technology in art--
Exhibitions. Machinery in art--
Exhibitions.
Series: Smart Art Press (Series); v. 6,
no. 56.
LC Classification: NB237.R34 A4 1999
Dewey Class No.: 709/.2 21

Raucci, Richard.
Personal robotics: real robots to
construct, program, and explore the
world / Richard Raucci.
Published/Created: Natick, Mass.: A K
Peters, c1999.
Description: xi, 190 p.: ill.; 23 cm.
ISBN: 156881089X (pbk.)
Notes: Includes index.
Subjects: Personal robotics.
LC Classification: TJ211.416 .R38 1999
Dewey Class No.: 629.8/92 21

Real-time object measurement and
classification / edited by Anil K. Jain.
Published/Created: Berlin; New York:
Springer-Verlag, c1988.
Related Authors: Jain, Anil Kumar.
North Atlantic Treaty Organization.
Scientific Affairs Division.
Description: viii, 407 p.: ill.; 25 cm.
ISBN: 0387187669 (U.S.)
Notes: "Proceedings of the NATO
Advanced Research Workshop on Real-
time Object and Environment
Measurement and Classification, held in
Maratea, Italy, August 31-September 3,
1987"--T.p. verso. "Published in
cooperation with NATO Scientific
Affairs Division." Includes
bibliographies.
Subjects: Robot vision--Congresses.
Real-time data processing--Congresses.
Robotics--Congresses.
Series: NATO ASI series. Series F,
Computer and systems sciences; no. 42.
LC Classification: TJ211.3 .N37 1987
Dewey Class No.: 629.8/92 19

Reasoning with uncertainty in robotics:
international workshop, RUR '95,
Amsterdam, The Netherlands,

December 4-6, 1995: proceedings / Leo
Dorst, Michiel van Lambalgen, Frans
Voorbraak (eds.).
Published/Created: Berlin; New York:
Springer, c1996.
Related Authors: Dorst, Leo, 1958-
Lambalgen, Michiel van, 1954-
Voorbraak, Frans, 1961-
Description: viii, 385 p.: ill.; 24 cm.
ISBN: 3540613765 (alk. paper)
Notes: Includes bibliographical
references and index.
Subjects: Robotics--Congresses.
Reasoning--Congresses. Uncertainty--
Congresses.
Series: Lecture notes in computer
science; 1093. Lecture notes in
computer science. Lecture notes in
artificial intelligence.
LC Classification: TJ210.3 .R87 1995
Dewey Class No.: 629.8/92633 20

Recent advances in robotics / edited by
Gerardo Beni, Susan Hackwood.
Published/Created: New York: J. Wiley,
c1985.
Related Authors: Beni, Gerardo.
Hackwood, Susan.
Description: xviii, 426 p.: ill.; 25 cm.
ISBN: 0471883832
Notes: Includes bibliographies and
index.
Subjects: Robotics.
Series: Advances in robotics, 0749-1603
LC Classification: TJ211 .R42 1985
Dewey Class No.: 629.8/92 19

Recent trends in mobile robots / editor,
Yuan F. Zheng.
Published/Created: Singapore; River
Edge, N.J.: World Scientific, c1993.
Related Authors: Zheng, Yuan-Fang,
1946-
Description: xiv, 364 p.: ill.; 23 cm.
ISBN: 9810215118
Notes: Includes bibliographical
references and index.
Subjects: Mobile robots.
Series: World Scientific series in
robotics and automated systems; vol. 11

LC Classification: TJ211.415 .R46 1993
Dewey Class No.: 629.8/92 20

Recent trends in robotics: modeling, control, and education: proceedings of the International Symposium on Robot Manipulators: Modeling, Control, and Education held ...
Published/Created: New York: North-Holland, c1986.
Description: 1 v.: ill.; 24 cm. Nov. 12-14, 1986.
Continued by: International Symposium on Robotics and Manufacturing.
Robotics and manufacturing 1052-4150 (DLC) 91641763 (OCoLC)22288200
Notes: SERBIB/SERLOC merged record
Subjects: Robotics--Congresses. Manufacturing processes--Automation--Congresses. Robots, Industrial--Congresses.
LC Classification: TJ210.3 .I6a
Dewey Class No.: 670.42/72/05 20

Redford, A. H.
Robots in assembly / A.H. Redford, E. Lo.
Published/Created: New York: Halsted Press, 1986.
Related Authors: Lo, E. (Eddie)
Description: x, 176 p.: ill.; 24 cm.
ISBN: 0470203269
Notes: Includes index. Bibliography: p. [170]-172.
Subjects: Robots, Industrial. Assembling machines.
Series: Open University Press robotics series
LC Classification: TS191.8 .R43 1986
Dewey Class No.: 670.42/7 19

Reed, Harold L.
Brains for animats / Harold Reed.
Published/Created: New York: Kroshka Books, 1996.
Description: p. cm.
ISBN: 1560723203
Notes: Includes bibliographical references.

Subjects: Robotics. Artificial intelligence. Brain.
LC Classification: TJ211 .R423 1996
Dewey Class No.: 003/.7 20

Rehg, James A.
Introduction to robotics in CIM systems / James A. Rehg.
Edition Information: 5th ed.
Published/Created: Upper Saddle River, NJ: Prentice Hall, 2003.
Description: p. cm.
ISBN: 0130602434
Notes: Includes bibliographical references and index.
Subjects: Robotics. Computer integrated manufacturing systems.
LC Classification: TJ211 .R422 2003
Dewey Class No.: 670.42/72 21

Rehg, James A.
Introduction to robotics in CIM systems / James A. Rehg.
Edition Information: 4th ed.
Published/Created: Upper Saddle River, N.J.: Prentice Hall, 2000.
Description: xxiii, 440 p.: ill.; 25 cm.
ISBN: 0139012087
Notes: Includes index.
Subjects: Robotics. Computer integrated manufacturing systems.
LC Classification: TJ211 .R43 2000
Dewey Class No.: 670.42/72 21

Rehg, James A.
Introduction to robotics in CIM systems / James A. Rehg.
Edition Information: 3rd ed.
Published/Created: Upper Saddle River, N.J.: Prentice Hall, c1997.
Description: xvii, 318 p.: ill.; 24 cm.
ISBN: 0132383950 (hbk.: alk. paper)
Notes: Includes index.
Subjects: Robotics. Computer integrated manufacturing systems.
LC Classification: TJ211 .R43 1997
Dewey Class No.: 670.42/72 20

Rehg, James A.
Introduction to robotics in CIM systems

/ James A. Rehg.
Edition Information: 2nd ed.
Published/Created: Englewood Cliffs,
N.J.: Prentice Hall, c1992.
Related Authors: Rehg, James A.
Introduction to robotics.
Description: xv, 218 p.: ill.; 24 cm.
ISBN: 0134891139
Notes: Rev. ed. of: Introduction to
robotics. c1985. Includes index.
Subjects: Robotics.
LC Classification: TJ211 .R43 1992
Dewey Class No.: 670.42/72 20

Rehg, James A.
Introduction to robotics: a systems
approach / James A. Rehg.
Published/Created: Englewood Cliffs,
N.J.: Prentice-Hall, c1985.
Description: x, 230 p.: ill.; 25 cm.
ISBN: 0134955811 :
Notes: Includes index.
Subjects: Robotics.
LC Classification: TJ211 .R43 1985
Dewey Class No.: 629.8/92 19

Reichardt, Jasia.
Robots: fact, fiction, and prediction /
Jasia Reichardt.
Published/Created: Harmondsworth,
Eng.; New York: Penguin Books,
c1978.
Description: 168 p.: ill.; 28 cm.
ISBN: 014004938X
Notes: Includes bibliographical
references and index.
Subjects: Robotics.
LC Classification: TJ211 .R44 1978b
Dewey Class No.: 629.8/92

Reichardt, Jasia.
Robots: fact, fiction, and prediction /
Jasia Reichardt.
Published/Created: New York: Viking
Press, 1978.
Description: 168 p.: ill.; 29 cm.
ISBN: 067060156X
Notes: Includes index. Bibliography: p.
167.
Subjects: Robotics.

LC Classification: TJ211 .R44 1978
Dewey Class No.: 629.8/92

RIA robotics glossary.
Published/Created: Dearborn, Mich.:
RIA, [c1984-
Related Authors: Robot Institute of
America.
Description: v.; 18 cm. [No. 1]-
ISSN: 0741-9473
Notes: Title from cover.
SERBIB/SERLOC merged record
Subjects: Robotics--Dictionaries.
Robots--Dictionaries.
LC Classification: TJ210.4 .R53
Dewey Class No.: 629.8/92 19

Rietman, Ed.
Genesis redux: experiments creating
artificial life / by Edward Rietman.
Edition Information: 1st ed.
Published/Created: New York:
Windcrest/McGraw-Hill, c1994.
Description: xvii, 347 p.: ill.; 24 cm. + 1
computer disk (3 1/2 in.)
ISBN: 0830645039 :
Notes: System requirements for
computer disk: IBM-compatible PC;
DOS. Source code in BASIC, C, and
Pascal. Includes bibliographical
references (p. 325-344) and index. LC
has 2 copies of disk.
Subjects: Artificial intelligence.
Robotics. Biological systems.
LC Classification: Q335 .R55 1994
Dewey Class No.: 003/.7 20

Robillard, Mark J.
Advanced robot systems / by Mark J.
Robillard.
Edition Information: 1st ed.
Published/Created: Indianapolis, Ind.:
H.W. Sams, c1984.
Description: 215 p.: ill.; 28 cm.
ISBN: 0672221667 (pbk.) :
Notes: Includes index.
Subjects: Robotics. Robots.
LC Classification: TJ211 .R525 1984
Dewey Class No.: 629.8/92 19

Robillard, Mark J.
    HERO 1, advanced programming and
    interfacing / by Mark J. Robillard.
    Edition Information: 1st ed.
    Published/Created: Indianapolis, Ind.:
    H.W. Sams, c1983.
    Description: 234 p.: ill.; 29 cm.
    ISBN: 0672221659 (pbk.) :
    Notes: Includes index.
    Subjects: Robotics. Robots.
    LC Classification: TJ211 .R526 1983
    Dewey Class No.: 629.8/92/02854 19

Robillard, Mark J.
    Microprocessor based robotics / by
    Mark J. Robillard; [illustrated by R.E.
    Lund].
    Edition Information: 1st ed.
    Published/Created: Indianapolis, Ind.,
    USA: H.W. Sams, c1983.
    Related Authors: Lund, R. E.
    Description: 220 p.: ill.; 28 cm.
    ISBN: 0672220504 (pbk.) :
    Notes: Includes index. Bibliography: p.
    215.
    Subjects: Robotics. Microprocessors.
    Series: Intelligent machine series; v. 1
    LC Classification: TJ211 .R53 1983
    Dewey Class No.: 629.8/92 19

Robinson, William S. (William Spencer),
    1940-
    Computers, minds & robots / William S.
    Robinson.
    Published/Created: Philadelphia:
    Temple University Press, 1992.
    Description: x, 281 p.: ill.; 24 cm.
    ISBN: 0877229155 (alk. paper)
    Notes: Includes bibliographical
    references (p. [265]-274) and indexes.
    Subjects: Computers. Robotics.
    Artificial intelligence. Philosophy of
    mind.
    LC Classification: QA76.5 .R497 1992
    Dewey Class No.: 006.3 20

RoboCup 2000: Robot Soccer World Cup
    IV / Peter Stone, Tucker Balch, Gerhard
    Kraetzschmar (eds.)
    Published/Created: New York: Springer,
    2001.
    Related Authors: Stone, Peter, 1971-
    Balch, Tucker. Kraetzschmar, Gerhard
    K.
    Description: p. cm.
    ISBN: 3540421858 (pbk.)
    Notes: Includes index.
    Subjects: Robotics--Congresses.
    Artificial intelligence--Congresses.
    Soccer--Computer simulation--
    Congresses.
    Series: Lecture notes in computer
    science; 2019. Lecture notes in
    computer science. Lecture notes in
    artificial intelligence.
    LC Classification: TJ210.3 .R6153 2000
    Dewey Class No.: 629.8/92 21

RoboCup-97: robot soccer World Cup I /
    Hiroaki Kitano (ed.).
    Published/Created: Berlin; New York:
    Springer, c1998.
    Related Authors: Kitano, Hiroaki, 1961-
    Description: xiv, 520 p.: ill.; 24 cm.
    ISBN: 3540644733 (pbk.: alk. paper)
    Notes: Includes bibliographical
    references and index.
    Subjects: RoboCup-97 (1997: Nagoya-
    shi, Japan) Robotics--Congresses.
    Artificial intelligence--Congresses.
    Robotics in sports--Congresses. Soccer-
    -Computer simulation--Congresses.
    Series: Lecture notes in computer
    science; 1395. Lecture notes in
    computer science. Lecture notes in
    artificial intelligence.
    LC Classification: TJ210.3 .R615 1998
    Dewey Class No.: 629.8/92 21

RoboCup-98: Robot Soccer World Cup II /
    Minoru Asada, Hiroaki Kitano (eds.).
    Published/Created: Berlin; New York:
    Springer, c1999.
    Related Authors: Asada, Minoru.
    Kitano, Hiroaki, 1961-
    Description: xi, 509 p.: ill.; 24 cm.
    ISBN: 3540663207 (pbk.: alk. paper)
    Notes: Includes bibliographical
    references and index.
    Subjects: Robotics--Congresses.

Artificial intelligence--Congresses.
Soccer--Computer simulation--
Congresses.
Series: Lecture notes in computer
science; 1604. Lecture notes in
computer science. Lecture notes in
artificial intelligence.
LC Classification: TJ210.3 .R6153 1998
Dewey Class No.: 629.8/92 21

RoboCup-99: Robo Soccer World Cup III /
Manuela Veloso, Enrico Pagello,
Hiroaki Kitano (eds.).
Published/Created: Berlin; New York:
Springer, c2000.
Related Authors: Veloso, Manuela M.
Pagello, Enrico, 1946- Kitano, Hiroaki,
1961-
Description: xiv, 802 p.: ill.; 23 cm.
ISBN: 3540410430 (softcover: alk.
paper)
Notes: Includes bibliographical
references and index.
Subjects: Robotics--Congresses.
Artificial intelligence--Congresses.
Soccer--Computer simulation--
Congresses.
Series: Lecture notes in computer
science; 1856. Lecture notes in
computer science. Lecture notes in
artificial intelligence.
LC Classification: TJ210.3 .R6153 1999
Dewey Class No.: 629.8/92 21

Robomatix reporter.
Published/Created: New York, N.Y.:
EIC/Intelligence, [1983-
Related Authors: EIC Intelligence Inc.
Environment Information Center.
Description: v.; 28 cm. Includes an
annual index called: Robomatix
reporter, annual index Vol. 1, no. 1
(Apr. 1983)-
ISSN: 0748-1624
Notes: Title from cover. Accompanied
by annual index. SERBIB/SERLOC
merged record Annual cumulation:
Robomatix reporter annual. Robomatix
reporter annual 0000-121X (DLC)sn
89007659

Subjects: Robots, Industrial--
Periodicals.
LC Classification: TS191.8 .R558
Dewey Class No.: 629.8/92 19

Robot colonies / edited by Ronald C. Arkin,
George A. Bekey.
Published/Created: Boston: Kluwer
Academic, c1997.
Related Authors: Arkin, Ronald C.,
1949- Bekey, George A., 1928-
Description: 153 p.: ill. (some col.); 27
cm.
ISBN: 0792399048 (acid-free paper)
Notes: "Reprinted from a special issue
of Autonomous robots, volume 4,
number 1, March 1997." Includes
bibliographical references and index.
Subjects: Robotics. Autonomous robots.
LC Classification: TJ211 .R533 1997
Dewey Class No.: 629.8/92 21

Robot components and systems / Francois
L'Hote ... [et al.].
Published/Created: London: Kogan
Page; Englewood Cliffs, NJ: Prentice-
Hall, c1983.
Related Authors: L'Hote, François.
Description: 346 p.: ill.; 24 cm.
ISBN: 0137821603
Notes: Translated from the French.
Includes bibliographical references and
index.
Subjects: Robotics. Robots, Industrial.
Series: Robots. English; v. 4.
LC Classification: TJ211 .R57313 1983
vol. 4
Dewey Class No.: 629.8/92 s 629.8/92
19

Robot control (SYROCO '85): proceedings
of the 1st IFAC symposium, Barcelona,
Spain, 6-8 November 1985 / edited by
L. Basañez, G. Ferraté, and G.N.
Saridis.
Edition Information: 1st ed.
Published/Created: Oxford
[Oxfordshire]; New York: Published for
the International Federation of
Automatic Control by Pergamon Press,

1986.
Related Authors: Basañez, L. Ferraté,
Gabriel A., 1932- Saridis, George N.,
1931- International Federation of
Automatic Control. Institut de
Cibernètica (Barcelona, Spain) IEEE
Robotics and Automation Council.
Description: xiii, 561 p.: ill.; 31 cm.
ISBN: 0080334466 (U.S.) :
Notes: Organized by Comité Español of
IFAC and the Institut de Cibernètica;
sponsored by various committees of
IFAC, with the cooperation of IEEE
Robotics and Automation Council.
Includes bibliographies and indexes.
Subjects: Robots, Industrial--
Congresses. Manipulators
(Mechanism)--Congresses. Robots--
Control systems--Congresses.
Series: IFAC proceedings series; 1986,
no. 9
LC Classification: TS191.8 .I34 1985
Dewey Class No.: 670.42/7 19

Robot control: theory and applications /
   edited by K. Warwick and A. Pugh.
Published/Created: London: P.
Peregrinus on behalf of the Institution of
Electrical Engineers, c1988.
Related Authors: Warwick, K. Pugh, A.
(Alan) Institution of Electrical
Engineers.
Description: xii, 238 p.: ill.; 24 cm.
ISBN: 0863411282
Notes: Based on papers from a
workshop held at the University of
Oxford in April 1988. Includes
bibliographical references and index.
Subjects: Robotics--Congresses. Robots,
Industrial--Congresses. Manipulators
(Mechanism)--Congresses.
Series: IEE control engineering series;
v. 36.
LC Classification: TJ211 .R534 1988
Dewey Class No.: 670.42/72 20

Robot design handbook / SRI International;
   Gerry B. Andeen, editor-in-chief.
Published/Created: New York:
McGraw-Hill, c1988.

Related Authors: Andeen, Gerry B. SRI
International.
Description: 1 v. (various pagings): ill.;
24 cm.
ISBN: 007060777X
Notes: Includes index. Bibliography: p.
R1-R7.
Subjects: Robotics. Manipulators
(Mechanism)--Design and construction.
LC Classification: TJ211 .R535 1988
Dewey Class No.: 629.8/92 19

Robot revolution: the implications of
   automation / [by M. Kenneth Boss ... et
   al.].
Published/Created: [s.l.]: Socialist
Society, USA, 1955.
Related Authors: Boss, M. Kenneth.
Socialist Society, USA.
Description: 47 p.; 23 cm.
Subjects: Robotics--Economic aspects--
United States. Machinery in the
workplace--United States.
LC Classification: HD6331.2.U5 R6

Robot teams: from diversity to
   polymorphism / edited by Tucker Balch,
   Lynne E. Parker.
Published/Created: Natick, MA: A K
Peters, 2001.
Related Authors: Balch, Tucker. Parker,
Lynne E.
Description: p. cm.
ISBN: 1568811551
Notes: Includes bibliographical
references and index.
Subjects: Robotics. Intelligent agents
(Computer software)
LC Classification: TJ211 .R542 2001
Dewey Class No.: 629.8/92 21

Robot tech talk / by Ruth and Ed Radlauer ...
   [et al.]; cartoons by Eileen Morris.
Published/Created: Chicago: Childrens
Press, c1985.
Related Authors: Radlauer, Ruth, 1926-
Morris, Eileen, ill.
Description: 63 p.: ill. (some col.); 24
cm.
ISBN: 051608254X

Summary: A dictionary of words and terms relating to robotics from "android" to "work envelope."
Notes: "An Elk Grove book." Includes index.
Subjects: Robotics--Dictionaries, Juvenile. Robotics--Dictionaries.
Series: Tech talk books
LC Classification: TJ210.4 .R63 1985
Dewey Class No.: 629.8/92 19

Robot technology and applications: proceedings of the 1st Robotics Europe Conference, Brussels, June 27-28, 1984 / editors, K. Rathmill ... [et al.].
Published/Created: Berlin; New York: Springer-Verlag, 1985.
Related Authors: Rathmill, K. (Keith), 1947-
Description: x, 199 p.: ill.; 24 cm.
ISBN: 0387139605 (U.S.)
Notes: Includes bibliographies.
Subjects: Robotics--Congresses. Robots, Industrial--Congresses.
LC Classification: TJ210.3 .R63 1984
Dewey Class No.: 629.8/92 19

Robot technology.
Published/Created: London: Kogan Page; Englewood Cliffs, N.J.: Prentice-Hall, c1983-
Description: v. <1-2, 3A-3B, 4-5, 7-8: ill.; 24 cm.
ISBN: 0137820941 (Prentice-Hall: v. 1) :
Notes: Translation of: Les Robots. Includes bibliographical references and indexes.
Subjects: Robotics. Robots, Industrial.
LC Classification: TJ211 .R57313 1983
Dewey Class No.: 629.8/92 19

Robot vision in Holland: proceedings of the seminar 'intelligent seeing robots in the Netherlands 1982' and the exposition 'robot vision,' Delft, June 18th, 1982 / N.J. Zimmerman, editor.
Published/Created: Pijnacker, Netherlands: D.E.B. Publishers, 1982.
Related Authors: Zimmerman, N. J.

Description: 257 p.: ill.; 24 cm.
ISBN: 9062311016 :
Notes: Includes bibliographies and indexes.
Subjects: Robot vision--Congresses. Robotics--Netherlands--Congresses.
LC Classification: TJ211.3 .R6 1982
Dewey Class No.: 629.8/92 19

Robot vision systems.
Published/Created: Norwalk, Conn., U.S.A. (6 Prowitt St., Norwalk 06855): International Resource Development, c1985.
Related Authors: International Resource Development, inc.
Description: v, 263 leaves: ill.; 28 cm.
Notes: "March 1985."
Subjects: Robot vision. Robotics.
Series: Report (International Resource Development, inc.); #645.
LC Classification: TJ211.3 .R63 1985
Dewey Class No.: 629.8/92 19

Robotic and semi-robotic ground vehicle technology: 15-16 April 1998, Orlando, Florida / Grant R. Gerhart, Ben Abbott, chairs/editors; sponsored ... by SPIE--the International Society for Optical Engineering.
Published/Created: Bellingham, Wash., USA: SPIE, c1998.
Related Authors: Gerhart, Grant R. Abbott, Ben Allen. Society of Photo-optical Instrumentation Engineers.
Description: vii, 274 p.: ill.; 28 cm.
ISBN: 0819428159
Notes: Includes bibliographical references and index.
Subjects: Robots--Control systems--Congresses. Robotics--Military applications--Congresses. Vehicles, Remotely piloted--Congresses. Vehicles, Military--Congresses. Remote control--Congresses.
Series: Proceedings of SPIE--the International Society for Optical Engineering; v. 3366.
LC Classification: TJ211.35 .R644 1998

Dewey Class No.: 629.8/92 21

Robotic education and training: meeting the educational challenge: August 20-22, 1984, Romulus, Michigan.
Published/Created: Dearborn, Mich. (1 SME Dr., Dearborn 48121): Society of Manufacturing Engineers, c1984.
Related Authors: Society of Manufacturing Engineers.
Description: 1 v. (various pagings): ill.; 28 cm.
Notes: Cover title. "For presentation at a creative manufacturing engineering program." Contains Technical papers MS84-560-574 of the Society of Manufacturing Engineers. Includes bibliographies.
Subjects: Robotics--Study and teaching--Congresses.
LC Classification: TS191.8 .R5592 1984
Dewey Class No.: 629.8/92/071 19

Robotic observatories / editor, Michael F. Bode.
Published/Created: Chichester; New York: Wiley; Chichester: Praxis Pub., 1995.
Related Authors: Bode, M. F.
Description: 158 p.: ill.; 25 cm.
ISBN: 0471956902
Notes: Includes bibliographical references and index.
Subjects: Astronomical observatories--Automation. Robotics--Scientific applications.
Series: Wiley-Praxis series in astronomy and astrophysics
LC Classification: QB81 .R63 1995
Dewey Class No.: 522/.2 20

Robotic solutions in aerospace manufacturing: conference, March 3-5, 1986, Orlando, Florida / sponsored and published by Robotics International of SME.
Edition Information: 1st ed.
Published/Created: Dearborn, Mich.: Robotics International of SME, c1986.

Related Authors: Robotics International of SME.
Description: 1 v. (various pagings): ill.; 28 cm.
ISBN: 0872632199 (pbk.)
Notes: Includes bibliographies.
Subjects: Aerospace industries--Automation--Congresses. Robots, Industrial--Congresses.
LC Classification: TL671.28 .R63 1986
Dewey Class No.: 629.134/2 19

Robotic systems education and training, August 11-13, 1986, Detroit, Michigan: creative manufacturing engineering program.
Published/Created: Dearborn, Mich. (1 SME Dr., Dearborn): Society of Manufacturing Engineers, c1986.
Related Authors: Society of Manufacturing Engineers.
Description: 1 v. (various pagings): ill.; 28 cm.
Notes: Cover title. "For presentation at a Creative manufacturing engineering program." Includes bibliographies.
Subjects: Robotics--Congresses. Robots, Industrial.
LC Classification: TJ210.3 .R616 1986
Dewey Class No.: 670.42/7 19

Robotic systems: advanced techniques and applications / edited by Spyros G. Tzafestas.
Published/Created: Dordrecht; Boston: Kluwer Academic, c1992.
Related Authors: Tzafestas, S. G., 1939-
Description: xi, 639 p.: ill.; 25 cm.
ISBN: 0792317491
Notes: "A selection of papers presented at the 'European Robotics and Intelligent Systems Conference' (EURISCON '91) held in Corfu, Greece (June 23-28, 1991)"--P. xi. Includes bibliographical references.
Subjects: Robotics.
Series: International series on microprocessor-based and intelligent systems engineering; v. 10
LC Classification: TJ211 .R548 1992

Dewey Class No.: 670.42/72 20

Robotic technology / edited by A. Pugh.
Published/Created: London, UK: P.
Peregrinus, c1983.
Related Authors: Pugh, A. (Alan)
Description: 168 p.: ill.; 24 cm.
ISBN: 0863410049
Notes: Series taken from jacket.
Includes bibliographical references.
Subjects: Robotics. Robots, Industrial.
Series: [IEE control engineering series;
23]
LC Classification: TJ211 .R55 1983
Dewey Class No.: 629.8/92 19

Robotic telescopes: current capabilities,
present developments, and future
prospects for automated astronomy:
proceedings of a symposium held as
part of the 106th annual meeting of the
Astronomical Society of the Pacific,
Flagstaff, Arizona, 28-30 June 1994 /
edited by Gregory W. Henry and Joel A.
Eaton.
Published/Created: San Francisco:
Astronomical Society of the Pacific,
1995.
Related Authors: Henry, Gregory W.
Eaton, Joel A. Astronomical Society of
the Pacific. Meeting (106th: 1994:
Flagstaff, Ariz.)
Description: xv, 270 p.: ill.; 24 cm.
ISBN: 0937707988
Notes: "A symposium on Robotic
Telescopes and Automated Astronomy
was held ..."--Preface. Includes
bibliographical references and index.
Subjects: Telescopes--Automatic
control--Congresses. Computerized
instruments--Congresses. Robotics--
Congresses.
Series: Astronomical Society of the
Pacific conference series; v. 79
LC Classification: QB88 .R615 1995
Dewey Class No.: 522/.2 20

Robotic welding: a guide to selection and
application / John A. Piotrowski III,
William T. Randolph, editors.

Edition Information: 1st ed.
Published/Created: Dearborn, Mich.:
Welding Division, Robotics
International of SME, Publications
Development Dept., Marketing
Division, c1987.
Related Authors: Piotrowski, John A.
Randolph, William T. Robotics
International of SME. Marketing
Services Division.
Description: 255 p.: ill.; 29 cm.
ISBN: 0872632660
Notes: Includes bibliographies and
index.
Subjects: Welding--Automation.
LC Classification: TS227.2 .R63 1987
Dewey Class No.: 671.5/2 19

Robotica.
Published/Created: Cambridge
[Cambridgeshire]; New York, NY:
Cambridge University Press, c1983-
Description: v.: ill.; 30 cm. Vol. 1, pt. 1
(Jan. 1983)-
ISSN: 0263-5747
Notes: Title from cover.
SERBIB/SERLOC merged record
Indx'd selectively by: Electronics and
communications abstracts journal
(Riverdale) 0361-3313 International
aerospace abstracts 0020-5842 1983-
ISMEC bulletin 0306-0039 Pollution
abstracts with indexes 0032-3624 Safety
science abstracts journal 0160-1342
Subjects: Robotics--Periodicals.
Artificial intelligence--Periodicals.
LC Classification: TJ210.2 .R6
Dewey Class No.: 629.8/92/05 19

Robotics / by the editors of Time-Life
Books.
Published/Created: Alexandria, Va.:
Time-Life Books, [1991]
Related Authors: Time-Life Books.
Description: 128 p.: col. ill.; 29 cm.
ISBN: 0809475820 0809475839 (lib.
bdg.)
Notes: Revised 1991. Includes
bibliographical references (p. 124-125)
and index.

Subjects: Robotics.
Series: Understanding computers
LC Classification: TJ211.15 .R62 1991
Dewey Class No.: 629.8/92 20

Robotics / edited by E.P. Popov and E.I.
  Yurevich; [translated from the Russian
  by Boris V. Kuznetsov].
  Published/Created: Moscow: Mir
  Publishers, 1987.
  Related Authors: Popov, E. P. (Evgenii
  Pavlovich) IUrevich, Evgenii Ivanovich.
  Description: 247 p.: ill.; 22 cm.
  Notes: Rev. translation of:
  Robototekhnika. Includes index.
  Bibliography: p. [243]-245.
  Subjects: Robotics. Robots, Industrial.
  LC Classification: TJ211 .R56913 1987
  Dewey Class No.: 670.42/7 19

Robotics / edited by Marvin Minsky; Omni
  editorial consultant, Douglas Colligan;
  photo editor, Robert Malone.
  Edition Information: 1st ed.
  Published/Created: Garden City, N.Y.:
  Anchor Press/Doubleday, 1985.
  Related Authors: Minsky, Marvin Lee,
  1927-
  Description: 317 p.: ill.; 24 cm.
  ISBN: 0385194145 :
  Notes: "An Omni Press book." Includes
  index.
  Subjects: Robotics.
  LC Classification: TJ211 .R557 1985
  Dewey Class No.: 629.8/92 19

Robotics / J. Baillieul ... [et al.].
  Published/Created: Providence, R.I.:
  American Mathematical Society, c1990.
  Related Authors: Baillieul, J. (John)
  American Mathematical Society.
  Description: x, 196 p.: ill.; 26 cm.
  ISBN: 0821801635 (acid-free paper)
  Notes: "Lecture notes prepared for the
  American Mathematical Society short
  course Robotics held in Louisville,
  Kentucky, January 16-17, 1990"--T.p.
  verso. Includes bibliographical
  references and index.
  Subjects: Robotics--Mathematics.

Series: Proceedings of symposia in
  applied mathematics; v. 41. Proceedings
  of symposia in applied mathematics.
  AMS short course lecture notes.
  LC Classification: TJ211 .R5572 1990
  Dewey Class No.: 629.8/92/0151 20

Robotics / research analyst, Neil
  DiGeronimo.
  Published/Created: Cleveland, Ohio
  (200 University Circle Research Center,
  11001 Cedar Ave., Cleveland 44106):
  Predicasts, 1982.
  Related Authors: DiGeronimo, Neil.
  Predicasts, inc.
  Description: iii, 85 leaves: ill.; 30 cm.
  Subjects: Robotics--United States.
  Market surveys--United States.
  Series: Industry study (Predicasts, inc.:
  1975); E70.
  LC Classification: HD9696.R623 U65
  1982
  Dewey Class No.: 338.4/7629892 19

Robotics 2000: proceedings of the Fourth
  International Conference and
  Exposition/Demonstration on Robotics
  for Challenging Situations and
  Environments: February 27-March 2,
  2000, Albuquerque. N.M. / edited by
  William C. Stone.
  Published/Created: Reston, VA:
  American Society of Civil Engineers,
  2000.
  Related Authors: Stone, William C.
  (William Curtis), 1952- American
  Society of Civil Engineers. Space
  (Conference) (7th: 2000: Albuquerque,
  N.M.)
  Description: x, 364 p.: ill.; 22 cm.
  ISBN: 0784404763
  Notes: Includes bibliographical
  references and index.
  Subjects: Robotics--Congresses.
  LC Classification: TJ210.3. I5768 2000
  Dewey Class No.: 629.8/92 21

Robotics 98: proceedings of the Third ASCE
  Specialty Conference on Robotics for
  Challenging Environments: April 26-30,

1998, Albuquerque, New Mexico /
edited by Laura A. Demsetz, Raymond
H. Bryne, John P. Wetzel.
Published/Created: Reston, Va.:
American Society of Civil Engineers,
c1998.
Related Authors: Demsetz, Laura A.
Bryne, Raymond H. Wetzel, John P.
Description: viii, 346 p.: ill.; 22 cm.
ISBN: 0784403376
Notes: Includes bibliographical
references and indexes.
Subjects: Robotics--Congresses.
LC Classification: TJ210.3 .A83 1998
Dewey Class No.: 629.8/92 21

Robotics abstracts annual.
Published/Created: New York, N.Y.:
Bowker A&I Pub., c1990-1991.
Related Authors: Bowker A&I
Publishing.
Description: 3 v.; 29 cm. Ceased
publication. Vol. 7 (1989)-v. 9 (1991).
ISSN: 1053-6051 Incorrect
ISSN: 0000-121X
Notes: SERBIB/SERLOC merged
record Annual cumulation of: Robotics
abstracts. Robotics abstracts 0000-1139
(DLC) 89646704 (OCoLC)18543950
Subjects: Robots, Industrial--Abstracts--
Periodicals. Robotics--Abstracts--
Periodicals. Robots--Abstracts--
Periodicals.
LC Classification: TS191.8 .R55925
Dewey Class No.: 670.42/72 20

Robotics abstracts.
Published/Created: [New York, NY]:
Bowker A&I Pub., -1992.
Related Authors: Bowker A&I
Publishing.
Description: v.; 28 cm. -v. 10, no. 8
(Aug. 1992).
ISSN: 0000-1139
Notes: Description based on: Vol. 7, no.
4 (Apr. 1989); title from cover.
SERBIB/SERLOC merged record
Cumulated by: Robotics abstracts
annual. Robotics abstracts annual 1053-
6051 (DLC) 90641147

(OCoLC)21477975
Subjects: Robotics--Abstracts--
Periodicals. Robots--Abstracts--
Periodicals. Robots, Industrial--
Abstracts--Periodicals.
LC Classification: TS191.8 .R558
Dewey Class No.: 670.42/72 20

Robotics age product guide: a sourcebook
for educators and experimentalists.
Published/Created: [U.S.]: North
American Technology; [Peterborough,
NH: Robotics Age, c1984]
Related Authors: Robotics age.
Description: 115 p.: ill.; 28 cm.
ISBN: 091686300X (pbk.) :
Notes: Cover title.
Subjects: Robotics--Catalogs.
LC Classification: TJ211 .R5644 1984
Dewey Class No.: 629.8/92/029473 19

Robotics age, in the beginning: selected
from Robotics age magazine / edited by
Carl Helmers.
Published/Created: Rochelle Park, N.J.:
Hayden, c1983.
Related Authors: Helmers, Carl.
Description: [x], 241 p.: ill.; 28 cm.
ISBN: 0810463253
Notes: Includes bibliographical
references.
Subjects: Automata.
LC Classification: TJ211 .R564 1983
Dewey Class No.: 629.8/92 19

Robotics age.
Published/Created: [Peterboro, N.H.,
etc., Robotics Age [etc.]
Description: 7 v. ill. 28 cm. v. 1-7;
summer 1979-Dec. 1985.
ISSN: 0197-1905 CODEN: ROAGD2
Notes: SERBIB/SERLOC merged
record Indexed entirely by: Applied
science & technology index 0003-6986
Indx'd selectively by: Computer &
control abstracts 0036-8113 July-Aug.
1982-1985 Electrical & electronics
abstracts 0036-8105 July-Aug. 1982-
1985 Physics abstracts. Science
abstracts. Series A 0036-8091 July-Aug.

1982-1985 Predicasts
Subjects: Automata--Periodicals.
Robots, Industrial--Periodicals.
LC Classification: TJ211 .R56
Dewey Class No.: 629.8/92/05

Robotics and artificial intelligence / edited
by Michael Brady, Lester A. Gerhardt,
Harold F. Davidson.
Published/Created: Berlin; New York:
Springer-Verlag: Published in
cooperation with NATO Scientific
Affairs Division, 1984.
Related Authors: Brady, Michael, 1945-
Gerhardt, Lester A. Davidson, Harold F.
North Atlantic Treaty Organization.
Description: xvii, 693 p.: ill.; 25 cm.
ISBN: 0387128883 (U.S.)
Notes: "Proceedings of the NATO
Advanced Study Institute on Robotics
and Artificial Intelligence held at
Castelvecchio Pascoli, Barga, Italy,
June 26-July 8, 1983"--T.p. verso.
Includes bibliographies.
Subjects: Robotics--Congresses.
Artificial intelligence--Congresses.
Series: NATO ASI series. Series F,
Computer and systems sciences; no. 11.
LC Classification: TJ210.3 .N38 1983
Dewey Class No.: 629.8/92 19

Robotics and artificial intelligence
applications series: overviews.
Published/Created: Sacramento, Calif.:
Business/Technology Books, c1984.
Description: 4 v. in 1: ill.; 29 cm.
ISBN: 0899341845
Notes: Includes bibliographies.
Subjects: Robotics. Artificial
intelligence.
LC Classification: TJ211 .R5657 1984
Dewey Class No.: 001.53/5 19

Robotics and automation: Second IASTED
International Symposium Robotics and
Automation, June 22-24, 1983, Lugano,
Switzerland / editor, M.H. Hamza.
Published/Created: Anaheim: Acta
Press, [1983]
Related Authors: Hamza, M. H.

International Association of Science and
Technology for Development.
Description: 173 p.: ill.; 28 cm.
ISBN: 0889860491 (pbk.)
Notes: Cover title. "A Publication of the
International Association of Science and
Technology for Development--
IASTED." Includes bibliographies.
Subjects: Robots, Industrial--
Congresses. Automation--Congresses.
LC Classification: TS191.8 .I27 1983
Dewey Class No.: 670.42/7 19

Robotics and autonomous systems.
Published/Created: Amsterdam:
Elsevier Science Publishers, c1988-
Description: v.: ill.; 26 cm. Vol. 4, no. 1
(Mar. 1988)-
ISSN: 0921-8890 Incorrect
ISSN: 0921-8830 0167-8493
Notes: Title from cover.
SERBIB/SERLOC merged record
Indx'd selectively by: Computer &
control abstracts 0036-8113 Nov. 1988-
Electrical & electronic abstracts 0036-
8105 Nov. 1988- Physics abstracts
0036-8091 Nov. 1988-
Subjects: Robotics--Periodicals. Robots-
-Periodicals. Robots, Industrial--
Periodicals. Automatic control--
Periodicals.
LC Classification: TJ210.2 .R62
Dewey Class No.: 629.8/92/05 20

Robotics and computer-integrated
manufacturing.
Published/Created: New York:
Pergamon, [c1984-
Description: v.: ill.; 28 cm. Vol. 1, no.
1-
ISSN: 0736-5845
Notes: Title from cover.
SERBIB/SERLOC merged record
Absorbed in 1999: Computer-integrated
manufacturing systems. Indexed entirely
by: Applied science & technology index
0003-6986 1991-
Subjects: Robots, Industrial--
Periodicals. Computer integrated
manufacturing systems--Periodicals.

Robotics--Periodicals.
LC Classification: TS191.8 .R5593
Dewey Class No.: 670/.285 19

Robotics and expert systems: proceedings of
ROBEXS / The ... annual Workshop on
Robotics and Expert Systems.
Published/Created: Research Triangle
Park, N.C.: Instrument Society of
America, c1985-
Related Authors: Instrument Society of
America. Robotics & Expert Systems
Division. Instrument Society of
America. Clear Lake--Galveston
Section. Instrument Society of America.
District 7.
Description: v.: ill.; 28 cm. 1st (1985)-
ISSN: 0891-4621
Notes: Sponsored by: Robotics &
Expert Systems Division, Clear Lake--
Galveston Section, and District 7 of the
Instrument Society of America.
SERBIB/SERLOC merged record
Subjects: Robotics--Congresses. Expert
systems (Computer science)--
Congresses.
LC Classification: TJ210.3 .W66a
Dewey Class No.: 629.8/92 20

Robotics and factories of the future:
proceedings of an international
conference, Charlotte, North Carolina,
USA, December 4-7, 1984 / editor,
Suren N. Dwivedi.
Published/Created: Berlin; New York:
Springer-Verlag, 1984.
Related Authors: Dwivedi, Suren N.
International Conference on Robotics
and Factories of the Future (1984:
Charlotte, N.C.)
Description: xxxii, 780 p.: ill.; 24 cm.
ISBN: 0387150153 (U.S.)
Notes: Includes bibliographies.
Subjects: Robots, Industrial--
Congresses. Automation--Congresses.
LC Classification: TS191.8 .R5594
1984
Dewey Class No.: 670.42/7 19

Robotics and factories of the future:
proceedings of the second International
Conference, San Diego, California,
USA, July 28-31, 1987 / R.
Radharamanan, editor.
Published/Created: Berlin; New York:
Springer-Verlag, c1988.
Related Authors: Radharamanan, R.
International Society for Productivity
Enhancement. International Conference
on Robotics and Factories of the Future
(2nd: 1987: San Diego, Calif.)
Description: xvi, 846 p.: ill.; 24 cm.
Notes: "The Second International
Conference on Robotics and Factories
of the Future ... sponsor, International
Society for Productivity Enhancement"-
-P. [v]. Includes bibliographies and
index.
Subjects: Robots, Industrial--
Congresses. Automation--Congresses.
LC Classification: TS191.8 .R5595
1988
Dewey Class No.: 670.42/7 19

Robotics and flexible manufacturing
systems: selected and revised papers
from the IMACS 13th World Congress,
Dublin, Ireland, July 1991, and the
IMACS Conference on Modelling and
Control of Technological Systems,
Lille, France, May 1991 / edited by
Jean-Claude Gentina, Spyros G.
Tzafestas.
Published/Created: Amsterdam: New
York: Elsevier, 1992.
Related Authors: Gentina, Jean-Claude.
Tzafestas, S. G., 1939- IMACS
Conference on Modelling and Control
of Technological Systems (1991: Lille,
France)
Description: viii, 436 p.: ill.; 25 cm.
ISBN: 0444897054 (acid-free paper)
Notes: Includes bibliographical
references.
Subjects: Robotics--Congresses.
Flexible manufacturing systems--
Congresses.
LC Classification: TJ210.3 .I48 1991

Dewey Class No.: 670.42/7 20

Robotics and flexible manufacturing technologies: assessment, impacts, and forecast / by Robert U. Ayres ... [et al.]. Published/Created: Park Ridge, N.J.: Noyes Publications, c1985. Related Authors: Ayres, Robert U. Description: xii, 443 p.: ill.; 25 cm. ISBN: 0815510438 : Notes: Bibliography: p. 430-436. Subjects: Robotics. Flexible manufacturing systems. Technology assessment. LC Classification: TJ211 .R56585 1985 Dewey Class No.: 670.42 19

Robotics and foreign affairs: a symposium / organized and co-edited by David T. Morrison; co-edited by Diane B. Bendahmane. Published/Created: Washington, D.C.: Center for the Study of Foreign Affairs, Foreign Service Institute, U.S. Dept. of State: For sale by the Supt. of Docs., U.S. G.P.O., 1985. Related Authors: Morrison, David T. Bendehmane, Diane B. Center for the Study of Foreign Affairs (U.S.) Description: x, 93 p.; 24 cm. Notes: Shipping list no.: 85-955-P. "Released August 1985"--T.p. verso. S/N 044-000-02061-1 Item 872-A Bibliography: p. 93. Subjects: Automation--United States-- Congresses. United States--Foreign relations--1977-1981--Congresses. LC Classification: T59.5 .R57 1985 Dewey Class No.: 338/.06 19 Govt. Doc. No.: S 1.114/3:R 57

Robotics and industrial control, RH 706-07 / NRI. Published/Created: Washington, D.C.: McGraw-Hill Continuing Education Center, [c1985] Related Authors: National Radio Institute (Washington, D.C.) Description: 28, 40 p.: col. ill.; 28 cm. Contents: Speech synthesis, RH 706 --

Robot applications, RH 707. Notes: Courses offered by NRI, the National Radio Institute. Cover title. Subjects: Speech synthesis. Robots, Industrial. LC Classification: TK7882.S65 R63 1985 Dewey Class No.: 006.5/4 19

Robotics and industrial engineering: selected readings / editors, Edward L. Fisher, Oded Z. Maimon. Published/Created: Atlanta, Ga.: Industrial Engineering and Management Press, Institute of Industrial Engineers, 1983-1986. Related Authors: Fisher, Edward L. Maimon, Oded Z. Institute of Industrial Engineers (1981-) Description: 2 v.: ill.; 28 cm. ISBN: 0898060451 (pbk.) Notes: Includes bibliographies and indexes. Subjects: Robots, Industrial. Robotics. LC Classification: TS191.8 .R56 1983 Dewey Class No.: 670.42/7 19

Robotics and industrial inspection: August 24-27, 1982 San Diego, California / David P. Casasent, chairman/editor, in cooperation with IEEE Computer Society. Published/Created: Bellingham, Wash.: SPIE--International Society for Optical Engineering, c1983. Related Authors: Casasent, David Paul. IEEE Computer Society. Society of Photo-optical Instrumentation Engineers. Description: x, 356 p.: ill.; 28 cm. ISBN: 0892523956 (pbk.) Notes: Includes bibliographical references and indexes. Subjects: Robotics--Congresses. Image processing--Congresses. Robots, Industrial--Congresses. Series: Proceedings of SPIE--the International Society for Optical Engineering; v. 360 LC Classification: TJ211 .R566 1983

Dewey Class No.: 629.8/92 19

Robotics and intelligent machines in agriculture: proceedings of the first International Conference on Robotics and Intelligent Machines in Agriculture, October 2-4, 1983, Curtis Hixon Convention Center, Tampa, Florida.
Published/Created: St. Joseph, Mich.: American Society of Agricultural Engineers, c1984.
Description: v, 155 p.: ill.; 23 cm.
ISBN: 0916150607 (pbk.)
Notes: Includes bibliographical references.
Subjects: Robotics--Congresses. Agricultural machinery--Congresses. Artificial intelligence--Congresses.
Series: ASAE publication; 4-84
LC Classification: S678.65 .I58 1983
Dewey Class No.: 631.3/7 19

Robotics and manufacturing automation: presented at the Winter Annual Meeting of the American Society of Mechanical Engineers, Miami Beach, Florida, November 17-22, 1985 / co-sponsored by the Production Engineering Division and the Dynamic Systems and Control Division, ASME; edited by Max Donath, Ming Leu.
Published/Created: New York, N.Y. (345 E. 47th St., New York 10017): ASME, c1985.
Related Authors: Donath, Max. Leu, M. C. American Society of Mechanical Engineers. Production Engineering Division. American Society of Mechanical Engineers. Dynamic Systems and Control Division.
Description: vi, 299 p.: ill.; 28 cm.
Notes: Includes bibliographies and index.
Subjects: Robotics--Congresses. Robots, Industrial--Congresses.
Series: PED (Series); vol. 15.
LC Classification: TJ210.3 .R618 1985
Dewey Class No.: 671.42/7 19

Robotics and manufacturing: proceedings of the Fourth IASTED International Conference, August 19-22, 1996, Honolulu, Hawaii / editor, R.V. Mayorga.
Published/Created: Anaheim, Calif.: IASTED: Acta Press, c1996.
Related Authors: Mayorga, R. V. (René V.) International Association of Science and Technology for Development. IASTED International Conference, Robotics and Manufacturing (4th: 1996: Honolulu, Hawaii)
Description: iv, 380 p.: ill.; 27 cm.
ISBN: 0889862095
Notes: "Proceedings of the Fourth IASTED International Conference, Robotics and Manufacturing." Includes bibliographical references and index.
Subjects: Robotics--Congresses. Robots, Industrial--Congresses. Flexible manufacturing systems--Congresses.
LC Classification: TJ210.3 .R619 1996
Dewey Class No.: 670.42/72 21

Robotics and manufacturing: recent trends in research, education, and applications: proceedings of the ... International Symposium on Robotics and Manufacturing: Research, Education, and Applications, held ...
Published/Created: New York: ASME Press, 1988-
Description: v.: ill.; 24 cm. 2nd (Nov. 16-18, 1988)-
ISSN: 1052-4150
Notes: Subtitle varies. SERBIB/SERLOC merged record
Subjects: Robotics--Congresses. Manufacturing processes--Automation--Congresses. Robots, Industrial--Congresses.
Series: ASME Press series--proceedings of the international symposia on robotics and manufacturing
LC Classification: TJ210.3 .I6a
Dewey Class No.: 670.42/72/05 20

Robotics and material flow / edited by S.Y. Nof.

Published/Created: Amsterdam; New York: Elsevier, 1986.
Related Authors: Nof, Shimon Y., 1946-
Description: 205 p.: ill.; 28 cm.
ISBN: 0444426213 :
Notes: "Published as a special issue of Material flow, vol. 3, issues 1-3."
Includes bibliographies.
Subjects: Robots, Industrial. Materials handling.
LC Classification: TS191.8 .R57 1986
Dewey Class No.: 670.42/7 19

Robotics and remote systems for hazardous environments / editors, Mo Jamshidi, Patrick J. Eicker.
Published/Created: Englewood Cliffs, N.J.: PTR Prentice Hall, c1993.
Related Authors: Jamshidi, Mohammad. Eicker, Patrick J.
Description: ix, 229 p.: ill.; 25 cm.
ISBN: 0137825900
Notes: Includes bibliographical references and indexes.
Subjects: Robotics. Manipulators (Mechanism) Hazardous waste sites.
Series: Prentice Hall series on environmental and intelligent manufacturing systems
LC Classification: TJ211 .R5663 1993
Dewey Class No.: 604.7 20

Robotics and remote systems: proceedings of the Fourth ANS Topical Meeting on Robotics and Remote Systems / editors, Mohammad Jamshidi, Patrick J. Eicker.
Published/Created: Albuquerque, N.M.: Dept. of Energy, Albuquerque Operations: Sandia National Laboratories: CAD Laboratory for Systems/Robotics, UNM, [1991]
Related Authors: Jamshidi, Mohammad. Eicker, Patrick J. American Nuclear Society.
Description: vii, 708 p.: ill.; 23 cm.
Notes: Includes bibliographical references and indexes.
Subjects: Robotics--Congresses.
Remote control--Congresses.

LC Classification: TJ210.3 .A58 1991

Robotics and robot sensing systems: August 25, 1983, San Diego, California / David Casasent, Ernest L. Hall, chairmen/editors.
Published/Created: Bellingham, Wash.: SPIE--the International Society for Optical Engineering, c1983.
Related Authors: Casasent, David Paul. Hall, Ernest L. Society of Photo-optical Instrumentation Engineers.
Description: vi, 143 p.: ill.; 28 cm.
ISBN: 0892524774 (pbk.)
Notes: Includes bibliographies and index.
Subjects: Image processing-- Congresses. Robotics--Congresses.
Series: SPIE critical reviews of technology series; 3rd. Proceedings of SPIE--the International Society for Optical Engineering; v. 442.
LC Classification: TA1632 .R627 1983
Dewey Class No.: 629.8/92 19

Robotics applications for industry: a practical guide / by L.L. Toepperwein ... [et al.].
Published/Created: Park Ridge, N.J., U.S.A.: Noyes Data Corp., 1983.
Related Authors: Toepperwein, L. L.
Description: x, 326 p.: ill.; 25 cm.
ISBN: 0815509626 :
Notes: Includes bibliographies.
Subjects: Robots, Industrial.
LC Classification: TS191.8 .R58 1983
Dewey Class No.: 629.8/92 19

Robotics crosscutting program: technology summary / [prepared by the Office of Science and Technology].
Published/Created: [Washington, D.C.?]: U.S. Dept. of Energy; Springfield, VA: Available to the public from the U.S. Dept. of Commerce, Technology Administration, National Technical Information Service, [1996]
Related Authors: United States. Office of Science and Technology.
Description: iv, 108 p.: ill.; 28 cm.

Notes: "August 1996." "DOE/EM-0299"--P. [2] of cover. Includes
bibliographical references.
Subjects: United States. Dept. of
Energy--Buildings. Radioactive wastes--United States--Management
Technological innovations. Hazardous
wastes--United States--Management--Technological innovations.
Environmental sampling--United States--Technological innovations. Robotics--United States.
LC Classification: TD898.118 .R63
1996
Dewey Class No.: 621.48/38 21

Robotics development and future
applications: seminar report / authors,
M. Illi ... [et al.].
Published/Created: Berlin: European
Centre for the Development of
Vocational Training; Washington, DC:
European Community Information
Service [distributor], 1985.
Related Authors: Illi, M. European
Centre for the Development of
Vocational Training.
Description: 100 p.: ill.; 20 cm.
ISBN: 9282549038 (pbk.) :
Notes: Includes bibliographical
references.
Subjects: Robotics--Congresses. Robots,
Industrial--Congresses.
LC Classification: TJ210.3 .R62 1985
Dewey Class No.: 670.42/7 19

Robotics engineering.
Published/Created: [Peterborough,
N.H.: Robotics Age Inc., 1986.
Description: 1 v.: ill.; 28 cm. Vol. 8, no.
1 (Jan. 1986)-v. 8, no. 12 (Dec. 1986).
ISSN: 0888-0816 Incorrect
ISSN: 0197-1905 CODEN: ROENEA
Notes: Title from cover.
SERBIB/SERLOC merged record
Indexed entirely by: Applied science &
technology index 0003-6986 Indx'd
selectively by: Computer & control
abstracts 0036-8113 1986 Electrical &
electronics abstracts 0036-8105 1986

Physics abstracts 0036-8091 1986
Predicasts
Subjects: Robotics--Periodicals. Robots,
Industrial--Periodicals.
LC Classification: TJ211 .R56
Dewey Class No.: 670.42/7 19

Robotics for bioproduction systems / edited
by Naoshi Kondo, K.C. Ting.
Published/Created: Saint Joseph, MI:
American Society of Agricultural
Engineers, c1998.
Related Authors: Kondo, Naoshi. Ting,
K. C.
Description: xvi, 325 p.: ill.; 24 cm.
ISBN: 0929355946
Notes: Includes bibliographical
references and index.
Subjects: Robotics. Agricultural
machinery. Artificial intelligence--
Agricultural applications.
Series: ASAE publication; 05-98
LC Classification: S678.65 .R63 1998
Dewey Class No.: 631.3 21

Robotics for challenging environments:
proceedings of the ASCE Specialty
Conference: Albuquerque, New Mexico
February 26-March 3, 1994 / sponsored
by Aerospace Division of the American
Society of Civil Engineers ... [et al.];
edited by Laura A. Demsetz and Paul R.
Klarer.
Published/Created: New York, N.Y.:
American Society of Civil Engineers,
c1994.
Related Authors: Demsetz, Laura A.
Klarer, Paul R. American Society of
Civil Engineers. Aerospace Division.
Description: x, 484 p.: ill.; 22 cm.
ISBN: 0872629139
Notes: Includes bibliographical
references and indexes.
Subjects: Robotics--Congresses.
LC Classification: TJ210.3 .A83 1994
Dewey Class No.: 629.8/92 20

Robotics for challenging environments:
proceedings of the RCE II, the second
conference, Albuquerque, New Mexico,

June 1-6, 1996 / sponsored by
Aerospace Division of the American
Society of Civil Engineers ... [et al.];
edited by Laura A. Demsetz.
Published/Created: New York: The
Society, c1996.
Related Authors: Demsetz, Laura A.
American Society of Civil Engineers.
Aerospace Division.
Description: x, 328 p.: ill.; 22 cm.
ISBN: 0784401780
Notes: Includes bibliographical
references and indexes.
Subjects: Robotics--Congresses.
LC Classification: TJ210.3 .A83 1996
Dewey Class No.: 629.8/92 20

Robotics for human resource professionals:
a research report / prepared by Bernard
Hodes Advertising.
Published/Created: New York, N.Y.
(555 Madison Ave., New York 10022):
B. Hodes Advertising, c1984.
Related Authors: Bernard Hodes
Advertising (New York, N.Y.)
Description: 30 p.: ill.; 28 cm.
Subjects: Manpower planning.
Technological unemployment. Robots,
Industrial--Planning. Occupational
retraining--Planning.
LC Classification: HF5549.5.M3 R63
1984
Dewey Class No.: 658.5/14 19

Robotics in Alpe-Adria region: proceedings
of the 2nd international workshop (RAA
'93), June 1993, Krems, Austria / P.
Kopacek, ed.
Published/Created: Vienna, Austria;
New York: Springer-Verlag, 1994.
Related Authors: Kopacek, Peter.
Description: p. cm.
ISBN: 0387825452 (U.S.)
Subjects: Robotics--Alps Region--
Congresses. Robotics--Adriatic Sea
Region--Congresses.
LC Classification: TJ210.3 .R633 1994
Dewey Class No.: 670.42/72/094947 20

Robotics in meat, fish, and poultry
processing / edited by K.
Khodabandehloo.
Edition Information: 1st ed.
Published/Created: London; New York:
Blackie Academic & Professional,
1993.
Related Authors: Khodabandehloo, K.
Description: xii, 214 p.: ill.; 24 cm.
ISBN: 0751400874 0442316615 (USA)
Notes: Includes bibliographical
references and index.
Subjects: Meat industry and trade--
Automation. Robotics. Poultry plants--
Automation. Fish processing--
Automation. Food Processing
Automation
LC Classification: TS1960 .R58 1993
Dewey Class No.: 664/.9/0028 20

Robotics industry directory.
Published/Created: [La Canada, Calif.:
Robotics Pub. Corp., c1981-
Description: v.: ill.; 28 cm. 1981-
ISSN: 0278-159X
Notes: Title from cover.
SERBIB/SERLOC merged record
Updated by: Computerized
manufacturing. Continued in 1984 by:
International robotics industry directory.
Computerized manufacturing 0746-
3405 (DLC) 90656036
(OCoLC)9934098
Subjects: Robot industry--United States-
-Directories.
LC Classification: TS191.8 .R6
Dewey Class No.: 629.8/92/09473

Robotics product database.
Published/Created: Orlando, Fla.:
TecSpec Division, Flora
Communications, c1988-1991.
Related Authors: Flora
Communications. TecSpec Division.
Description: 4 v.; 23 cm. 5th ed. (1988)-
Ceased with 1991 ed.
Notes: "International."
SERBIB/SERLOC merged record
Subjects: Robot industry--Directories.
Robots--Catalogs--Periodicals. Robots--

Catalogs. Robots, Industrial--Catalogs.
LC Classification: TS191.8 .R6
Dewey Class No.: 629.8/92 19

Robotics research and advanced applications
/ presented at the Winter Annual
Meeting of the American Society of
Mechanical Engineers, Phoenix,
Arizona, November 14-19, 1982;
sponsored by the Dynamic Systems and
Controls Division, ASME; edited by
Wayne J. Book.
Published/Created: New York (345 East
47th St., New York 10017): American
Society of Mechanical Engineers,
c1982.
Related Authors: Book, Wayne J.
American Society of Mechanical
Engineers. Dynamic Systems and
Control Division.
Description: vi, 287 p.: ill.; 26 cm.
Notes: Includes bibliographical
references.
Subjects: Robotics--Congresses. Robots,
Industrial--Congresses.
LC Classification: TJ211 .A48 1982
Dewey Class No.: 629.8/92 19

Robotics research laboratories directory.
Published/Created: Dearborn, MI:
Research and Development Division of
RI/SME,
Related Authors: Robotics International
of SME. Research and Development
Division.
Description: v.; 28 cm.
Notes: Description based on: 1984-
1985. SERBIB/SERLOC merged record
Subjects: Robotics laboratories--
Directories.
LC Classification: TJ210.5 .R637
Dewey Class No.: 629.8/92 19

Robotics research, 1989: presented at the
Winter Annual Meeting of the
American Society of Mechanical
Engineers, San Francisco, California,
December 10-15, 1989 / sponsored by
the Robotics Technical Panel of the
Dynamic Systems and Control Division,

ASME; edited by Y. Youcef-Toumi, H.
Kazerooni.
Published/Created: New York, N.Y.:
ASME, c1989.
Related Authors: Youcef-Toumi,
Kamal. Kazerooni, H. (Homayoon)
American Society of Mechanical
Engineers. Winter Meeting (1989: San
Francisco, Calif.) American Society of
Mechanical Engineers. Dynamic
Systems and Control Division. Robotics
Technical Panel. Symposium on
Robotics (1989: San Francisco, Calif.)
Description: vi, 307 p.: ill.; 28 cm.
ISBN: 0791804135
Notes: "Papers presented in the
Symposium on Robotics held ... in San
Francisco during the period of
December 11 to December 15, 1989"--
P. iii. Includes bibliographical
references.
Subjects: Robotics--Research--
Congresses.
Series: DSC (Series); vol. 14.
LC Classification: TJ210.3 .R6338 1989
Dewey Class No.: 629.8/92 20

Robotics research, 1990: presented at the
Winter Annual Meeting of the
American Society of Mechanical
Engineers, Dallas, Texas, November 25-
30, 1990 / sponsored by the Dynamic
Systems and Control Division, ASME;
edited by K. Youcef-Toumi, H.
Kazerooni.
Published/Created: New York, N.Y.:
ASME, c1991.
Related Authors: Youcef-Toumi,
Kamal. Kazerooni, H. (Homayoon)
American Society of Mechanical
Engineers. Winter Meeting (1990:
Dallas, Tex.) American Society of
Mechanical Engineers. Dynamic
Systems and Control Division.
Symposium on Robotics (1990: Dallas,
Tex.)
Description: vi, 247 p.: ill.; 28 cm.
ISBN: 0791807444
Notes: "Papers presented in the
Symposium of Robotics held ... in

Dallas, Texas during the week of November 25-30, 1990"--P. iii. Includes bibliographical references and index.
Subjects: Robotics--Congresses.
Series: DSC (Series); vol. 26.
LC Classification: TJ210.3 .R634 1991
Dewey Class No.: 629.8/92 20

Robotics research, the next five years and beyond / the First World Conference on Robotics Research: conference, August 14-16, 1984, Bethlehem, Pennsylvania; sponsored and published by Robotics International of SME.
Edition Information: 1st ed.
Published/Created: Dearborn, Mich.: RI/SME, c1984.
Related Authors: Robotics International of SME.
Description: 1 v. (various pagings): ill.; 28 cm.
ISBN: 0872631524 (pbk.)
Notes: Contains Technical papers MS84-480-499; 418, 500-506, 355, of the Society of Manufacturing Engineers. Includes bibliographies.
Subjects: Robotics--Research--Congresses.
LC Classification: TJ210.3 .W67 1984
Dewey Class No.: 629.8/92 19

Robotics research: the eighth international symposium / Yoshiaki Shirai and Shigeo Hirose (eds.).
Published/Created: London; New York: Springer, c1998.
Related Authors: Shirai, Yoshiaki. Hirose, Shigeo, 1947- International Symposium on Robotics Research (8th: 1997: Sh⁻onan-machi, Japan)
Description: x, 458 p.: ill.; 28 cm.
ISBN: 3540762442 (casebound: alk. paper)
Notes: "Eight International Symposium of Robotics Research was held in Shonan near Kamakura, Japan from October 4 to 7, 1997, organized by the International Foundation of Robotics Research (IFRR)"--Pref. Includes bibliographical references.

Subjects: Robotics--Research--Congresses.
LC Classification: TJ211.3 .R64 1998
Dewey Class No.: 629.8/92/072 21

Robotics research: the fifth international symposium / edited by Hirofumi Miura and Suguru Arimoto.
Published/Created: Cambridge, Mass.: MIT Press, c1990.
Related Authors: Miura, Hirofumi, 1938- Arimoto, Suguru, 1936- International Symposium on Robotics Research (5th: 1989: University of Tokyo)
Description: vii, 468 p.: ill.; 29 cm.
ISBN: 0262132532
Notes: Papers of the Fifth International Symposium on Robotics Research, held at the University of Tokyo, Aug. 28-31, 1989. Includes bibliographical references.
Subjects: Robotics--Research--Congresses.
Series: Artificial intelligence (Cambridge, Mass.)
LC Classification: TJ210.3 .R635 1990
Dewey Class No.: 629.8/92 20

Robotics research: the first international symposium / edited by Michael Brady and Richard Paul.
Published/Created: Cambridge, Mass.: MIT Press, c1984.
Related Authors: Brady, Michael, 1945- Paul, Richard P. International Symposium on Robotics Research (1st: 1983: Bretton Woods, N.H.)
Description: xiv, 1001 p.: ill.; 27 cm.
ISBN: 0262022079
Notes: Papers presented at the First International Symposium on Robotics Research held at Bretton Woods, N.H., Aug. 25-Sept. 2, 1983. Includes bibliographies.
Subjects: Robotics--Research--Congresses. Artificial intelligence--Congresses.
Series: The MIT Press series in artificial intelligence

LC Classification: TJ211 .R568 1984
Dewey Class No.: 629.8/92 19

Robotics research: the fourth international
symposium / edited by Robert C. Bolles
and Bernard Roth.
Published/Created: Cambridge, Mass.:
MIT Press, c1988.
Related Authors: Bolles, Robert C.
Roth, Bernard. National Science
Foundation (U.S.) System Development
Foundation (Palo Alto, Calif.)
International Symposium on Robotics
Research (4th: 1987: University of
California at Santa Cruz)
Description: xii, 520 p.: ill.; 29 cm.
ISBN: 0262022729 :
Notes: Papers of the Fourth
International Symposium on Robotics
Research, held at the University of
California at Santa Cruz on Aug. 9-14,
1987; sponsored by the National
Science Foundation and the System
Development Foundation. Includes
bibliographies
Subjects: Robotics--Research--
Congresses.
Series: The MIT Press series in artificial
intelligence
LC Classification: TJ210.3 .R636 1988
Dewey Class No.: 629.8/92 19

Robotics research: the ninth international
symposium / John M. Hollerbach and
Daniel E. Koditschek (eds.).
Published/Created: London; New York:
Springer, 2000.
Related Authors: Hollerbach, John M.
Koditschek, Daniel E. International
Symposium on Robotics Research (9th:
1999: Snowbird, Utah)
Description: viii, 460 p.: ill.; 27 cm.
ISBN: 1852332921 (alk. paper)
Notes: The 9th International
Symposium of Robotics Research
(ISRR'99) was held from October 9-12,
1999, at Snowbird, Utah. Includes
bibliographical references and index.
Subjects: Robotics--Research--
Congresses.

LC Classification: TJ210.3. R638 2000
Dewey Class No.: 629.8/92 21

Robotics research: the second international
symposium / edited by Hideo Hanafusa
and Hirochika Inoue.
Published/Created: Cambridge, Mass.:
MIT Press, c1985.
Related Authors: Hanafusa, Hideo,
1923- Inoue, Hirochika, 1942-
International Symposium on Robotics
Research (2nd: 1984: Kyoto, Japan)
Description: xx, 530 p.: ill.; 29 cm.
ISBN: 0262081512
Notes: Papers presented at the Second
International Symposium on Robotics
Research, held Aug. 20-23, 1984, in
Kyoto, Japan. Includes bibliographies.
Subjects: Robotics--Research--
Congresses.
Series: The MIT Press series in artificial
intelligence
LC Classification: TJ210.3 .R64 1985
Dewey Class No.: 629.8/92 19

Robotics research: the seventh international
symposium / Georges Giralt and
Gerhard Hirzinger, eds.
Published/Created: New York: Springer,
c1996.
Related Authors: Giralt, Georges.
Hirzinger, Gerd.
Description: xii, 645 p.: ill.; 28 cm.
ISBN: 3540760431 (hardcover: alk.
paper)
Notes: Includes bibliographical
references.
Subjects: Robotics--Research--
Congresses.
LC Classification: TJ211 .R56817 1996
Dewey Class No.: 629.8/92 20

Robotics research: the third international
symposium / edited by O.D. Faugeras
and Georges Giralt.
Published/Created: Cambridge, Mass.:
MIT Press, c1986.
Related Authors: Faugeras, Olivier,
1949- Giralt, Georges. International
Symposium on Robotics Research (3rd:

1985: Gouvieux, France)
Description: vii, 404 p.: ill.; 29 cm.
ISBN: 0262061015
Notes: Proceedings of the Third
International Symposium on Robotics
Research held at the Château
Monvillargenes in Gouvieux, France,
10/7-11/85. "3"--Cover. Includes
bibliographies.
Subjects: Robotics--Research--
Congresses.
Series: The MIT Press series in artificial
intelligence
LC Classification: TJ210.3 .R643 1986
Dewey Class No.: 629.8/92 19

Robotics science / edited by Michael Brady.
Published/Created: Cambridge, Mass.:
MIT Press, c1989.
Related Authors: Brady, Michael, 1945-
Description: vii, 617 p.: ill.; 24 cm.
ISBN: 0262022842
Notes: Includes bibliographies and
indexes.
Subjects: Robotics.
Series: System Development
Foundation benchmark series; [1]
LC Classification: TJ211 .R5683 1989
Dewey Class No.: 629.8/92 19

Robotics technical directory, 1986 / William
M. Rowe, editor.
Published/Created: Research Triangle
Park, NC: Instrument Society of
America, c1986.
Related Authors: Rowe, William M.
Description: v, 170 p.: ill.; 28 cm.
ISBN: 0876649177 (pbk.)
Subjects: Robotics--Directories.
LC Classification: TJ210.5 .R65 1986
Dewey Class No.: 629.8/92/025 19

Robotics technical directory.
Published/Created: Research Triangle
Park, NC: Instrument Society of
America, c1986-
Related Authors: Instrument Society of
America.
Description: v.: ill.; 28 cm. 1986-
ISSN: 0891-4400

Notes: SERBIB/SERLOC merged
record
Subjects: Robotics--Directories. Robots-
-Catalogs.
LC Classification: TJ210.5 .R64
Dewey Class No.: 629.8/92 19

Robotics technology abstracts.
Published/Created: Bedford
[Bedfordshire]: Cranfield Press, c1982-
Description: v.; 21 cm. Vol. 1, no. 1
(Jan. 1982)-
ISSN: 0262-4004
Notes: Title from cover.
SERBIB/SERLOC merged record
Subjects: Robotics--Abstracts--
Periodicals. Robots--Abstracts--
Periodicals. Robotics--Bibliography--
Periodicals. Robots--Bibliography--
Periodicals. Robots, Industrial--
Abstracts--Periodicals.
LC Classification: TJ211 .R5684
Dewey Class No.: 629.8/92 19

Robotics technology and its varied uses:
hearing before the Subcommittee on
Science, Research, and Technology of
the Committee on Science, Space, and
Technology, U.S. House of
Representatives, One Hundred First
Congress, first session, September 25,
1989.
Published/Created: Washington: U.S.
G.P.O.: For Sale by the Supt. of Docs.,
Congressional Sales Office, U.S.
G.P.O., 1989.
Description: iii, 238 p.: ill.; 24 cm.
Notes: "[No. 56]." Includes
bibliographical references.
Subjects: Robotics--United States.
Robotics--Economic aspects--United
States.
LC Classification: KF27 .S399 1989w
Dewey Class No.: 338.4/7629892/0973
20

Robotics today.
Edition Information: Annual edition
Published/Created: Dearborn, Mich.:
Robotics International, Society of

Manufacturing Engineers, c1982-
Related Authors: Robotics International
of SME.
Description: v.: ill.; 29 cm. '82-
ISSN: 0734-287X
Notes: SERBIB/SERLOC merged
record Robotics today 0193-6913
(OCoLC)5228354
Subjects: Robots, Industrial--
Periodicals.
LC Classification: TS191.8 .R63
Dewey Class No.: 629.8/92/05

Robotics today.
Published/Created: [Dearborn, Mich.,
Society of Manufacturing Engineers]
Related Authors: Society of
Manufacturing Engineers.
Description: v. ill. 28 cm. summer
1979- Ceased with issue for Dec. 1987.
ISSN: 0193-6913 CODEN: ROTODJ
Notes: Vols. for summer 1979-
published in cooperation with the Robot
Institute of America. SERBIB/SERLOC
merged record Robotics today. Annual
edition 0734-287X Indx'd selectively
by: Predicasts Computer & control
abstracts 0036-8113 April 1982-
Electrical & electronics abstracts 0036-
8105 April 1982- Physics abstracts.
Science abstracts. Series A 0036-8091
April 1982- Metals abstracts 0026-0924
World aluminum abstracts 0002-6697
Subjects: Robots, Industrial--
Periodicals.
LC Classification: TS191 .R62
Dewey Class No.: 629.8/92/05

Robotics update.
Published/Created: Boca Raton, FL:
Robotics Newsletter Associates, c1983-
Related Authors: Robotics Newsletter
Associates.
Description: v.; 28 cm. Vol. 1, no. 1
(Apr. 1983)-
ISSN: 0737-5700
Notes: Title from caption.
SERBIB/SERLOC merged record
Subjects: Robotics--Periodicals. Robots-
-Periodicals.

LC Classification: TJ210.2 .R63
Dewey Class No.: 629.8/92/05 19

Robotics world.
Published/Created: [Atlanta, GA]:
Communication Channels, [c1983-
Description: v.: ill.; 28 cm. Vol. 1, no. 1
also called premiere issue. [Vol. 1, no.
1] (Jan. 1983)-
ISSN: 0737-7908
Notes: Title from cover. Publisher:
Argus Business, SERBIB/SERLOC
merged record Indexed by: Applied
science & technology index 0003-6986
Engineering index annual (1968) 0360-
8557 Indx'd selectively by: Predicasts
Subjects: Robots, Industrial--
Periodicals.
LC Classification: TS191.8 .R635
Dewey Class No.: 629.8/92/05 19

Robotics worldwide: market study &
forecast / DMS; Seth B. Golbey,
Thomas P. Lydon, project directors.
Published/Created: Greenwich, CT,
U.S.A. (100 Northfield St., Greenwich
06830): Defense Marketing Services,
c1985.
Related Authors: Golbey, Seth B.
Lydon, Thomas. Defense Marketing
Service, inc.
Description: 637, [133] p.: ill.; 29 cm.
Subjects: Robotics industry. Robotics
industry--Directories. Market surveys.
LC Classification: HD9696.R622 R625
1985
Dewey Class No.: 380.1/45629892 19

Robotics, CAD/CAM market place, 1985 /
with an introduction by Ken Susnjara.
Published/Created: New York: Bowker,
c1985.
Related Authors: R.R. Bowker
Company.
Description: xix, 239 p.; 28 cm.
ISBN: 0835218201 (pbk.) :
Notes: Includes index.
Subjects: Robotics--Directories.
Computer-aided design--Directories.
LC Classification: TJ210.5 .R63 1985

Dewey Class No.: 670/.28/54 19

Robotics, control, and society: essays in
honor of Thomas B. Sheridan / edited
by N. Moray, W.R. Ferrell, W.B. Rouse.
Published/Created: London; New York:
Taylor & Francis, c1990.
Related Authors: Moray, Neville.
Ferrell, William R. Rouse, William B.
Sheridan, Thomas B.
Description: viii, 268 p.: ill.; 25 cm.
ISBN: 0850668506
Notes: Includes bibliographical
references and index.
Subjects: Robotics. Human-machine
systems. Technology--Social aspects.
LC Classification: TJ211 .R5748 1990
Dewey Class No.: 629.8/92 20

Robotics, its technology, applications, and
impacts / [by Barry Brownstein ... et
al.].
Published/Created: Columbus, Ohio
(505 King Ave., Columbus 43201):
Battelle Memorial Institute, c1982.
Related Authors: Brownstein, Barry.
Description: 40 p.: ill. (some col.); 28
cm.
Notes: Cover title.
Subjects: Robotics.
Series: Battelle technical inputs to
planning; rept. no. 27
LC Classification: TS191.8 .R62 1982
Dewey Class No.: 629.8/92 19

Robotics, mechatronics and manufacturing
systems: transactions of the
IMACS/SICE International Symposium
on Robotics, Mechatronics, and
Manufacturing Systems, Kobe, Japan,
16-20 September, 1992 / edited by
Toshi Takamori, Kazuo Tsuchiya.
Published/Created: Amsterdam; New
York: North-Holland: IMACS, 1993.
Related Authors: Takamori, Toshi,
1940- Tsuchiya, Kazuo, 1943-
International Association for
Mathematics and Computers in
Simulation Keisoku Jid⁻o Seigyo
Gakkai (Japan)

Description: xv, 960 p.: ill.; 25 cm.
ISBN: 0444897003 (acid-free paper)
Notes: Includes bibliographical
references.
Subjects: Robotics--Congresses.
Mechatronics--Congresses. Computer
integrated manufacturing systems--
Congresses.
LC Classification: TJ210.3 .I586 1992
Dewey Class No.: 629.8/92 20

Robotics, spatial mechanisms, and
mechanical systems: presented at the
1992 ASME design technical
conferences, 22nd Biennal Mechanisms
Conference, Scottsdale, Arizona,
September 13-16, 1992 / sponsored by
the Design Engineering Division,
ASME; edited by G. Kinzel ... [et al.].
Published/Created: New York:
American Society of Mechanical
Engineers, c1992.
Related Authors: Kinzel, G. (Gary)
American Society of Mechanical
Engineers. Design Engineering
Division. Mechanisms Conference
(22nd: 1992: Scottsdale, Ariz.)
Description: viii, 644 p.: ill.; 28 cm.
ISBN: 0791809390
Notes: Includes bibliographical
references and index.
Subjects: Robotics--Congresses.
Mechanical movements--Congresses.
Series: DE (Series) (American Society
of Mechanical Engineers. Design
Engineering Division); vol. 45.
LC Classification: TJ210.3 .R644 1992
Dewey Class No.: 629.8/92 20

Robotics.
Published/Created: Amsterdam: North-
Holland: Elsevier Science Publishers,
c1985-c1987.
Description: 3 v.: ill.; 26 cm. Vol. 1, no.
1 (May 1985)-v. 3, no. 3 & 4 (Sept.-
Dec. 1987).
ISSN: 0167-8493
Notes: Title from cover.
SERBIB/SERLOC merged record
Indx'd selectively by: Computer &

control abstracts 0036-8113 1985-
Electrical & electronics abstracts 0036-
8105 1985- Physics abstracts. Science
abstracts. Series A 0036-8091 1985-
Subjects: Robotics--Periodicals. Robots-
-Periodicals. Robots, Industrial--
Periodicals.
LC Classification: TJ210.2 .R62
Dewey Class No.: 629.8/92 19

Robotics.
   Published/Created: New York, N.Y.,
   USA: Engineering Information, 1984,
   c1983.
   Related Authors: Engineering
   Information, Inc.
   Description: x, 167 p.; 28 cm.
   ISBN: 0911820434 (pbk.)
   Notes: Includes indexes.
   Subjects: Robotics--Bibliography.
   Series: Technical bulletin (Engineering
   Information, Inc.)
   LC Classification: Z5853.R58 R6 1983
   TJ211
   Dewey Class No.: 016.6298/92 19

Robotics: an overview.
   Published/Created: Sacramento, CA:
   Business/Technology Books, c1984.
   Related Authors: Business/Technology
   Books (Firm)
   Description: x, 90 p.: ill.; 28 cm.
   ISBN: 0899341764 (unb.)
   Notes: Bibliography: p. 89-90.
   Subjects: Robotics. Artificial
   intelligence.
   Series: Robotics and artificial
   intelligence applications series; v. 1
   LC Classification: TJ211 .R565 1984
   Dewey Class No.: 629.8/92 19

Robotics: applied mathematics and
   computational aspects: based on the
   proceedings of a conference on robotics,
   applied mathematics and computational
   aspects / organized by the Institute of
   Mathematics and Its Applications and
   held at Loughborough University of
   Technology in July 1989; edited by
   Kevin Warwick.

Published/Created: Oxford: Clarendon
Press; New York: Oxford University
Press, 1993.
Related Authors: Warwick, K. Institute
of Mathematics and Its Applications.
Description: xvii, 596 p.: ill.; 25 cm.
ISBN: 0198536496 :
Notes: Includes bibliographical
references.
Subjects: Robotics.
Series: The Institute of Mathematics and
Its Applications conference series; new
ser., 41
LC Classification: TJ211 .R5665 1993
Dewey Class No.: 629.8/92 20

Robotics: hearings before the Subcommittee
   on Investigations and Oversight of the
   Committee on Science and Technology,
   U.S. House of Representatives, Ninety-
   seventh Congress, second session, June
   2, 23, 1982.
   Published/Created: Washington: U.S.
   G.P.O., 1983.
   Description: iii, 449 p.: ill.; 24 cm.
   Notes: "No. 148." Item 1025-A-1, 1025-
   A-2 (microfiche) Bibliography: p. 333-
   334.
   Subjects: Robots, Industrial--
   Government policy--United States.
   Robot industry--United States. Robot
   industry.
   LC Classification: KF27 .S3975 1982f
   Dewey Class No.: 338.4/7629892/0973
   19
   Govt. Doc. No.: Y 4.Sci 2:97/148

Robotics: kinematics, dynamics and
   controls: presented at the 1994 ASME
   design technical conferences, 23rd
   Biennial Mechanisms Conference,
   Minneapolis, Minnesota, September 11-
   14, 1994 / sponsored by the Design
   Engineering Division, ASME; edited by
   G.R. Pennock; associate editors, J.
   Angeles ... [et al.].
   Published/Created: New York:
   American Society of Mechanical
   Engineers, c1994.
   Related Authors: Pennock, G. R.

Angeles, Jorge, 1943- American Society of Mechanical Engineers. Design Engineering Division.
Description: viii, 510 p.: ill.; 28 cm.
ISBN: 0791812863
Notes: Includes bibliographical references and index.
Subjects: Robotics--Congresses. Mechanical movements--Design and construction--Congresses.
Series: DE (Series) (American Society of Mechanical Engineers. Design Engineering Division); vol. 72.
LC Classification: TJ210.3 .M43 1994
Dewey Class No.: 629.8/92 20

Robotics: proceedings of national workshop, April 1987 / organised by Department of Science & Technology, Government of India, in collaboration with Confederation of Engineering Industry.
Published/Created: New Delhi: Oxford & IBH Pub. Co., c1988.
Related Authors: India. Dept. of Science and Technology Confederation of Engineering Industry (India)
Description: vii, 307 p.: ill.; 25 cm.
ISBN: 8120403495 :
Notes: Includes bibliographies.
Subjects: Robotics--Congresses. Robotics--India--Congresses.
LC Classification: TJ211 .R5574 1988
Dewey Class No.: 629.8/92 20

Robotics: proceedings of the IASTED International Symposium, Davos, March 2-5, 1982 / [editor, M.H. Hamza].
Published/Created: Anaheim: Acta Press, [1982?]
Related Authors: Hamza, M. H.
Description: 61 p.: ill.; 28 cm.
ISBN: 0889860319 (pbk.)
Notes: English and French. Cover title. "An official publication of the International Association of Science and Technology for Development." Includes bibliographies.
Subjects: Robotics--Congresses.
LC Classification: TJ210.3 .I27 1982

Dewey Class No.: 629.8/92 19

Robotics: the algorithmic perspective: the Third Workshop on the Algorithmic Foundations of Robotics / edited by Pankaj K. Agarwal, Lydia E. Kavraki, Matthew T. Mason.
Published/Created: Wellesley, Mass.: A K Peters, c1998.
Related Authors: Agarwal, Pankaj K. Kavraki, Lydia. Mason, Matthew T.
Description: x, 389 p.: ill.; 29 cm.
ISBN: 1568810814
Notes: Includes bibliographical references and index.
Subjects: Robotics--Congresses. Algorithms--Congresses.
LC Classification: TJ210.3 .W664 1998
Dewey Class No.: 629.8/92 21

Robotics: theory and applications: presented at the Winter Annual Meeting of the American Society of Mechanical Engineers, Anaheim, California, December 7-12, 1986 / sponsored by the Dynamic Systems and Controls Division, ASME; edited by F.W. Paul, K. Youcef-Toumi.
Published/Created: New York, N.Y. (345 E. 47th St., New York 10017): ASME, c1986.
Related Authors: Paul, Frank W. Youcef-Toumi, Kamal. American Society of Mechanical Engineers. Winter Meeting (1986: Anaheim, Calif.) American Society of Mechanical Engineers. Dynamic Systems and Control Division. Symposium on Robotics (1986: Anaheim, Calif.)
Description: vi, 207 p.: ill.; 28 cm.
Notes: "Presented in the Symposium on Robotics ... 1986"--P. iii. Includes bibliographies.
Subjects: Robotics--Congresses.
Series: DSC (Series); vol. 3.
LC Classification: TJ210.3 .R645 1986
Dewey Class No.: 629.8/92 19

Robotics: understanding computers / by the editors of Time-Life Books.

Published/Created: Alexandria, Va.:
Time-Life Books, c1986.
Related Authors: Time-Life Books.
Description: 128 p.: col. ill.; 29 cm.
ISBN: 0809456966 0809456974 (lib.
bdg.)
Notes: Includes index. Bibliography: p.
124-125.
Subjects: Robotics--Popular works.
Series: Understanding computers
LC Classification: TJ211.15 .R63 1986
Dewey Class No.: 629.8/92 19

Robots 10: conference proceedings, April
20-24, 1986, Chicago, Illinois.
Edition Information: 1st ed.
Published/Created: Dearborn, Mich.:
Robotics International of SME, c1986.
Related Authors: Robotics International
of SME.
Description: 1 v. (various pagings): ill.;
28 cm.
ISBN: 0872632237 (pbk.)
Notes: Includes bibliographies and
index.
Subjects: Robots, Industrial--
Congresses.
LC Classification: TS191.8 .R6365
1986
Dewey Class No.: 670.42/7 19

Robots 12 and Vision '88 Conference.
Published/Created: Dearborn, Mich.:
Society of Manufacturing Engineers,
c1988-
Related Authors: Society of
Manufacturing Engineers.
Description: v. <1-4: ill.; 28 cm.
Notes: "For presentation at a Creative
Manufacturing Engineering Program."
Includes bibliographies.
Subjects: Robots, Industrial--
Congresses. Robot vision--Congresses.
Robotics--Congresses.
LC Classification: TS191.8 .R6366
1988
Dewey Class No.: 670.42/72 20

Robots 13: conference proceedings, May 7-
11, 1989, Gaithersburg, Maryland /

sponsored by Society of Manufacturing
Engineers, Robotics International of
SME.
Published/Created: Dearborn, Mich.:
Society of Manufacturing Engineers,
c1989.
Related Authors: Society of
Manufacturing Engineers. Robotics
International of SME.
Description: 2 v.: ill.; 28 cm.
ISBN: 0872633608
Notes: "Creative manufacturing
engineering program": Book 2. Book 2
contains SME technical papers MS89-
310--MS89-334. Includes
bibliographical references.
Subjects: Robotics--Congresses.
Robots--Congresses.
LC Classification: TJ210.3 .R65 1989
Dewey Class No.: 670.42/72 20

ROBOTS 14: November 12-15, 1990,
Detroit, Michigan.
Published/Created: Dearborn, Mich.:
Society of Manufacturing Engineers,
c1990.
Related Authors: Society of
Manufacturing Engineers. Robotics
International of SME.
Description: 1 v. (various pagings): ill.;
28 cm.
ISBN: 087263390X
Notes: Cover title. "Sponsored by
Robotics International of the Society of
Manufacturing Engineers"--P. [5] "For
presentation at a Creative
Manufacturing Engineering Program."
Contains SME technical papers MS90-
12, etc. Includes bibliographical
references.
Subjects: Robotics--Congresses.
LC Classification: TJ210.3 .R65 1990
Dewey Class No.: 670.42/72 20

Robots 8: conference proceedings, June 4-7,
1984, Detroit, Michigan / sponsored by
Robotics International of SME.
Edition Information: 1st ed.
Published/Created: Dearborn, Mich.:
Society of Manufacturing Engineers,

c1984.
Related Authors: Robotics International
of SME.
Description: 2 v.: ill.; 28 cm.
ISBN: 0872631478 (pbk.: set)
Contents: v. 1. Applications for today --
v. 2. Future considerations.
Notes: Includes bibliographies.
Subjects: Robots, Industrial--
Congresses.
LC Classification: TS191.8 .R637 1984
Dewey Class No.: 629.8/92 19

Robots 9: conference proceedings, June 2-6,
1985, Detroit, Michigan / sponsored by
Robotics International of SME.
Edition Information: 1st ed.
Published/Created: Dearborn, Mich.:
Robotics International of SME, c1985.
Related Authors: Robotics International
of SME.
Description: 2 v.: ill.; 28 cm.
ISBN: 0872631893 (pbk.: set)
Contents: v. 1. Advancing applications -
- v. 2. Current issues, future concerns.
Notes: Includes bibliographies and
index.
Subjects: Robots, Industrial--
Congresses. Robotics--Congresses.
LC Classification: TS191.8 .R637 1985
Dewey Class No.: 629.8/92 19

Robots and biological systems: towards a
new bionics? / edited by Paolo Dario,
Giulio Sandini, Patrick Aebischer.
Published/Created: Berlin; New York:
Springer-Verlag, c1993.
Related Authors: Dario, Paolo, 1951-
Sandini, G. (Giulio) Aebischer, Patrick.
North Atlantic Treaty Organization.
Scientific Affairs Division. NATO
Advanced Workshop on Robots and
Biological Systems (1989: Il Ciocco,
Italy)
Description: xii, 786 p.: ill.; 25 cm.
ISBN: 3540561587: 0387561587 :
Notes: "Proceedings of the NATO
Advanced Workshop on Robots and
Biological Systems, held at Il Ciocco,
Toscana, Italy, June 26-30, 1989"--

Verso t.p. "Published in cooperation
with NATO Scientific Affairs
Division." Includes bibliographical
references.
Subjects: Robotics--Congresses.
Bionics--Congresses.
Series: NATO ASI series. Series F,
Computer and systems sciences; no.
102.
LC Classification: TJ210.3 .R648 1993
Dewey Class No.: 629.8/92 20

Robots for competitive industries:
proceedings of an international
conference of the Australian Robot
Association and the International
Federation of Robotics, Brisbane,
Queensland, Australia, 14- 16 July
1993.
Published/Created: Sydney: Australian
Robot Association; London: Mechanical
Engineering Publications, c1993.
Related Authors: Australian Robot
Association. International Federation of
Robotics.
Description: xvii, 541 p.: ill.; 29 cm.
ISBN: 0646142194 0852988869
(Mechanical Engineering Publications)
Notes: Includes bibliographical
references and index.
Subjects: Robotics--Congresses. Robots,
Industrial--Congresses.
LC Classification: TJ210.3 .R655 1993
Dewey Class No.: 670.42/72 20

Robots for kids: exploring new technologies
for learning / edited by Allison Druin
and James Hendler.
Published/Created: San Francisco:
Morgan Kaufmann, 2000.
Related Authors: Druin, Allison, 1963-
Hendler, James A.
Description: xvii, 377 p.: ill. (some
col.); 24 cm.
ISBN: 1558605975
Notes: Includes bibliographical
references and index.
Subjects: Robotics.
LC Classification: TJ211 .R5749 2000

Dewey Class No.: 629.8/92 21

Robots in inspection / Jay Lee, editor;
    Robert E. King, manager.
    Edition Information: 1st ed.
    Published/Created: Dearborn, Mich.:
    Society of Manufacturing Engineers,
    Machine Vision Association of SME,
    Robotics International of SME,
    Publications Development Dept., c1987.
    Related Authors: Lee, Jay, 1957- King,
    Robert E. (Robert Edward), 1949-
    Description: 171 p.: ill.; 29 cm.
    ISBN: 0872632865
    Notes: Includes bibliographies and
    index.
    Subjects: Engineering inspection--
    Automation. Robots, Industrial.
    Series: Advanced manufacturing series
    LC Classification: TS156.2 .R63 1987
    Dewey Class No.: 620/.0044 19

Robots in unstructured environments: Fifth
    International Conference on Advanced
    Robotics: '91 ICAR, June 19-22, 1991,
    Pisa, Italy.
    Published/Created: New York, N.Y.:
    IEEE; Piscataway, NJ: Available from
    IEEE Service Center, c1991.
    Description: 2 v. (1827 p.): ill.; 28 cm.
    ISBN: 0780300785 0780300793
    (microfiche)
    Notes: Includes bibliographical
    references and index.
    Subjects: Robotics--Congresses.
    LC Classification: TJ210.3 .I5765 1991
    Dewey Class No.: 629.8/92 20

Robots V Conference, October 28-30, 1980,
    Dearborn, Michigan.
    Published/Created: Dearborn, Mich.
    (One SME Drive, Dearborn, 48128):
    Society of Manufacturing Engineers,
    c1980.
    Related Authors: Robotics International
    of SME.
    Description: 1 v. (various pagings): ill.;
    22 cm.
    Notes: Cover title. Contains Technical
    papers MS80-690-693; 695, 697, 699-

700, 703-705, 708, 712, of the Society
    of Manufacturing Engineers. "For
    presentation at a creative manufacturing
    engineering program." Sponsored by
    Robotics International of SME.
    Subjects: Robotics--Congresses.
    LC Classification: TS191 .R63 1980
    Dewey Class No.: 629.8/92 19

Robots VI: conference proceedings, March
    2-4, 1982, Detroit, Michigan /
    sponsored by Robotics International of
    SME; in cooperation with Society of
    Manufacturing Engineers and Robot
    Institute of America.
    Published/Created: Dearborn, Mich.:
    Robotics International, c1982.
    Related Authors: Robotics International
    of SME. Society of Manufacturing
    Engineers. Robot Institute of America.
    Description: 603 p.: ill.; 28 cm.
    ISBN: 0872630781 (pbk.)
    Notes: Includes bibliographies and
    index.
    Subjects: Robotics--Congresses.
    LC Classification: TS191 .R63 1982
    Dewey Class No.: 629.8/92 19

Robust vision for vision-based control of
    motion / edited by Markus Vincze,
    Gregory D. Hager.
    Published/Created: Bellingham, Wash.;
    SPIE Optical Engineering Press; New
    York: IEEE Press, c2000.
    Related Authors: Vincze, Markus, 1965-
    Hager, Gregory D., 1961- IEEE
    Robotics and Automation Society.
    Description: xxiv, 237 p.: ill.; 26 cm.
    ISBN: 0780353781 (IEEE)
    Notes: "IEEE Robotics and Automation
    Society, sponsor." Includes
    bibliographical references and index.
    Subjects: Robot vision. Robots--Control
    systems. Robots--Motion.
    Series: SPIE/IEEE series on imaging
    science & engineering
    LC Classification: TJ211.3 .R57 2000
    Dewey Class No.: 629.8/92637 21

Rosenberg, Jerry Martin.
Dictionary of artificial intelligence and robotics / Jerry M. Rosenberg.
Published/Created: New York: Wiley, c1986.
Description: xi, 203 p.; 24 cm.
ISBN: 0471849820 0471849812 (pbk.)
Subjects: Artificial intelligence--Dictionaries. Robotics--Dictionaries.
LC Classification: Q334.6 .R67 1986
Dewey Class No.: 629.8/92/0321 19

Rosenschein, Stanley J.
The synthesis of digital machines with provable epistemic properties / Stanley J. Rosenschein and Leslie Pack Kaelbling.
Published/Created: Menlo Park, CA.: Center for the Study of Language and Information/SRI International, c1987.
Related Authors: Kaelbling, Leslie Pack.
Description: 38 p.: ill.; 28 cm.
Notes: "March 1987." Bibliography: p. 36-38.
Subjects: Robotics. Artificial intelligence. Logic design.
Series: Report (Center for the Study of Language and Information (U.S.)); no. CSLI-87-83.
LC Classification: TJ211 .R665 1987
Dewey Class No.: 006.3 19

Rosheim, Mark E.
Robot evolution: the development of anthrobotics / Mark E. Rosheim.
Published/Created: New York, N.Y.: Wiley, c1994.
Description: xvi, 423 p.: ill.; 29 cm.
ISBN: 0471026220 (alk. paper)
Notes: Includes index.
Subjects: Robotics.
LC Classification: TJ211 .R67 1994
Dewey Class No.: 629.8/92 20

Ross, Dave, 1949-
Making robots / written and illustrated by Dave Ross.
Published/Created: New York: F. Watts, 1980.

Description: 32 p.: col. ill.; 26 cm.
ISBN: 0531041425
Summary: Gives instructions for making toys resembling robots using materials found around the house.
Subjects: Robotics--Juvenile literature. Toys--Juvenile literature. Handicraft--Juvenile literature. Robots--Models. Models and modelmaking. Toy making.
LC Classification: TJ211 .R68
Dewey Class No.: 745.592

Russell, R. Andrew.
Odour detection by mobile robots / R. Andrew Russell.
Published/Created: Singapore; River Edge, NJ: World Scientific, c1999.
Description: xiv, 217 p.: ill.; 23 cm.
ISBN: 981023791X
Notes: Includes bibliographical references and index.
Subjects: Mobile robots. Gas detectors.
Series: World Scientific series in robotics and intelligent systems; vol. 22
LC Classification: TJ211.415 .R87 1999
Dewey Class No.: 629.8/9264 21

Russo, Mark F.
Automating science and engineering laboratories with visual basic / Mark F. Russo, Martin M. Echols.
Published/Created: New York: Wiley, c1999.
Related Authors: Echols, Martin M.
Description: xx, 355 p.: ill.; 24 cm.
ISBN: 0471254932 (pbk.: alk. paper)
Notes: Includes bibliographical references and index.
Subjects: Laboratories--Data processing. Engineering laboratories--Data processing. Laboratories--Automation. Engineering laboratories--Automation. Robotics. Robots, Industrial.
Series: Wiley-Interscience series on laboratory automation
LC Classification: Q183.A1 R88 1999
Dewey Class No.: 660/.28/002855268 21

Rutland, Jonathan.
> The world of robots / Jonathan Rutland.
> Published/Created: New York: Warwick
> Press, 1979.
> Description: 22 p.: col ill.; 23 cm.
> ISBN: 0531091309: 0531091155 (lib.
> bdg.)
> Summary: Discusses the history of
> robots and describes the many different
> types of machines that can "think" for
> themselves and their uses.
> Notes: Cover Exploring the world of
> robots.
> Subjects: Robotics--Juvenile literature.
> Robots. Robotics.
> LC Classification: TJ211 .R87
> Dewey Class No.: 629.8/91

Ryan, Daniel L., 1941-
> Robotic simulation / Daniel L. Ryan.
> Published/Created: Boca Raton: CRC
> Press, c1994.
> Description: 317 p.: ill.; 25 cm.
> ISBN: 0849344689 (acid-free paper)
> Notes: Includes index.
> Subjects: Robotics--Computer
> simulation.
> LC Classification: TJ211.47 .R93 1994
> Dewey Class No.: 629.8/92 20

Ryder, Joanne.
> C-3PO's book about robots / by Joanne
> Ryder; illustrated by John Gampert.
> Published/Created: New York: Random
> House, c1983.
> Related Authors: Gampert, John.
> Description: [32] p.: col. ill.; 13 cm.
> ISBN: 0394856902 (pbk.) :
> Summary: Text and illustrations portray
> a whole range of robots now in
> existence--some in factories, some who
> work underwater, and some who travel
> in space, and others who do jobs too
> dangerous for man.
> Subjects: Robots--Juvenile literature.
> Robots, Industrial--Juvenile literature.
> Robotics. Robots.
> Series: Star wars
> LC Classification: TJ211 .R93 1983

Dewey Class No.: 629.8/92 19

Safford, Edward L.
> Handbook of advanced robotics / by
> Edward L. Safford, Jr.
> Edition Information: 1st ed.
> Published/Created: Blue Ridge Summit,
> Pa.: Tab Books, c1982.
> Description: xii, 468 p.: ill.; 22 cm.
> ISBN: 0830625216: 0830614214 (pbk.)
> :
> Notes: Includes index.
> Subjects: Robotics.
> LC Classification: TJ211 .S24 1982
> Dewey Class No.: 629.8/92 19

Safford, Edward L.
> The complete handbook of robotics / by
> Edward L. Safford.
> Edition Information: 1st ed.
> Published/Created: Blue Ridge Summit,
> Pa.: Tab Books, c1978.
> Description: 358 p.: ill.; 22 cm.
> ISBN: 0830698728: 0830610715
> Notes: Includes index.
> Subjects: Robotics.
> LC Classification: TJ211 .S23
> Dewey Class No.: 629.8/92

Salant, Michael A. (Michael Alan)
> Introduction to robotics / Michael A.
> Salant.
> Published/Created: New York:
> McGraw-Hill Book Co., c1988.
> Description: vi, 121 p.: ill.; 28 cm.
> ISBN: 0070544689 (pbk.)
> Notes: Bibliography: p. 120-121.
> Subjects: Robotics.
> LC Classification: TJ211 .S34 1988
> Dewey Class No.: 629.8/92 19

SAMA 83: Internationale Fachmesse von
> Spitzentechniken: Produktion,
> Automation, industrielle Robotik,
> Oberflächenbehandlung, in den Hallen
> der Schweizer Mustermesse
> Basel/Schweiz, 6.-10. Sept. 1983 =
> Salon international des techniques
> avancées: production, automation,
> robotique industrielle, traitements de

surface, dans les halles de la Foire suisse d'echantillons Bâle/Suisse, du 6 au 10 sept. 1983 = International Exhibition on Advanced Techniques: Production, Automation, Industrial Robotics, Surface Treatment, in the halls of the Swiss Industries Fair, Basle/Switzerland, 6-10 Sept. 1983. Published/Created: Basel: Schweizer Mustermesse, [1983] Related Authors: Schweizer Mustermesse. Description: 196 p.: ill.; 21 cm. Notes: English, French, and German. Organized by SAMA International S.A. in collaboration with the Schweizer Mustermesse. Includes indexes. Subjects: Production engineering--Exhibitions. LC Classification: TS176 .I652 1983 Dewey Class No.: 670/.42 20

Sanders, David A. (David Adrian), 1958- Making complex machinery move: automatic programming and motion planning / David A. Sanders. Published/Created: Taunton, Somerset, England: Research Studies Press; New York: Wiley, c1993. Description: xvi, 276 p.: ill.; 24 cm. ISBN: 0863801412 (Research Studies Press) 0471937932 (Wiley) Notes: Includes bibliographical references (p. 259-265) and index. Subjects: Robots--Motion. Robots--Programming. Series: Robotics and mechatronics series; 1 LC Classification: TJ211.4 .S26 1993 Dewey Class No.: 629.8/92 20

Sandhu, Harpit Singh. An introduction to robotics / Harpit Singh Sandhu. Published/Created: London: Nexus Special Interests, 1997. Description: 198 p.; cm. ISBN: 1854861530 LC Classification: IN PROCESS

Sandler, B. Z., 1932- Robotics: designing the mechanisms for automated machinery / Ben-Zion Sandler. Edition Information: 2nd ed. Published/Created: San Diego: Academic Press, 1999. Description: x, 433 p.: ill.; 26 cm. ISBN: 0126185204 Notes: Includes bibliographical references (p. 423-424) and index. Subjects: Automatic machinery--Design and construction. LC Classification: TJ213 .S1157 1999 Dewey Class No.: 670.42/72 21

Sandler, B. Z., 1932- Robotics: designing the mechanisms for automated machinery / Ben-Zion Sandler. Published/Created: Englewood Cliffs, N.J.: Prentice Hall, c1991. Description: x, 479 p.: ill.; 25 cm. ISBN: 0137816006 : Notes: "A Solomon Press book." Includes bibliographical references (p. 471-472) and index. Subjects: Automatic machinery--Design and construction. LC Classification: TJ213 .S1157 1991 Dewey Class No.: 670.42/72 20

Sayers, Craig. Remote control robotics / Craig Sayers. Published/Created: New York: Springer, 1998. Description: xix, 224 p.: ill. (some col.); 24 cm. ISBN: 0387985972 (alk. paper) Notes: Based on the author's Ph.D. dissertation, University of Pennsylvania. Includes bibliographical references (p. [205]-214) and index. Subjects: Robots--Control systems. Remote control. LC Classification: TJ211.35 .S29 1998 Dewey Class No.: 620/.46 21

Schilling, Robert J. (Robert Joseph), 1947- Fundamentals of robotics: analysis and

control / Robert J. Schilling.
Published/Created: Englewood Cliffs,
N.J.: Prentice Hall, 1990, c1988.
Description: p. cm.
ISBN: 0133444333
Notes: Includes index.
Subjects: Robotics.
LC Classification: TJ211 .S38 1990
Dewey Class No.: 629.8/92 19

Schilling, Robert J. (Robert Joseph), 1947-
Robotic manipulation: programming
and simulation studies / Robert J.
Schilling, Robert B. White.
Published/Created: Englewood Cliffs,
N.J.: Prentice-Hall, c1990.
Related Authors: White, Robert B.
(Robert Bruce), 1944-
Description: x, 182 p.: ill.; 24 cm. + 1
computer disk (5 1/4 in.)
ISBN: 0133444414
Notes: System requirements for
computer disk: IBM PC/XT/AT, or pure
compatible computer; 512K; MS-DOS
2.18 through 3.30; a COM1 serial port
and a graphics adaptor card (CGA,
EGA, Hercules) Includes
bibliographical references (p. 169-170)
and index.
Subjects: Robotics--Computer
simulation.
LC Classification: TJ211.47 .S35 1990
Dewey Class No.: 629.8/92 20

Schodt, Frederik L., 1950-
Inside the robot kingdom: Japan,
mechatronics, and the coming robotopia
/ Frederik L. Schodt.
Edition Information: 1st ed.
Published/Created: Tokyo; New York:
Kodansha International; New York,
N.Y.: Distributed in the U.S. by
Kodansha International/USA through
Harper & Row, 1988.
Description: 256 p.: ill.; 24 cm.
ISBN: 0870118544 (U.S.) :
Notes: Includes index. Bibliography: p.
247-251.
Subjects: Mechatronics--Japan.
Robotics--Japan.

LC Classification: TJ211 .S39 1988
Dewey Class No.: 338/.06 19

Schraft, R. D. (Rolf-Dieter), 1942-
Service robots / Rolf Dieter Schraft,
Gernot Schmierer.
Published/Created: Natick, MA: A K
Peters, c2000.
Related Authors: Schmierer, Gernot,
1969-
Description: 216 p.: col. ill.; 26 cm.
ISBN: 1568811098 (alk. paper)
Notes: Translation of: Serviceroboter.
Subjects: Robotics. Robots, Industrial.
LC Classification: TJ211 .S42 2000
Dewey Class No.: 629.8/92 21

Schuler, Charles A.
Industrial electronics and robotics /
Charles A. Schuler, William L.
McNamee.
Published/Created: New York:
McGraw-Hill, c1986.
Related Authors: McNamee, William L.
Description: v, 474 p.: ill.; 29 cm.
ISBN: 0070556253
Notes: Includes index.
Subjects: Industrial electronics. Robots,
Industrial.
LC Classification: TK7881 .S38 1986
Dewey Class No.: 621.3815 19

Schuler, Charles A.
Modern industrial electronics / Charles
A. Schuler, William L. McNamee.
Published/Created: New York: Glencoe,
c1993.
Related Authors: McNamee, William L.
Schuler, Charles A. Industrial
electronics and robotics.
Description: vii, 534 p.: ill.; 29 cm.
ISBN: 0028008626
Notes: Rev. and expanded ed. of:
Industrial electronics and robotics.
c1986. Includes index.
Subjects: Industrial electronics. Robots,
Industrial.
LC Classification: TK7881 .S4 1993
Dewey Class No.: 621.381 20

Science and ethics: papers presented at a symposium held under the aegis of the Australian Academy of Science, University of New South Wales, November 7, 1980 / edited by David Oldroyd.
Published/Created: Kensington, NSW, Australia: New South Wales University Press, 1982.
Related Authors: Oldroyd, D. R. (David Roger) Australian Academy of Science. University of New South Wales.
Description: vii, 120 p.; 22 cm.
ISBN: 0868403520 (pbk.) :
Contents: Scientific neutrality versus normative learning / Clifford Hooker -- Drawing the limits / Ron Johnston -- Science under social and political pressures / Mark Diesendorf -- Ethical aspects of robotics / Michael Kassler -- There are few hazards in recombinant DNA research abstract / Robert Wake -- Social engineering / Lloyd Reinhardt -- When politics is harder than physics / Julius Stone.
Notes: Includes bibliographical references.
Subjects: Science--Moral and ethical aspects.
LC Classification: BJ57 .S29 1982
Dewey Class No.: 174/.95 19

Scott, Peter B.
The robotics revolution: the complete guide for managers and engineers / Peter B. Scott.
Published/Created: Oxford, OX, U.K.; New York, N.Y., USA: B. Blackwell, 1984.
Description: xv, 345 p.: ill.; 24 cm.
ISBN: 0631131620 :
Notes: Includes index. Bibliography: p. [310]-311.
Subjects: Robots, Industrial.
LC Classification: TS191.8 .S36 1984
Dewey Class No.: 670.42/7 19

Second Annual International Robotic Education and Training Conference, August 12-14, 1985, Plymouth,
Michigan / sponsored and published by Robotics International of SME.
Edition Information: 1st ed.
Published/Created: Dearborn, Mich.: RI/SME, c1985.
Related Authors: Robotics International of SME.
Description: 1 v. (various pagings): ill.; 28 cm.
ISBN: 0872632040 (pbk.)
Notes: Papers presented at the Second Annual International Robotics Education and Training Conference.
Subjects: Robotics--Study and teaching--Congresses.
LC Classification: TJ211.26 .I58 1985
Dewey Class No.: 629.8/92 19

Second European In-Orbit Operations Technology Symposium: proceedings of a symposium held in Toulouse, France, 12-14 September 1989 / sponsord by European Space Agency; symposium co-chairmen, I. Braga & J-J. Runavot; symposium editor, E.R. Rolfe.
Published/Created: Paris, France: The Agency, c1989.
Related Authors: Rolfe, Erica. European Space Agency.
Description: xxxi, 451 p.: ill.; 30 cm.
Notes: "December 1989." Includes bibliographical references.
Subjects: Space vehicles--Maintenance and repair--Congresses. Space robotics--Congresses.
Series: ESA SP, 0379-6566; 297
LC Classification: TL915 .E97 1989
Dewey Class No.: 629.47 20

Segre, Alberto Maria.
Machine learning of robot assembly plans / by Alberto Maria Segre.
Published/Created: Boston: Kluwer Academic Publishers, c1988.
Description: xvi, 233 p.: ill.; 25 cm.
ISBN: 0898382696 :
Notes: Includes index. Bibliography: p. [219]-227.
Subjects: Robotics. Robots, Industrial. Knowledge representation (Information

theory)
Series: Kluwer international series in
engineering and computer science.
Knowledge representation, learning, and
expert systems
LC Classification: TJ211 .S43 1988
Dewey Class No.: 670.42/7 19

Selected papers on smart structures for
spacecraft / editors, Alok Das, Ben
Wada.
Published/Created: Bellingham, Wash.:
SPIE Optical Engineering Press, 2001.
Related Authors: Das, Alok, 1954-
Wada, B. K. (Ben K.)
Description: p. cm.
ISBN: 0819440507
Notes: Includes indexes.
Subjects: Space robotics. Smart
structures. Space vehicles--Equipment
and supplies. Large space structures
(Astronautics)
Series: SPIE milestone series; v. MS
167
LC Classification: TL1097 .S45 2001
Dewey Class No.: 629.47 21

Self-organization, computational maps, and
motor control / edited by Pietro Morasso
and Vittorio Sanguineti.
Published/Created: Amsterdam; New
York: Elsevier, 1997.
Related Authors: Morasso, P. (Pietro)
Sanguineti, Vittorio.
Description: xvii, 635 p.: ill.; 24 cm.
ISBN: 0444823239 (acid-free paper)
Notes: Includes bibliographical
references and indexes.
Subjects: Sensorimotor cortex.
Locomotion--Computer simulation.
Self-organizing systems. Robotics.
Series: Advances in psychology
(Amsterdam, Netherlands); 119.
LC Classification: QP383.15 .S45 1997
Dewey Class No.: 612.8/252 21

Selig, J. M.
Geometrical methods in robotics / J.M.
Selig.
Published/Created: New York: Springer,

c1996.
Description: xiii, 269 p.: ill.; 24 cm.
ISBN: 0387947280 (Berlin: hardcover:
acid-free paper)
Notes: Includes bibliographical
references (p. 251-256) and index.
Subjects: Robotics. Geometry. Lie
groups.
Series: Monographs in computer science
LC Classification: TJ211 .S433 1996
Dewey Class No.: 629.8/92 20

Selig, J. M.
Introductory robotics / J.M. Selig.
Published/Created: New York: Prentice
Hall, c1992.
Description: viii, 152 p.: ill.; 24 cm.
ISBN: 0134888758 (pbk.) :
Notes: Includes index.
Subjects: Robotics.
LC Classification: TJ211 .S434 1992
Dewey Class No.: 629.8/92 20

Sensor based intelligent robots: international
workshop, Dagstuhl Castle, Germany,
September 28-October 2, 1998: selected
papers / Hendrik I. Christensen, Horst
Bunke, Hartmut Noltemeier eds.
Published/Created: Berlin; New York:
Springer, c1999.
Related Authors: Christensen, H. I.
(Henrik I.), 1962- Bunke, Horst.
Noltemeier, Hartmut.
Description: viii, 325 p.: ill.; 24 cm.
ISBN: 3540669337 (softcover: alk.
paper)
Notes: Includes bibliographical
references and index.
Subjects: Robotics. Intelligent control
systems.
Series: Lecture notes in computer
science; 1724. Lecture notes in
computer science. Lecture notes in
artificial intelligence.
LC Classification: TJ211 .S437 1999
Dewey Class No.: 629.8/92 21

Sensor based intelligent robots: international
workshop, Dagstuhl Castle, Germany,
October 2000: selected revised papers /

Gregory D. Hager ... [et al.
Published/Created: Berlin; New York:
Springer, 2002.
Related Authors: Hager, Gregory D.,
1961-
Description: p. cm.
ISBN: 3540433996 (softcover: alk.
paper)
Notes: Includes bibliographical
references and index.
Subjects: Robotics--Congresses.
Artificial intelligence--Congresses.
Intelligent control systems--Congresses.
Series: Lecture notes in computer
science; 2238
LC Classification: TJ210.3 .I64 2000
Dewey Class No.: 629.8/92 21

Sensor devices and systems for robotics /
edited by Alícia Casals.
Published/Created: Berlin; New York:
Springer-Verlag, c1989.
Related Authors: Casals, Alícia, 1955-
North Atlantic Treaty Organization.
Scientific Affairs Division.
Description: ix, 362 p.: ill.; 25 cm.
ISBN: 0387508856 (U.S.: alk. paper)
Notes: "Proceedings of the NATO
Advanced Research Workshop on
Sensor Devices and Systems for
Robotics, held in Catalonia, Spain,
October 13-16, 1987"--T.p. verso.
"Published in cooperation with NATO
Scientific Affairs Division." Includes
bibliographical references.
Subjects: Robotics--Congresses.
Detectors--Congresses.
Series: NATO ASI series. Series F,
Computer and systems sciences; no. 52.
LC Classification: TJ210.3 .N3755 1987
Dewey Class No.: 629.8/92 19

Sensor fusion and networked robotics.
Published/Created: Bellingham, Wash.,
USA: SPIE, c1995-
Related Authors: Society of Photo-
optical Instrumentation Engineers.
Description: v.: ill.; 28 cm. 8 (23-24
Oct. 1995)-
Subjects: Optical data processing--

Congresses. Optical detectors--
Congresses. Robotics--Congresses.
Multisensor data fusion--Congresses.
Three-dimensional display systems--
Congresses.
Series: Proceedings of SPIE--the
International Society for Optical
Engineering
LC Classification: TA1632 .S46

Sensor fusion.
Published/Created: Bellingham, Wash,:
SPIE, c1988-c1994.
Related Authors: Society of Photo-
optical Instrumentation Engineers.
Description: v.: ill.; 28 cm. 4-6 Apr.
1988-7 (31 Oct.-1 Nov. 1994).
ISSN: 1084-3175
Notes: Nov. issue each year 1988-1991
has also distinctive thematic title.
SERBIB/SERLOC merged record
Subjects: Optical data processing--
Congresses. Optical detectors--
Congresses. Robotics--Congresses.
Multisensor data fusion--Congresses.
Three-dimensional display systems--
Congresses.
Series: Proceedings of SPIE--the
International Society for Optical
Engineering
LC Classification: TA1632 .S46
Dewey Class No.: 621 12

Sensors and control for automation:
[conference] 22-24 June, Frankfort,
FRG / Marcus Becker, R.W. Daniel,
Otmar Loffeld, chairs/editors; sponsored
by the Commission of the European
Communities, Directorate General for
Sciences, Research and Development,
EOS--the European Optical Society,
SPIE-the International Society for
Optical Engineering.
Published/Created: Bellington, Wash.:
SPIE--the International Society for
Optical Engineering, c1994.
Related Authors: Becker, Markus.
Daniel, R. W. (Ron W.) Loffeld, Otmar.
Commission of the European
Communities. Directorate-General for

Science, Research, and Development.
European Optical Society. Society of
Photo-optical Instrumentation
Engineers.
Description: vii, 312 p.: ill.; 28 cm.
ISBN: 0819415537
Notes: Includes bibliographical
references and index.
Subjects: Engineering inspection--
Automation--Congresses. Quality
control--Optical methods--Congresses.
Optical detectors--Industrial
applications--Congresses. Automatic
control--Congresses. Robotics--
Congresses.
Series: Proceedings EurOpt series.
Proceedings of SPIE--the International
Society for Optical ENgineering; v.
2247.
LC Classification: TS156.2 .S46 1994
Dewey Class No.: 670.42/7 20

Sensors and controls for automated
manufacturing and robotics: presented
at the Winter Annual Meeting of the
American Society of Mechanical
Engineers, New Orleans, Louisiana,
December 9-14, 1984 / sponsored by the
Dynamic Systems and Control Division,
ASME; edited by K.A. Stelson, L.M.
Sweet.
Published/Created: New York, N.Y.
(345 E. 47th St., New York 10017):
ASME, c1984.
Related Authors: Stelson, K. A. (Kim
A.) Sweet, L. M. (Larry M.) American
Society of Mechanical Engineers.
Dynamic Systems and Control Division.
Description: vi, 308 p.: ill.; 26 cm.
Notes: Includes bibliographies.
Subjects: Robots, Industrial--
Congresses. Automation--Congresses.
Detectors--Congresses.
LC Classification: TS191.8 .A47 1984
Dewey Class No.: 629.8/92 19

Sensors and sensory systems for advanced
robots / edited by Paolo Dario.
Published/Created: Berlin; New York:
Springer-Verlag, c1988.

Related Authors: Dario, Paolo, 1951-
North Atlantic Treaty Organization.
Scientific Affairs Division.
Description: xi, 597 p.: ill.; 25 cm.
ISBN: 0387190899 (U.S.)
Notes: "Proceedings of the NATO
Advanced Research Workshop on
Sensors and Sensory Systems for
Advanced Robots, held in Maratea,
Italy, April 28-May 3, 1986"--
"Published in cooperation with NATO
Scientific Affairs Division." Includes
bibliographies.
Subjects: Robotics--Congresses.
Detectors--Congresses.
Series: NATO ASI series. Series F,
Computer and systems sciences; no. 43.
LC Classification: TJ210.3 .N376 1986
Dewey Class No.: 629.8/92 19

Sensory robotics for the handling of limp
materials / edited by Paul M. Taylor.
Published/Created: Berlin; New York:
Springer-Verlag, c1990.
Related Authors: Taylor, Paul M., 1951-
Description: ix, 342 p.: ill.; 25 cm.
ISBN: 3540522999 (Berlin)
0387522999 (New York)
Notes: Workshop held in Il Ciocco,
Italy, Oct. 16-22, 1988. "Published in
cooperation with the NATO Scientific
Affairs Division." Includes
bibliographical references.
Subjects: Robots, Industrial--
Congresses. Materials handling--
Automation--Congresses. Detectors--
Congresses.
Series: NATO ASI series. Series F,
Computer and systems sciences; no. 64.
LC Classification: TS191.8 .N37 1988
Dewey Class No.: 670.42/72 20

Service robot: an international journal.
Published/Created: Bradford, West
Yorkshire, England: MCB University
Press: ISRA, 1995-
Related Authors: MCB University
Press. ISRA (Association)
Description: v.: ill.; 30 cm. Vol. 1, no. 1
(1995)-

ISSN: 1356-3378 CODEN: SEROFK
Notes: Title from cover.
SERBIB/SERLOC merged record
Subjects: Robots, industrial--
Periodicals. Robotics--Periodicals.
LC Classification: TS191.8 .S475
Dewey Class No.: 670.42/72 20

Shafer, Steven A.
Shadows and silhouettes in computer
vision / by Steven A. Shafer.
Published/Created: Boston: Kluwer
Academic Publishers, c1985.
Description: xv, 198 p.: ill.; 25 cm.
ISBN: 0898381673
Notes: "This research was sponsored by
the Defense Advanced Research
Projects Agency (DOD), ARPA order
no. 3597, monitored by the Air Force
Avionics Laboratory under contract
F33615-81-K-1539"--T.p. verso.
Includes indexes. Bibliography: p.
[163]-167.
Subjects: Shades and shadows.
Geometry, Descriptive. Cylinders.
Series: Kluwer international series in
engineering and computer science;
SECS 3. Kluwer international series in
engineering and computer science.
Robotics and vision.
LC Classification: QA519 .S43 1985
Dewey Class No.: 516.2 19

Shahinpoor, Mohsen.
A robot engineering textbook / Mohsen
Shahinpoor.
Published/Created: New York: Harper
& Row, c1987.
Description: xvi, 480 p.: ill.; 24 cm.
ISBN: 006045931X
Notes: Includes bibliographies and
indexes.
Subjects: Robotics.
LC Classification: TJ211 .S45 1987
Dewey Class No.: 629.8/92 19

Shape memory alloys / edited by Hiroyasu
Funakubo; translated from the Japanese
by J.B. Kennedy.
Published/Created: New York: Gordon

and Breach Science Publishers, c1987.
Related Authors: Funakubo, Hiroyasu,
1927-
Description: xii, 275 p.: ill.; 24 cm.
ISBN: 2881241360
Notes: Translation of: Keij⁻o kioku
g⁻okin. Includes bibliographies and
index.
Subjects: Alloys--Thermomechanical
properties. Martensite--
Thermomechanical properties. Shape
memory effect. Shape memory alloys.
Series: Precision machinery and
robotics, 0889-860X; v. 1
LC Classification: TN690 .K4213 1987
Dewey Class No.: 620.1/692 19

Sharp, Mike.
Robot world / by Mike Sharp; editor,
Jacqui Bailey; designer, David Jefferis.
Edition Information: A Warwick Press
library ed.
Published/Created: New York: Warwick
Press, 1985, c1984.
Related Authors: Bailey, Jacqui.
Jefferis, David, ill.
Description: 93 p., [2] p. of plates: ill.
(some col.); 19 cm.
ISBN: 0531190005 (lib. bdg.)
Notes: "Packed with color pictures,
diagrams, and hundreds of fascinating
facts"--Cover. Includes index.
Subjects: Robotics--Juvenile literature.
Robots--Juvenile literature. Robotics.
Robots.
LC Classification: TJ211.2 .S53 1985
Dewey Class No.: 629.8/92 20

Sheridan, Thomas B.
Telerobotics, automation, and human
supervisory control / Thomas B.
Sheridan.
Published/Created: Cambridge, Mass.:
MIT Press, c1992.
Description: xx, 393 p.: ill.; 24 cm.
ISBN: 0262193167
Notes: Includes bibliographical
references (p. [365]-381) and indexes.
Subjects: Supervisory control systems.
Robotics. Human-machine systems.

LC Classification: TJ213 .S45522 1992
Dewey Class No.: 620/.46 20

Sheu, Phillip C.-Y.
Intelligent robotic planning systems /
Phillip C.-Y. Sheu, Q. Xue.
Published/Created: Singapore; River
Edge, N.J.: World Scientific, c1993.
Related Authors: Xue, Q. (Qing)
Description: xi, 265 p.: ill.; 23 cm.
ISBN: 9810207581 981020759X (pbk.)
Notes: Includes bibliographical
references (p. 253-261) and index.
Subjects: Robots--Motion. Robots--
Programming. Mobile robots.
Series: World Scientific series in
robotics and automated systems; vol. 3
LC Classification: TJ211.4 .S44 1993
Dewey Class No.: 629.8/92 20

Shircliff, David R.
Build a remote-controlled robot / David
R. Shircliff.
Published/Created: New York:
McGraw-Hill, 2002.
Description: p. cm.
ISBN: 0071385436
Subjects: Robotics--Popular works.
Remote control--Popular works.
LC Classification: TJ211.15 .S44 2002
Dewey Class No.: 629.8/92 21

Shircliff, David R.
Build a remote-controlled robot for
under $300 / David R. Shircliff.
Edition Information: 1st ed.
Published/Created: Blue Ridge Summit,
PA: Tab Books, c1986.
Description: xxi, 119 p.: ill.; 22 cm.
ISBN: 0830604170: 0830615172 (pbk.)
:
Notes: Includes index. Bibliography: p.
112-113.
Subjects: Robotics--Popular works.
Remote control--Popular works.
LC Classification: TJ211.15 .S45 1986
Dewey Class No.: 629.8/92 19

Shoham, Moshe.
Robotics training program.

Published/Created: [Tel Aviv, Israel]:
Eshed Robotec, [c1984-
Description: v. <1, 4: ill.; 24 cm.
ISBN: 965291004X (pbk.: set)
Notes: Cover title.
Subjects: Robotics. Robots, Industrial.
LC Classification: TJ211 .S484 1984
Dewey Class No.: 629.8/92 19

Silverstein, Alvin.
The robots are here / Alvin Silverstein
& Virginia B. Silverstein.
Published/Created: Englewood Cliffs,
N.J.: Prentice-Hall, c1983.
Related Authors: Silverstein, Virginia
B.
Description: 128 p.: ill.; 24 cm.
ISBN: 0137821859 (lib. bdg.) :
Summary: Describes various kinds of
robots and their uses and considers the
relationship of robots and humans and
the implications of a world increasingly
run by robots.
Notes: "10 & up; gr. 5 & up"--Jacket.
Includes index. Bibliography: p. 124-
125.
Subjects: Robots--Juvenile literature.
Robotics. Robots.
LC Classification: TJ211.2 .S54 1983
Dewey Class No.: 629.8/92 19

Simons, G. L. (Geoffrey Leslie), 1939-
Are computers alive?: evolution and
new life forms / Geoff Simons.
Published/Created: Boston, [Mass.]:
Birkhäuser, [1983]
Description: xi, 212 p.; 22 cm.
ISBN: 0817631429 0817631445 (pbk.)
3764331429 0817631429
Notes: Includes index. Bibliography: p.
196-206.
Subjects: Artificial intelligence.
Computers. Robotics.
LC Classification: Q335 .S53 1983
Dewey Class No.: 001.53/5 19

Simons, G. L. (Geoffrey Leslie), 1939-
Is man a robot? / Geoff Simons.
Published/Created: Chichester [West
Sussex]; New York: Wiley, c1986.

Description: xvi, 316 p.; 23 cm.
ISBN: 0471911062 :
Notes: Includes index. Bibliography: p.
302-307.
Subjects: Robotics.
LC Classification: TJ211 .S56 1986
Dewey Class No.: 629.8/92 19

Simons, G. L. (Geoffrey Leslie), 1939-
Robots: the quest for living machines /
Geoff Simons.
Published/Created: London: Cassell;
New York, NY: Distributed in the
United States by Sterling Pub. Co.,
1992.
Description: 224 p.: ill.; 24 cm.
ISBN: 0304340863
Notes: Includes bibliographical
references (p. 209-221) and index.
Subjects: Robotics.
LC Classification: TJ211 .S563 1992
Dewey Class No.: 629.8/92 20

Simulation software for robotics / edited by
J.D. Lee.
Published/Created: Oxford; New York:
Pergamon Press, c1989.
Related Authors: Lee, J. D. (James D.)
Description: p. 279-380, iii, iii p.: ill.;
29 cm.
ISBN: 0080371965
Notes: Published also as v. 5, no. 4 of
the Robotics & computer-integrated
manufacturing. Includes bibliographical
references.
Subjects: Robotics--Congresses.
LC Classification: TJ210.3 .S56 1989
Dewey Class No.: 670.42/72 20

SIRS'98: proceedings of the 6th
International Symposium on Intelligent
Robotic Systems, 21-23 July 1998, the
University of Edinburgh / edited by
Robert B. Fisher, Gillian M. Hayes.
Published/Created: [Edinburgh:
University of Edinburgh, Department of
Artificial Intelligence, 1998]
Related Authors: Fisher, R. B. Hayes,
Gillian M.
Description: 319 p.: ill.; 30 cm.

ISBN: 1901725022 (pbk)
Notes: Includes bibliographical
references and index.
Subjects: Robotics--Congresses.
Artificial intelligence--Congresses.
Intelligent control systems--Congresses.
LC Classification: TJ210.3 .I5867 1998
Dewey Class No.: 629.8/92 21

Sixth European Space Mechanisms &
Tribology Symposium: an international
symposium organised by the European
Space Agency and the Swiss Space
Industries and held at the Technopark,
Zürich, Switzerland on 4-6 October,
1995 / [compiled by W.R. Burke].
Published/Created: Paris, France:
European Space Agency, c1995.
Related Authors: Burke, W. R.
European Space Agency. Swiss Space
Industries.
Description: xv, 436 p.: ill.; 28 cm.
ISBN: 929092179X
Notes: "August 1995." Includes
bibliographical references.
Subjects: Space vehicles--Design and
construction--Congresses. Space
vehicles--Lubrication--Congresses.
Tribology--Congresses. Space robotics--
Congresses. Mechanical engineering--
Congresses.
Series: ESA SP, 03796566; 374
LC Classification: TL875 .E97 1995
Dewey Class No.: 629.47 20

Skowro'nski, Janislaw M.
Control dynamics of robotic
manipulators / J.M. Skowronski.
Published/Created: Orlando [Fla.]:
Academic Press, 1986.
Description: ix, 268 p.: ill.; 24 cm.
ISBN: 012648130X (alk. paper)
Notes: Includes index. Bibliography: p.
247-261.
Subjects: Robotics. Manipulators
(Mechanism)
LC Classification: TJ211 .S59 1986
Dewey Class No.: 629.8/92 19

Skurzynski, Gloria.
    Robots: your high-tech world / by
    Gloria Skurzynski.
    Edition Information: 1st American ed.
    Published/Created: New York:
    Bradbury Press, c1990.
    Description: 64 p.: col. ill.; 21 cm.
    ISBN: 0027829170 :
    Summary: An introduction to robotics,
    focusing on such topics as the robot
    revolution, bionics, and avionics.
    Notes: Includes index.
    Subjects: Robots--Juvenile literature.
    Robots. Robotics.
    LC Classification: TJ211.2 .S58 1990
    Dewey Class No.: 629.8/92 20

Smart materials and structures: proceedings
    of the 4th European Conference on
    Smart Structures and Materials in
    conjunction with the 2nd International
    Conference on Micromechanics,
    Intelligent Materials, and Robotics,
    Harrogate, UK, 6-8 July, 1998 / edited
    by G.R. Tomlinson and W.A. Bullough.
    Published/Created: Bristol;
    Philadelphia: Institute of Physics
    Publishing, c1998.
    Related Authors: Tomlinson, Geoffrey
    R. Bullough, W. A. International
    Conference on Micromechanics,
    Intelligent Materials, and Robotics (2nd:
    1998: Harrogate, England)
    Description: xvii, 834 p.: ill.; 31 cm.
    ISBN: 0750305479
    Notes: Includes bibliographical
    references and author index.
    Subjects: Smart structures--Congresses.
    Smart materials--Congresses. Intelligent
    control systems--Congresses.
    LC Classification: TA418.9.S62 E95
    1998
    Dewey Class No.: 624.1 21

SME technical reports, December 1983,
    Dearborn, Michigan.
    Published/Created: Dearborn, Mich.
    (One SME Dr., Dearborn 48128):
    Society of Manufacturing Engineers,
    c1983.

    Related Authors: Society of
    Manufacturing Engineers.
    Description: [224] p.: ill.; 28 cm.
    Contents: High speed steels for
    broaching / Per Hellman -- If
    productivity is a problem-- / Kenneth F.
    Huddleston -- Health and safety aspects
    of office automation--the European
    scene / Graham Briscoe -- Substrate
    preparation / edited by Emery P. Miller
    -- Robotics education and national
    needs / Barry Irvin Soroka -- A
    taxonomy of computer-aided
    manufacturing and its educational
    implications / R.P. Sadowski -- How
    should we teach engineering design? /
    Larry G. Richards -- Robotics and
    artificial intelligence in lay and
    handicapped education / Abby Gelles.
    Notes: Cover title. Contains Technical
    reports MRR 83-01, etc., of the Society
    of Manufacturing Engineers. "For
    presentation at a creative manufacturing
    engineering program." Includes
    bibliographies.
    Subjects: Production engineering.
    LC Classification: TS176 .S596 1983
    Dewey Class No.: 670 19

Smith, Penelope Probert.
    Active sensors for local planning in
    mobile robotics / Penelope Probert
    Smith.
    Published/Created: River Edge, NJ:
    World Scientific, 2001.
    Description: xvii, 317 p.: ill.; 22 cm.
    ISBN: 9810246811
    Notes: Includes bibliographical
    references (p. 291-305) and index.
    Subjects: Mobile robots. Detectors.
    Signal processing.
    Series: World Scientific series in
    robotics and intelligent systems; v. 26
    LC Classification: TJ211.415 .S65 2001

Sobey, Edwin J. C., 1948-
    How to build your own prize-winning
    robot / Ed Sobey.
    Published/Created: Berkeley Heights,
    NJ: Enslow Publishers, 2002.

ISBN: 0766016277
Summary: Teaches the fundamentals of robotics, from motors to wheel alignment, and including the construction of a personal robot.
Notes: Includes bibliographical references and index.
Subjects: Robotics--Juvenile literature. Robots. Robotics--Experiments. Experiments.
Series: Science fair success
LC Classification: TJ211.2 .S62 2002
Dewey Class No.: 629.8/92 21

Sonenklar, Carol.
Robots rising / Carol Sonenklar; with illustrations by John Kaufmann.
Edition Information: 1st ed.
Published/Created: New York: Henry Holt, 1999.
Related Authors: Kaufmann, John, ill.
Description: 103 p.: ill.; 20 cm.
ISBN: 0805060960 (hc.: alk. paper)
Summary: Simple text and illustrations describe technological advancements in the field of robotics.
Notes: Includes bibliographical references (p. 97) and index.
Subjects: Robots--Juvenile literature. Robots.
LC Classification: TJ211.2 .S66 1999
Dewey Class No.: 629.8/92 21

Song, Shin-Min.
Machines that walk: the adaptive suspension vehicle / Shin-Min Song, Kenneth J. Waldron.
Published/Created: Cambridge, Mass.: MIT Press, c1989.
Related Authors: Waldron, Kenneth J.
Description: xiv, 314 p.: ill.; 24 cm.
ISBN: 0262192748
Notes: Includes index. Bibliography: p. [301]-308.
Subjects: Robotics. Robots. Human locomotion.
Series: The MIT Press series in artificial intelligence.
LC Classification: TJ211 .S66 1989

Dewey Class No.: 629.8/92 19

Space robotics (SPRO'98): a proceedings volume from the IFAC workshop, St-Hubert, Quebec, Canada, 19-22 October 1998 / edited by S. Rondeau.
Edition Information: 1st ed.
Published/Created: New York: Published for the International Federation of Automatic Control by Pergamon, 1999.
Related Authors: Rondeau, S.
Description: viii, 191 p.: ill.; 30 cm.
ISBN: 0080430503 (pbk.)
Notes: Includes bibliographical references and index.
Subjects: Space robotics--Congresses.
LC Classification: TL1097 .S67 1998
Dewey Class No.: 629.47 21

Space robotics: dynamics and control / edited by Yangsheng Xu, Takeo Kanade.
Published/Created: Boston: Kluwer Academic Publishers, c1993.
Related Authors: Xu, Yangsheng. Kanade, Takeo.
Description: viii, 284 p.: ill.; 25 cm.
ISBN: 0792392663 (acid-free)
Notes: Includes bibliographical references and index.
Subjects: Space stations--Automation. Space robotics.
Series: The Kluwer international series in engineering and computer science; SECS 188
LC Classification: TL797 .S6445 1993
Dewey Class No.: 629.47 20

Spatial reasoning and multi-sensor fusion: proceedings of the 1987 workshop, October 5-7, 1987, Pheasant Run Resort, St. Charles, Illinois; sponsored by AAAI / [edited by Avi Kak and Su-shing Chen].
Published/Created: Los Altos, CA: M. Kaufmann, c1987.
Related Authors: Kak, Avinash C. Chen, Su-shing. American Association for Artificial Intelligence. Workshop on

Spatial Reasoning and Multi-Sensor
Fusion (1987: Saint Charles, Ill.)
Description: xiv, 441 p.; 29 cm.
ISBN: 0934613591
Notes: Proceedings of the Workshop on
Spatial Reasoning and Multi-Sensor
Fusion, held at Pheasant Run (a resort
hotel) in St. Charles, Ill.; sponsored by
the American Association for Artificial
Intelligence. Includes bibliographies and
index.
Subjects: Artificial intelligence--
Congresses. Robotics--Congresses.
Computer vision--Congresses.
LC Classification: Q334 .S635 1987
Dewey Class No.: 006.3 19

Spiteri, Charles J.
Robotics technology / Charles J. Spiteri.
Published/Created: Philadelphia:
Saunders College Pub., c1990.
Description: xii, 393 p.: ill.; 24 cm.
ISBN: 0030208580
Notes: Includes index. Includes
bibliographical references (p. [378]-
383).
Subjects: Robotics.
LC Classification: TJ211 .S68 1990

Stadler, Wolfram.
Analytical robotics and mechatronics /
Wolfram Stadler.
Published/Created: New York:
McGraw-Hill, 1995.
Description: p. cm.
ISBN: 0070606080
Notes: Includes index.
Subjects: Robotics. Mechatronics.
Series: McGraw-Hill series in electrical
and computer engineering
LC Classification: TJ211 .S728 1995
Dewey Class No.: 629.8/92 20

Staugaard, Andrew C.
Robotics and AI: an introduction to
applied machine intelligence / Andrew
C. Staugaard, Jr.
Published/Created: Englewood Cliffs,
N.J.: Prentice-Hall, c1987.
Description: x, 373 p.: ill.; 24 cm.

ISBN: 0137822693
Notes: Includes index. Bibliography: p.
337-339.
Subjects: Robotics. Artificial
intelligence.
LC Classification: TJ211 .S73 1987
Dewey Class No.: 006.3 19

Stevens, Ray, 1939- prf
Surely you joust [sound recording] /
Ray Stevens.
Published/Created: Universal City,
Calif.: MCA Records, p1986.
Description: 1 sound disc (42 min.):
analog, 33 1/3 rpm; 12 in. Publisher
Number: MCA-5795 MCA Records
Contents: Southern air (5:25) -- People's
court (4:43) -- Bionie and the Robotics
(4:28) -- Makin' the best of a bad
situation (2:56) -- Fat (3:29) --Can he
love you half as much as I (2:50) --
Smokey Mountain rattlesnake retreat
(3:21) -- The camping trip (5:50) --
Camp Werthahekahwee (3:53) --
Dudley Dorite (of the highway patrol)
(4:35).
Notes: Humorous country songs. Cast:
Ray Stevens, vocals, keyboards,
synthesizers, tympano; with
instrumental and vocal acc.
Subjects: Country music--1981-1990.
Humorous songs.
LC Classification: MCA Records MCA-
5795

Stone, Henry W.
Kinematic modeling, identification, and
control of robotic manipulators / Henry
W. Stone.
Published/Created: Boston: Kluwer
Academic, c1987.
Description: xviii, 224 p.: ill.; 25 cm.
Notes: Based on the author's
dissertation--Carnegie Mellon
University, 1986. Includes index.
Bibliography: p. [219]-222.
Subjects: Robotics. Manipulators
(Mechanism)
Series: Kluwer international series in
engineering and computer science;

SECS 29. Kluwer international series in engineering and computer science. Robotics.
LC Classification: TJ211 .S76 1987
Dewey Class No.: 629.8/92 19

Stone, Peter, 1971-
Layered learning in multiagent systems: a winning approach to robotic soccer / Peter Stone.
Published/Created: Cambridge, Mass.: MIT Press, c2000.
Description: xii, 272 p.: ill.; 24 cm.
ISBN: 0262194384
Notes: Includes bibliographical references (p. [261]-272).
Subjects: Intelligent agents (Computer software) Robotics.
Series: Intelligent robotics and autonomous agents.
LC Classification: QA76.76.I58 S76 2000
Dewey Class No.: 006.3 21

Stonecipher, Ken.
Industrial robotics: a handbook of automated systems design / Ken Stonecipher.
Published/Created: Hashbrouck Heights, NJ: Hayden Book Co., c1985.
Description: 364 p.: ill.; 26 cm.
ISBN: 0810463288
Notes: Includes index. Bibliography: p. 347-350.
Subjects: Robots, Industrial.
LC Classification: TS191.8 .S76 1985
Dewey Class No.: 670.42/7 19

Storrs, Graham.
The robot age / Graham Storrs.
Published/Created: New York: Bookwright Press, 1985.
Description: 48 p.: ill. (some col.); 24 cm.
ISBN: 0531180204 (lib. bdg.)
Summary: Discusses the history and uses of robots and how they are programmed to work in dangerous places.
Notes: Includes index. Bibliography: p.

46.
Subjects: Robotics--Juvenile literature. Robots--Juvenile literature. Robots. Robotics.
Series: Tomorrow's world (Bookwright Press)
LC Classification: TJ211.2 .S76 1985
Dewey Class No.: 629.8/92 19

Sullivan, George, 1927-
Rise of the robots. Illustrated with photos.
Published/Created: New York, Dodd, Mead [1971]
Description: ix, 114 p. illus. 25 cm.
ISBN: 039606292X
Summary: Describes various types of electro-mechanical devices that are man-like in the way they perform tasks.
Subjects: Robots--Juvenile literature. Robots. Robotics.
LC Classification: TJ211 .S85
Dewey Class No.: 629.8/92

Symposium on Robotics: presented at the Winter Annual Meeting of the American Society of Mechanical Engineers, Chicago, Illinois, November 27-December 2, 1988 / sponsored by the Dynamic Systems and Controls Division, ASME; edited by K. Youcef-Toumi, H. Kazerooni.
Published/Created: New York, N.Y. (345 E. 47th St., New York 10017): ASME, c1988.
Related Authors: Youcef-Toumi, Kamal. Kazerooni, H. (Homayoon) American Society of Mechanical Engineers. Dynamic Systems and Control Division.
Description: vii, 470 p.: ill.; 26 cm.
Notes: Includes bibliographies.
Subjects: Robotics--Congresses.
Series: DSC (Series); vol. 11.
LC Classification: TJ210.3 .S95 1988
Dewey Class No.: 629.8/92 20

Szeliski, Richard, 1958-
Bayesian modeling of uncertainty in low-level vision / by Richard Szeliski;

with a foreword by Takeo Kanade.
Published/Created: Boston: Kluwer
Academic Publishers, c1989.
Description: xvii, 198 p.: ill.; 25 cm.
ISBN: 0792390393
Notes: Includes bibliographical
references (p. [155]-165).
Subjects: Computer vision--
Mathematical models.
Series: Kluwer international series in
engineering and computer science;
SECS 79. Kluwer international series in
engineering and computer science.
Robotics.
LC Classification: TA1632 .S94 1989

Tactile sensors for robotics and medicine /
edited by John G. Webster.
Published/Created: New York: Wiley,
c1988.
Related Authors: Webster, John G.,
1932-
Description: xiv, 365 p.: ill.; 24 cm.
ISBN: 0471606073
Notes: "A Wiley-Interscience
publication." Includes bibliographies
and index.
Subjects: Tactile sensors. Robotics.
Biomedical Engineering. Robotics.
Touch.
LC Classification: R857.T32 T32 1988
Dewey Class No.: 629.8/92 19

Tanner, William R.
A manager's guide to robotics / William
A. [i.e. R.] Tanner.
Published/Created: New York, N.Y.:
American Management Association,
Extension Institute, c1986.
Description: xii, 196 p.: ill.; 30 cm.
Notes: Bibliography: p. 145-146.
Subjects: Robotics. Robots, Industrial.
Series: The Manufacturing management
skills series
LC Classification: TJ211 .T36 1986
Dewey Class No.: 670.42/7 19

Taylor, Paul M., 1951-
Understanding robotics / P.M. Taylor.
Published/Created: Boca Raton, Fla.:

CRC Press, 1990.
Description: x, 148 p.: ill.; 24 cm.
ISBN: 0849371457
Notes: Includes bibliographical
references (p. 139-146) and index.
Subjects: Robotics.
LC Classification: TJ211 .U53 1989
Dewey Class No.: 629.8/92 20

Teaching & learning with robots / edited by
Colin Terry and Peter Thomas.
Published/Created: London; New York:
Croom Helm, c1988.
Related Authors: Terry, Colin. Thomas,
Peter.
Description: 227 p.: ill.; 23 cm.
ISBN: 0709943180
Notes: Includes bibliographies.
Subjects: Robotics. Robots. Teaching--
Aids and devices.
LC Classification: TJ211 .T42 1988
Dewey Class No.: 629.8/92 19

Technology and innovation for
manufacturing: proceedings of a
Conference, May 4,5, 1979 / prepared
by the Committee on Science and
Technology, U.S. House of
Representatives, Ninety-sixth Congress,
second session.
Published/Created: Washington: U.S.
Govt. Print. Off., 1980.
Related Authors: United States.
Congress. House. Committee on
Science and Technology. University of
Florida. Center for Intelligent Machines
and Robotics.
Description: iv, 350 p.: ill.; 24 cm.
Notes: At head of Committee print.
"Serial FF." "Organized by: Center for
Intelligent Machines and Robotics,
College of Engineering, University of
Florida." Includes bibliographical
references.
Subjects: Manufacturing processes--
Technological innovations Congresses.
Technology and state--United States--
Congresses.
LC Classification: TS183 .C66 1979

Dewey Class No.: 338.4/767/0973 19

Technology for small spacecraft / Panel on
  Small Spacecraft Technology,
  Committee on Advanced Space
  Technology, Aeronautics and Space
  Engineering Board, Commission on
  Engineering and Technical Systems,
  National Research Coouncil.
  Published/Created: Washington, D.C.:
  National Academy Press, 1994.
  Related Authors: Adams, Laurence J.
  Description: xiv, 139 p.; 26 cm.
  ISBN: 0309050758
  Contents: Introduction -- Systems
  engineering and operations -- Spacecraft
  propulsion technology -- Spacecraft
  electric power -- Spacecraft structures
  and materials -- Small spacecraft
  communications technology --
  Guidance and control technology --
  Sensors for small spacecraft -- Robotics,
  automation, and artificial intelligence --
  Launch vehicle technology for small
  spacecraft -- Overall findings and
  recommendations.
  Notes: Includes appendices. Panel
  chairman: Laurence J. Adams. Study
  was supported by a contract from the
  National Aeronautics and Space
  Administration. Includes bibliographical
  references (p. 103-109).
  Subjects: Microspacecraft.
  LC Classification: TL795.4 .N38 1994
  Dewey Class No.: 629.46 21

Telemanipulator and telepresence
  technologies V: 4-5 November 1998,
  Boston, Massachusetts / Matthew R.
  Stein, chair/editor.
  Published/Created: Bellingham, Wash.:
  SPIE, c1998.
  Related Authors: Stein, Matthew R.
  Society of Photo-optical
  Instrumentation Engineers.
  Description: vii, 240 p.: ill.; 28 cm.
  ISBN: 0819429856
  Notes: Includes bibliographical
  references and index.
  Subjects: Manipulators (Mechanism)--

Congresses. Robotics--Congresses.
Remote control--Congresses.
Series: Proceedings of SPIE--the
International Society for Optical
Engineering; v. 3524.
LC Classification: TJ210.3 .T3343 1998
Dewey Class No.: 620/.46 21

Telemanipulator technology and space
  telerobotics: 7-9 September 1993,
  Boston, Massachusetts / Won S. Kim,
  chair/editor; sponsored and published by
  SPIE--the International Society for
  Optical Engineering in cooperation with
  Automated Imaging Association ... [et
  al.].
  Published/Created: Bellingham, Wash.,
  USA: SPIE, c1993.
  Related Authors: Kim, Won S. Society
  of Photo-optical Instrumentation
  Engineers.
  Description: ix, 464 p.: ill.; 28 cm.
  ISBN: 0819413224 (pbk.)
  Notes: Includes bibliographical
  references and author index.
  Subjects: Manipulators (Mechanism)--
  Congresses. Space robotics--
  Congresses. Remote control--
  Congresses.
  Series: Proceedings of SPIE--the
  International Society for Optical
  Engineering; v. 1833.
  LC Classification: TJ210.3 .T344 1993
  Dewey Class No.: 620/.46/0919 20

Telematics applications in automation and
  robotics (TA 2001): a proceedings
  volume from the IFAC Conference,
  Weingarten, Germany, 24-26 July 2001
  / edited by K. Schilling and H. Roth.
  Published/Created: New York:
  Pergamon, 2001.
  Related Authors: Schilling, Klaus,
  1956- Roth, H. (Hubert)
  Description: p. cm.
  ISBN: 0080438563
  Subjects: Automatic control--
  Congresses. Robotics--Congresses.
  Telematics--Congresses.
  LC Classification: TJ212.2 .T33 2001

Dewey Class No.: 629.8 21

Teleoperated robotics in hostile
environments / H. Lee Martin, editor;
Daniel P. Kuban, editor.
Edition Information: 1st ed.
Published/Created: Dearborn, Mich.:
Robotics International of SME,
Publications Development Dept.,
Marketing Services Division, c1985.
Related Authors: Martin, H. Lee.
Kuban, Daniel P.
Description: 273 p.: ill.; 29 cm.
ISBN: 0872631850
Notes: Includes bibliographies and
index.
Subjects: Robotics. Manipulators
(Mechanism)
Series: Manufacturing update series
LC Classification: TJ211 .T44 1985
Dewey Class No.: 629.8/92 19

Teleoperation: numerical simulation and
experimental validation / edited by
Marc C. Becquet.
Published/Created: Dordrecht; Boston:
Kluwer Academic, c1992.
Related Authors: Becquet, Marc C.,
1960- Commission of the European
Communities. Joint Research Centre.
Ispra Establishment. Commission of the
European Communities. Directorate
General for Telecommunications,
Information Industries, and Innovation.
Scientific and Technical
Communications Service.
Description: vii, 258 p.: ill. (some col.);
25 cm.
ISBN: 0792315847 (alk. paper)
Notes: "Based on the lectures given
during the Eurocourse on Teleoperation:
Numerical Simulation and Experimental
Validation, held at the Joint Research
Centre, Ispra, Italy, 18-22 November
1991"--T.p. verso. "Published for the
Commission of the European
Communities, Directorate-General
Telecommunications, Information
Industries and Innovation, Scientific and
Technical Communications Service"--P.

opp. t.p. Includes bibliographical
references.
Subjects: Manipulators (Mechanism)
Remote control. Robotics.
Series: Euro courses. Computer and
information science; v. 4
LC Classification: TJ211 .T443 1992
Dewey Class No.: 620/.46 20

Telerobotic applications / edited by T.
Schilling.
Published/Created: London:
Professional Engineering Pub., 2000.
Related Authors: Schilling, T. (Tyler)
Description: 162 p.: ill.; 25 cm.
ISBN: 1860582354
Notes: Includes bibliographical
references.
Subjects: Manipulators (Mechanism)
Remote handling (Radioactive
substances) Robotics.
LC Classification: TJ211 .T445 2000
Dewey Class No.: 620/.46 21

The 14th IEEE International Conference on
Micro Electro Mechanical Systems:
MEMS 2001: Interlaken, Switzerland,
January 21-25, 2001 / sponsored by the
IEEE Robotics and Automation Society.
Published/Created: Piscataway N.J.:
Institute of Electrical and Electronics
Engineers, c2001.
Related Authors: IEEE Robotics and
Automation Society.
Description: xxxviii, 610 p.: ill.; 28 cm.
ISBN: 0780359984 0780359992
(casebound) 0780362500 (microfiche)
Notes: "IEEE Catalog Number:
01CH37090"--T.p. verso. Ahead of
"Technical Digest". Includes
bibliographical references.
Subjects: Microelectromechanical
systems--Congresses.
LC Classification: TK7875 .I35 2001
Dewey Class No.: 621.381 21

The 3rd World Congress on Intelligent
Control and Automation: proceedings,
June 28-July 2, 2000, Hefei, China /
sponsored by University of Science and

Technology of China (USTC); technical cosponsored by IEEE Robotics and Automation Society, IEEE Control System Society, Beijing Chapter, Chinese Association of Automation, Chinese Association of Artificial Intelligence, Anhui Association for Science and Technology, China.
Published/Created: Piscataway, NJ: IEEE, c2000.
Description: 5 v.; 29 cm.
ISBN: 078035995X (set)

The Automatic control systems/robotics problem solver: a complete solution guide to any textbook / Staff of Research and Education Association, M. Fogiel, chief editor.
Edition Information: Rev. ed.
Published/Created: Piscataway, NJ: Research and Education Association, 2000.
Related Authors: Fogiel, M. (Max) Research and Education Association.
Description: xi, 1076 p.: ill.; 26 cm.
ISBN: 0878915427 (pbk.)
Subjects: Automatic control--Problems, exercises, etc. Robotics--Problems, exercises, etc.
Series: REA's problem solvers
LC Classification: TJ213.8 .A98 2000
Dewey Class No.: 629.8/076 21

The cyborg experiments: the extensions of the body in the media age / edited by Joanna Zylinska.
Published/Created: New York: Continuum, 2002.
Related Authors: Zylinska, Joanna, 1971-
Description: p. cm.
ISBN: 0826459021 082645903X (pbk.)
Notes: Includes bibliographical references and index.
Subjects: Human-machine systems. Cyborgs. Robotics--Social aspects.
Series: Technologies (London, England)
LC Classification: TA167 .C93 2002
Dewey Class No.: 303.48/3 21

The digital control of systems: applications to vehicles and robots / edited by C. Fargeon.
Published/Created: New York: Van Nostrand Reinhold, 1989.
Related Authors: Fargeon, C.
Description: xv, 449 p.: ill.; 24 cm.
ISBN: 0442239424
Notes: "First published in French under the Command Numérique des systèmes"--T.p. verso. Includes bibliographical references and index.
Subjects: Digital control systems. Robotics. Aeronautical instruments. Astronautical instruments.
LC Classification: TJ223.M53 C6513 1989
Dewey Class No.: 629.8 20

The Directory of consultants in robotics and mechanics.
Published/Created: Woodbridge, Conn.: Research Publications, c1985-
Description: v.; 28 cm. 1st ed.-
ISSN: 0882-603X
Notes: SERBIB/SERLOC merged record
Subjects: Robotics consultants--Directories. Mechanics consultants--Directories.
LC Classification: TJ210.5 .D57
Dewey Class No.: 629/92 19

The Eleventh Annual International Workshop on Micro Electro Mechanical Systems: an investigation of micro structures, sensors, actuators, machines and robots: proceedings, January 25-29, 1998, Heidelberg, Germany / sponsored by the IEEE Robotics and Automation Society in cooperation with the ASME Dynamic Systems and Control Division and the Ministry of Economic Affairs of Baden-Württemberg.
Published/Created: [New York]: Institute of Electrical and Electronics Engineers, c1998.
Related Authors: IEEE Robotics and Automation Society. American Society of Mechanical Engineers. Dynamic

Systems and Control Division. Baden-Württemberg (Germany). Wirtschaftsministerium.
Description: xxx, 666 p.: ill.; 30 cm.
ISBN: 078034412X (softbound) 0780344138 (casebound) 0780344146 (microfiche)
Notes: "IEEE catalog number 98CH36176." Includes bibliographical references and indexes.
Subjects: Microelectromechanical systems--Congresses. Detectors--Congresses. Microactuators--Congresses.
LC Classification: TK7875 .I35 1998
Dewey Class No.: 621.3 21

The essentials of automatic control systems/robotics I / Max Fogiel, editor.
Published/Created: Piscataway, NJ: Research and Education Association, 1997.
Description: p. cm.
ISBN: 0878915710 (pbk.)
LC Classification: 9709 BOOK NOT YET IN LC

The Essentials of automatic control systems/robotics I.
Published/Created: N.Y.: Research & Education Ass., 1987.
Description: p.
ISBN: 0878915710
LC Classification: 87-- BOOK NOT IN LC

The Future of the semiconductor, computer, robotics, and telecommunications industries / compiled by the editorial staff of Petrocelli Books.
Published/Created: Princeton, N.J.: Petrocelli, c1984.
Related Authors: Petrocelli Books (Firm)
Description: xxii, 244 p.; 24 cm.
ISBN: 0894332597
Subjects: Semiconductor industry--United States--Congresses. Computer industry--United States--Congresses. Robot industry--United States--

Congresses. Telecommunication equipment industry--United States Congresses.
LC Classification: HD9696.S43 U4817 1984
Dewey Class No.: 338.4/762138/0973 19

The impact of robotics on employment: hearing before the Subcommittee on Economic Goals and Intergovernmental Policy of the Joint Economic Committee, Congress of the United States, Ninety-eighth Congress, first session, March 18, 1983.
Published/Created: Washington: U.S. G.P.O., 1983.
Description: iii, 31 p.; 24 cm.
Notes: Item 1000-B, 1000-C (microfiche)
Subjects: Robots, Industrial--Economic aspects--United States. Robots, Industrial--Social aspects--United States. Manpower policy--United States.
Series: United States. Congress. Senate. S. hrg.; 98-131.
LC Classification: KF25 .E2314 1983a
Dewey Class No.: 331.12 19
Govt. Doc. No.: Y 4.Ec 7:R 57/2

The International journal of robotics research.
Published/Created: [Cambridge, MA]: MIT Press, [c1982-
Description: v.: ill.; 28 cm. Vol. 1, no. 1 ([spring 1982])-
ISSN: 0278-3649
Notes: Title from cover. Published: Thousand Oaks, CA: Sage Science Press, SERBIB/SERLOC merged record Indexed entirely by: Applied science & technology index 0003-6986 1991- Indx'd selectively by: International aerospace abstracts 0020-5842 1983- Mathematical reviews 0025-5629
Subjects: Automata--Periodicals. Robots, Industrial--Periodicals.
LC Classification: TJ211 .I485
Dewey Class No.: 629.8/92/05 19

The International robotics yearbook.
  Published/Created: Cambridge, Mass.:
  Ballinger Pub. Co., 1984-
  Description: v.; 29 cm. First ed.
  preceded by an unnumbered ed.
  published in 1983 in London, by Kogan
  Page Ltd. 1st ed.-
  ISSN: 0739-1595
  Notes: SERBIB/SERLOC merged
  record
  Subjects: Robotics--Directories.
  LC Classification: TJ210.5 .I58
  Dewey Class No.: 629.8/92/05 19

The McGraw-Hill illustrated encyclopedia
  of robotics & artificial intelligence /
  Stan Gibilisco, editor in chief.
  Published/Created: New York:
  McGraw-Hill, c1994.
  Related Authors: Gibilisco, Stan.
  Description: ix, 420 p.: ill., map; 25 cm.
  ISBN: 0070236135 (hbk.): 0070236143
  (pbk.) :
  Notes: Includes bibliographical
  references (p. 409) and index.
  Subjects: Robotics--Encyclopedias.
  Artificial intelligence--Encyclopedias.
  LC Classification: TJ210.4 .G53 1994
  Dewey Class No.: 629.892/03 20

The promotion of robotics and CAD/CAM
  in Sweden: report from the Computers
  and Electronics Commission, Ministry
  of Industry.
  Published/Created: Stockholm:
  Industridepartementet, 1981.
  Description: iv, 35 p.; 21 cm.
  ISBN: 9138066394 (pbk.)
  Notes: "Ds I 1981:26."
  Subjects: Robotics--Sweden.
  CAD/CAM systems--Sweden.
  LC Classification: TS191.8 .S98 1981
  Dewey Class No.: 338.948506 19

The robot in the garden: telerobotics and
  telepistemology in the age of the
  Internet / edited by Ken Goldberg.
  Published/Created: Cambridge, Mass.:
  MIT Press, c2000.
  Related Authors: Goldberg, Ken.

Description: xix, 366 p.: ill.; 24 cm.
  ISBN: 0262072033 (alk. paper)
  Notes: Includes bibliographical
  references and index.
  Subjects: Robotics. Knowledge, Theory
  of.
  Series: Leonardo (Series) (Cambridge,
  Mass.)
  LC Classification: TJ211 .R537 2000
  Dewey Class No.: 121 21

The Robotics handbook / [writer/editor,
  Lucia Corduan Luce].
  Published/Created: [Pittsburgh, Pa.]:
  Westinghouse Electric Corp., c1984.
  Related Authors: Luce, Lucia Corduan.
  Westinghouse Electric Corporation.
  Description: 129 p. in various pagings:
  ill.; 28 cm.
  Notes: Cover title. Includes
  bibliography.
  Subjects: Robotics--Handbooks,
  manuals, etc.
  LC Classification: TJ211 .R567 1984
  Dewey Class No.: 629.8/92 19

The Robotics industry of Japan: today and
  tomorrow / prepared by The Japan
  Industrial Robot Association.
  Published/Created: Tokyo: Fuji Corp.,
  c1982.
  Related Authors: Nihon Sangyⁿo
  Robotto Kⁿogyⁿokai.
  Description: vii, 581 p.: ill.; 29 cm.
  Subjects: Robot industry--Japan. Market
  surveys.
  LC Classification: HD9696.R623 J369
  1982
  Dewey Class No.: 338.4/7629892 19

The Robotics review.
  Published/Created: Cambridge, Mass.:
  MIT Press, c1989-
  Description: v.: ill.; 24 cm. 1-
  ISSN: 1049-5207
  Notes: SERBIB/SERLOC merged
  record
  Subjects: Robotics--Periodicals.
  LC Classification: TJ210.2 .R625

Dewey Class No.: 629 11

The role of automation and robotics in
advancing United States
competitiveness: hearing before the
Subcommittee on Science, Research,
and Technology of the Committee on
Science and Technology, House of
Representatives, Ninety-ninth Congress,
first session, October 7, 1985.
Published/Created: Washington: U.S.
G.P.O., 1986.
Description: iii, 173 p.: ill., 1 map; 24
cm.
Notes: Distributed to some depository
libraries in microfiche. Shipping list no.:
86-226-P. "No. 62." Item 1025-A-1,
1025-A-2 (microfiche) Includes
bibliographies.
Subjects: Robotics--Economic aspects--
United States. Automation--Economic
aspects--United States. Automatic
machinery--Economic aspects--United
States.
LC Classification: KF27 .S399 1985h
Dewey Class No.: 338.4/54/0973 19
Govt. Doc. No.: Y 4.Sci 2:99-62

The social implications of robotics and
advanced industrial automation:
proceedings of the IFIP TC 9
International Working Conference on
the Social Implications of Robotics and
Advanced Industrial Automation, Tel-
Aviv, Israel, 14-16 December 1987 /
edited by D. Millin, B.H. Raab.
Published/Created: Amsterdam; New
York: North-Holland; New York, N.Y.,
U.S.A.: Sole distributors for the U.S.A.
and Canada, Elsevier Science Pub. Co.,
1989.
Related Authors: Millin, D. (Daniel)
Raab, B. H. IFIP Technical Committee
9.
Description: xviii, 289 p.: ill.; 23 cm.
ISBN: 0444873201
Notes: Includes bibliographies.
Subjects: Automation--Social aspects--
Congresses. Robotics--Social aspects--
Congresses. Industries--Social aspects--

Congresses. Technology and
civilization--Congresses.
LC Classification: HC79.A9 I45 1987
Dewey Class No.: 303.4/83 19

The Tenth Annual International Workshop
on Micro Electro Mechanical Systems:
an investigation of micro structures,
sensors, actuators, machines and robots:
proceedings, Nagoya, Japan, January
26-30, 1997 / sponsored by the IEEE
Robotics and Automation Society in
cooperation with the ASME Dynamic
Systems and Control Division and the
Micromachine Center.
Published/Created: [New York]:
Institute of Electrical and Electronics
Engineers, c1997.
Related Authors: IEEE Robotics and
Automation Society. American Society
of Mechanical Engineers. Dynamic
Systems and Control Division.
Micromachine Center (Japan)
Description: xxiv, 545 p.: ill.; 30 cm.
ISBN: 0780337441 (softbound)
078033745X (casebound) 0780337468
(microfiche)
Notes: "IEEE catalog number
97CH36021." Includes bibliographical
references and indexes.
Subjects: Microelectronics--Congresses.
Microelectromechanical systems--
Congresses. Detectors--Congresses.
Actuators--Congresses.
LC Classification: TK7874 .I32747
1997
Dewey Class No.: 621.381 21

The Third Annual Applied Machine Vision
Conference proceedings, February 27-
March 1, 1984, Schaumburg (Chicago)
Illinois / sponsored by Robotics
International of SME.
Edition Information: 1st ed.
Published/Created: Dearborn, Mich.:
Robotics International of SME, c1984.
Related Authors: Robotics International
of SME.
Description: 1 v. (various pagings): ill.;
28 cm.

ISBN: 0872631400 (pbk.)
Notes: Includes bibliographies and index.
Subjects: Image processing--Industrial applications--Congresses. Robots, Industrial--Congresses.
LC Classification: TA1632 .A69 1984
Dewey Class No.: 621.36/7 19

The World yearbook of robotics research and development.
Published/Created: London, England: Kogan Page; Detroit, Mich.: Gale Research Co., 1985-
Related Authors: Gale Research Company.
Description: v.; 24 cm. 1st ed.-
Notes: "The concept of the book is based on a section of The International Robotics Yearbook (Kogan Page, London; Ballinger, Boston -1983)."
SERBIB/SERLOC merged record
Subjects: Robotics--Research--Periodicals. Robots--Periodicals. Robotics--Directories. Robots, Industrial--Periodicals.
LC Classification: TJ210.2 .W67
Dewey Class No.: 629.8/92 19

Theory and practice of robots and manipulators: proceedings of RoManSy '84, the Fifth CISM-IFToMM Symposium / edited by A. Morecki, G. Bianchi, and K. Kedzior.
Edition Information: 1st MIT Press ed.
Published/Created: Cambridge, Mass.: MIT Press, 1985.
Related Authors: Morecki, Adam. Bianchi, G. (Giovanni), 1924- Kedzior, K.
Description: 443 p.: ill.; 24 cm.
ISBN: 0262132087
Notes: "Sponsored by the CISM-Centre International des Sciences Mécaniques IFToMM-International Federation for the Theory of Machines and Mechanisms in association with the IVth Technical Division of the Polish Academy of Sciences." Includes bibliographies.

Subjects: Robotics--Congresses. Manipulators (Mechanism)--Congresses.
Series: The MIT Press series in artificial intelligence
LC Classification: TJ210.3 .R66 1984
Dewey Class No.: 629.8/92 19

Theory and practice of robots and manipulators: Second International CISM-IFToMM Symposium, Warsaw, Poland, September 14-17, 1976: proceedings / edited by A. Morecki and K. Kedzior: sponsored by CISM--Centre international des sciences mécaniques, IFToMM--International Federation for the Theory of Machines and Mechanisms, in association with the Technical Division of the Polish Academy of Sciences.
Published/Created: Amsterdam; New York: Elsevier Scientific Pub. Co.; New York: distribution for the U.S.A. and Canada, Elsevier/North-Holland, 1977.
Related Authors: Morecki, Adam. Kedzior, K. International Centre for Mechanical Sciences. International Federation for the Theory of Machines and Mechanisms. Polska Akademia Nauk. Wydzial IV-Nauk Technicznych.
Description: x, 500 p.: ill.; 24 cm.
ISBN: 0444998128
Notes: Includes bibliographical references.
Subjects: Robotics--Congresses. Manipulators (Mechanism)--Congresses.
LC Classification: TJ211 .R64 1976
Dewey Class No.: 629.8/92

Theory of robot control / Carlos Canudas de Wit, Bruno Siciliano, and Georges Bastin, eds.
Published/Created: Berlin; New York: Springer, c1996.
Related Authors: Canudas de Wit, Carlos A. Siciliano, Bruno, 1959- Bastin, G. (Georges), 1947-
Description: xvi, 392 p.; 24 cm.
ISBN: 3540760547 (hc: alk. paper)

Notes: At head of The Zodiac. Includes bibliographical references and index.
Subjects: Robots--Control systems. Robotics.
Series: Communications and control engineering
LC Classification: TJ211.35 .T37 1996
Dewey Class No.: 629.8/92 20

Theory of robots: selected papers from the IFAC/IFIP/IMACS Symposium, Vienna, Austria, 3-5 December 1986 / edited by P. Kopacek, I. Troch, and K. Desoyer.
Edition Information: 1st ed.
Published/Created: Oxford, England; New York: Published for the International Federation of Automatic Control by Pergamon Press, 1988.
Related Authors: Kopacek, Peter. Troch, I. (Inge) Desoyer, Kurt. International Federation of Automatic Control. IFAC/IFIP/IMACS Symposium on Theory of Robots (1986: Vienna, Austria)
Description: xii, 352 p.: ill.; 31 cm.
ISBN: 0080348033 :
Notes: "IFAC/IFIP/IMACS Symposium on Theory of Robots"--Prelim. p. Includes bibliographies and indexes.
Subjects: Robotics--Congresses.
Series: IFAC proceedings series; 1988, no. 3
LC Classification: TJ210.3 .T44 1988
Dewey Class No.: 629.8/92 19

Third National Conference on Robotics: Monday 4th-Wednesday 6th June 1990, Melbourne exhibition building with Automate Australia Exhibition / edited by K.P. Dabke and C.S. Berger.
Published/Created: [Melbourne?, Australia]: Australian Robot Association, [1990?]
Related Authors: Dabke, K. P. (Kishor P.) Berger, C. S. (Clive S.) Australian Robot Association.
Description: 417 p.: ill.; 30 cm.
Notes: Proceedings of the conference organized by the Australian Robot

Association and co-sponsored by Institution of Engineers, Australia ... [et al.]. Includes bibliographical references.
Subjects: Robotics--Congresses.
LC Classification: TJ210.3 .N35 1990
Dewey Class No.: 629.8/92 20

Three-dimensional machine vision / edited by Takeo Kanade.
Published/Created: Boston: Kluver Academic Publishers, c1987.
Related Authors: Kanade, Takeo.
Description: ix, 609 p.: ill.; 25 cm.
ISBN: 0898381886
Subjects: Robot vision.
Series: The Kluwer international series in engineering and computer science. Robotics
LC Classification: TJ211.3 .T47 1987
Dewey Class No.: 629.8/92 19

Thro, Ellen.
Robotics / Ellen Thro.
Edition Information: New ed.
Published/Created: New York: Facts on File, 2003.
Description: p. cm.
ISBN: 0816047014 (hardcover)
Subjects: Robotics--Popular works.
Series: Science and technology in focus
LC Classification: TJ211.15 .T457 2003
Dewey Class No.: 629.8/92 21

Thro, Ellen.
Robotics careers / Ellen Thro.
Published/Created: New York: F. Watts, 1987.
Description: 111 p.: ill.; 24 cm.
ISBN: 0531104257
Summary: Describes the duties, salaries, and training required for a variety of occupations related to the design, sales, and servicing of robots.
Notes: Includes index. Bibliography: p. 105-106.
Subjects: Robotics--Vocational guidance--Juvenile literature. Robotics--Vocational guidance. Robots. Vocational guidance.
Series: High-tech careers

LC Classification: TJ211.25 .T47 1987
Dewey Class No.: 629.8/92/023 19

Thro, Ellen.
　Robotics: the marriage of computers
　and machines / Ellen Thro.
　Published/Created: New York: Facts on
　File, 1993.
　Description: vii, 119 p.: ill.; 23 cm.
　ISBN: 0816026289
　Summary: Introduces the science of
　robotics, discussing the nature of
　artificial intelligence, the history of
　robotics, the different kinds of robots,
　and their uses.
　Notes: Includes bibliographical
　references and index.
　Subjects: Robotics--History--Juvenile
　literature. Robotics. Robots.
　Series: Facts on File science
　sourcebooks
　LC Classification: TJ211.2 .T47 1993
　Dewey Class No.: 629.8/92 20

Thuillard, Marc.
　Wavelets in soft computing / Marc
　Thuillard.
　Published/Created: Singapore; River
　Edge, N.J.: World Scientific, c2001.
　Description: xvii, 225 p.: ill.; 26 cm.
　ISBN: 9810246099
　Notes: Includes bibliographical
　references (p. 209-219) and index.
　Subjects: Wavelets (Mathematics) Soft
　computing.
　Series: World Scientific series in
　robotics and intelligent systems; vol. 25

Timms, Howard.
　Living in the future / Howard Timms.
　Published/Created: New York:
　Gloucester Press, 1990.
　Description: 36 p.: col. ill.; 30 cm.
　ISBN: 0531172244 (lib. bdg.)
　Summary: Speculates on living
　conditions of the future and shows how
　current technology, such as robotics,
　computers, and alternative energy
　sources, can affect future health care,
　shopping, leisure, and work.

　Notes: "A Watts/Gloucester book"--
　Cover. Includes index.
　Subjects: Technological innovations--
　Juvenile literature. Technological
　innovations. Forecasting.
　Series: Today's world (New York, N.Y.)
　LC Classification: T173.8 .T56 1990
　Dewey Class No.: 600 20

Todd, D. J.
　Walking machines: an introduction to
　legged robots / D.J. Todd.
　Published/Created: New York:
　Chapman and Hall, c1985.
　Description: 190 p.: ill.; 24 cm.
　ISBN: 041201131X
　Notes: Includes index. Bibliography: p.
　[179]-185.
　Subjects: Robotics. Robots.
　LC Classification: TJ211 .T63 1985
　Dewey Class No.: 629.8/92 19

Toward a practice of autonomous systems:
　proceedings of the first European
　Conference on Artificial Life / edited by
　Francisco J. Varela and Paul Bourgine.
　Published/Created: Cambridge, Mass.:
　MIT Press, c1992.
　Related Authors: Varela, Francisco J.,
　1946- Bourgine, Paul.
　Description: xvii, 515 p.: ill.; 28 cm.
　ISBN: 0262720191
　Notes: "A Bradford book." Selected
　papers from the first European
　Conference on Artificial Life, held in
　Paris, France in December 1991.
　Includes bibliographical references and
　index.
　Subjects: Robotics--Congresses.
　Artificial intelligence--Congresses.
　Series: Complex adaptive systems
　LC Classification: TJ210.3 .E87 1991
　Dewey Class No.: 629.8/92 20

Toward learning robots / edited by Walter
　Van de Velde.
　Edition Information: 1st MIT Press ed.
　Published/Created: Cambridge, Mass.:
　MIT Press, 1993.
　Related Authors: Velde, Walter Van de.

Description: 165 p.: ill.; 26 cm.
ISBN: 0262720175 (acid-free paper)
Notes: "A Bradford book." "Reprinted
from Robotics and autonomous systems,
volume 8, numbers 1-2 (1991)"--T.P.
verso. Includes bibliographical
references and index.
Subjects: Robots--Control systems--
Design and construction. Knowledge
representation (Information theory)
Series: Special issues of Robotics and
autonomous systems
LC Classification: TJ211.35 .T68 1993
Dewey Class No.: 006.3 20

Towards third generation robotics:
    proceedings of the 3rd International
    Conference on Advanced Robotics,
    ICAR '87: 13-15 October 1987,
    Versailles, France / edited by B. Espiau.
    Published/Created: Kempston, Bedford,
    UK: IFS Publications; Berlin; New
    York: Springer, c1987.
    Related Authors: Espiau, Bernard.
    Association française de la robotique
    industrielle.
    Description: 615 p.: ill.; 31 cm.
    ISBN: 038718404X (U.S.)
    Notes: "Sponsored by: Association
    française de robotique industrielle"--
    T.p. verso. Includes bibliographies and
    index.
    Subjects: Robotics--Congresses.
    LC Classification: TJ210.3 .I5765 1987
    Dewey Class No.: 629.8/92 20

Townsend, Carl, 1938-
    How to make money with your
    microcomputer / Carl Townsend and
    Merl Miller.
    Published/Created: Beaverton, Or.:
    Dilithium Press, c1981.
    Related Authors: Miller, Merl K., 1942-
    Description: x, 154 p.; 22 cm.
    ISBN: 0918398746 (pbk.) :
    Notes: Reprint. Originally published:
    Portland, Or.: Robotics Press, 1979.
    Includes index. Bibliography: p. [143]-
    [145]
    Subjects: Computer service industry.

Data processing service centers.
Microcomputers. Small business.
LC Classification: HD9696.C62 T677
1981
Dewey Class No.: 001.64 19

Townsend, Carl, 1938-
    How to make money with your
    microcomputer / Carl Townsend and
    Merl Miller.
    Published/Created: Portland, Or.:
    Robotics Press, c1979.
    Related Authors: Miller, Merl K., 1942-
    joint author.
    Description: x, 154 p.; 22 cm.
    ISBN: 0896610012
    Notes: Includes index. Bibliography: p.
    [143]-[145]
    Subjects: Computer service industry.
    Data processing service centers.
    Microcomputers. Small business.
    LC Classification: HD9696.C62 T677
    1979
    Dewey Class No.: 001.64 19

Traditional and non-traditional robotic
    sensors / edited by Thomas C.
    Henderson.
    Published/Created: Berlin; New York:
    Springer-Verlag, c1990.
    Related Authors: Henderson, Thomas C.
    Description: viii, 468 p.: ill.; 25 cm.
    ISBN: 354053007X (Berlin)
    038753007X (New York)
    Notes: "Proceedings of the NATO
    Advanced Research Workshop on
    Traditional and Non-Traditional
    Robotic Sensors, held in Maratea, Italy,
    August 28-September 2, 1989"--T.p.
    verso. Includes bibliographical
    references.
    Subjects: Robotics--Congresses.
    Detectors--Congresses.
    Series: NATO ASI series. Series F,
    Computer and systems sciences; no. 63.
    LC Classification: TJ210.3 .N377 1989
    Dewey Class No.: 629.8/92 20

Transactions: robotics research /
    Bartholomew O. Nnaji, editor; Robert

Hoekstra, associate editor.
Published/Created: Dearborn, Mich.:
Robotics International Association of
the Society of Manufacturing Engineers,
<1992
Related Authors: Nnaji, Bartholomew
O., 1956- Hoekstra, Robert L. Robotics
International of SME.
Description: v. <2: ill.; 28 cm.
ISBN: 0872634299 (v. 2)
Notes: Includes bibliographical
references.
Subjects: Robotics--Research--
Congresses.
LC Classification: TJ210.3 .T73 1992
Dewey Class No.: 629.8/92 20

Transactions: the best papers of 1987 /
sponsored by Society of Manufacturing
Engineers.
Published/Created: Dearborn, Mich.:
The Society, c1988.
Related Authors: Society of
Manufacturing Engineers.
Description: 1 v. (various pagings): ill.;
28 cm.
ISBN: 0872633276
Notes: On cover: Best technical papers.
28 papers, most relating to robotics.
Subjects: Production engineering.
LC Classification: TS176 .T73 1988
Dewey Class No.: 670.42 20

Trends & applications, 1983: automating
intelligent behavior, applications and
frontiers: proceedings, May 25-26,
1983, National Bureau of Standards,
Gaithersburg, Maryland / sponsored by
IEEE Computer Society, Washington
Chapter, Mideastern Area Committee
[and] IEEE Washington Section [and]
National Bureau of Standards, U.S.
Department of Commerce.
Published/Created: Silver Spring, MD
(1109 Spring St., Suite 300, Silver
Spring 20910): IEEE Computer Society
Press, c1983.
Related Authors: United States.
National Bureau of Standards. IEEE
Computer Society. Mideastern Area

Committee. Institute of Electrical and
Electronics Engineers. Washington
Section.
Description: vi, 285 p.: ill.; 28 cm.
ISBN: 0818604727 (pbk.) 0818684720
(hard)
Notes: "IEEE catalog no. 83CH1887-9."
"Computer Society no. 472." Includes
bibliographical references and index.
Subjects: Artificial intelligence--Data
processing--Congresses. Robotics--
Congresses.
LC Classification: Q336 .T74 1983
Dewey Class No.: 001.53/5 19

Trends and developments in mechanisms,
machines, and robotics, 1988: presented
at the 1988 ASME design technology
conferences, 20th Biennial Mechanisms
Conference, Kissimmee, Florida,
September 25-28, 1988 / sponsored by
the Mechanisms Committee of the
Design Engineering Division, ASME;
edited by A. Midha.
Published/Created: New York, N.Y.
(345 E. 47th St., New York 10017):
American Society of Mechanical
Engineers, c1988.
Related Authors: Midha, A. (Ashok)
American Society of Mechanical
Engineers. Mechanisms Committee.
Description: 3 v.: ill.; 28 cm.
Notes: Includes bibliographies.
Subjects: Mechanical movements--
Congresses. Machinery--Congresses.
Robotics--Congresses.
Series: DE (Series) (American Society
of Mechanical Engineers. Design
Engineering Division); vol. 15.
LC Classification: TJ181 .M37 1988
Dewey Class No.: 621 20

Trends in information technology / edited by
D.A. Linkens & R.I. Nicolson.
Published/Created: London, U.K.: P.
Peregrinus Ltd. on behalf of the
Institution of Electrical Engineers,
c1990.
Related Authors: Linkens, D. A.
Nicolson, R. I. Institution of Electrical

Engineers.
Description: 301 p.: ill.; 26 cm.
ISBN: 0863412319
Notes: Includes bibliographical
references and index.
Subjects: Computers. Computer
software. Robotics.
Series: IEE control engineering series;
43
LC Classification: QA76.5 .T688 1990
Dewey Class No.: 004 20

Trevelyan, James P.
Robots for shearing sheep: shear magic /
James P. Trevelyan.
Published/Created: Oxford; New York:
Oxford University Press, 1992.
Description: xii, 398 p.: ill.; 24 cm.
ISBN: 0198562527 :
Notes: Includes bibliographical
references (p. [383]-389) and index.
Subjects: Sheep-shearing. Robotics.
Series: Oxford science publications
LC Classification: SF379 .T74 1992
Dewey Class No.: 636.3/145 20

Tsai, Lung-Wen.
Robot analysis: the mechanics of serial
and parallel manipulators / Lung-Wen
Tsai.
Published/Created: New York: Wiley,
c1999.
Description: xiii, 505 p.: ill.; 25 cm.
ISBN: 0471325937 (cloth: alk. paper)
Notes: Includes index.
Subjects: Robotics. Manipulators
(Mechanism)
LC Classification: TJ211 .T75 1999
Dewey Class No.: 629.8/92 21

Tutorial on robotics / [edited by] C.S.G.
Lee, R.C. Gonzalez, K.S. Fu.
Edition Information: 2nd ed.
Published/Created: Washington, D.C.:
IEEE Computer Society Press, c1986.
Related Authors: Lee, C. S. G. (C. S.
George) Gonzalez, Rafael C. Fu, K. S.
(King Sun), 1930-
Description: xii, 731 p.: ill.; 28 cm.
ISBN: 0818606584 (pbk.) 0818646586

(microfiche)
Notes: "IEEE Computer Society order
number 658." "IEEE catalog number
EH0241-0." Includes bibliographies.
Subjects: Robotics. Robots, Industrial.
LC Classification: TJ211 .T88 1986
Dewey Class No.: 629.8/92 19

Tutorial on robotics / [edited by] C.S.G.
Lee, R.C. Gonzalez, K.S. Fu.
Published/Created: Silver Spring, MD:
IEEE Computer Society Press, c1983.
Related Authors: Lee, C. S. G. (C. S.
George) Gonzalez, Rafael C. Fu, K. S.
(King Sun), 1930-
Description: x, 573 p.: ill.; 28 cm.
ISBN: 0818605154 (pbk.) 0818645156
(microfiche)
Notes: "IEEE catalog number EHO207-
1." "IEEE Computer Society order
number 515." Includes bibliographical
references.
Subjects: Robotics. Robots, Industrial.
LC Classification: TJ211 .T88 1983
Dewey Class No.: 629.8/92 19

Tver, David F.
Robotics sourcebook and dictionary /
David F. Tver and Roger W. Bolz.
Published/Created: New York, N.Y.:
Industrial Press, c1983.
Related Authors: Bolz, Roger William.
Description: 258 p.: ill.; 25 cm.
ISBN: 0831111526 :
Subjects: Robot industry--Directories.
Robotics--Dictionaries.
LC Classification: HD9696.R622 T83
1983
Dewey Class No.: 629.8/92 19

UKACC International Conference on
Control '96, 2-5 September 1996, venue,
University of Exeter, UK / conference
co-sponsors, Institute of Chemical
Engineers (Process Control Subject
Group) ... [et al.] in association with the
British Computer Society ... [et al.].
Published/Created: London: Institution
of Electrical Engineers, c1996.
Related Authors: Institute of Chemical

Engineers (Great Britain). Process Control Subject Group.
Description: 2 v. (xxxxiii, 1489 p.): ill.; 30 cm.
ISBN: 0852966687 (set) 0852966660 (v. 1)
Notes: Includes bibliographical references.
Subjects: Automatic control--Congresses. Robotics--Congresses. Neural networks (Computer science)--Congresses. System analysis--Congresses.
Series: Conference publication (Institution of Electrical Engineers); no. 427.
LC Classification: TJ212.2 .U35 1996
Dewey Class No.: 629.8 21

Ullrich, Robert A.
The robotics primer: the what, why, and how of robots in the workplace / Robert A. Ullrich.
Published/Created: Englewood Cliffs, N.J.: Prentice-Hall, c1983.
Description: xii, 121 p.: ill.; 24 cm.
ISBN: 0137821441 0137821360 (pbk.)
Notes: "A Spectrum book"--T.p. verso. Includes index. Bibliography: p. 107-116.
Subjects: Robotics. Robots, Industrial.
LC Classification: TJ211 .U44 1983
Dewey Class No.: 629.8/92 19

Valavanis, K. (Kimon)
Intelligent robotic systems: theory, design, and applications / by Kimon P. Valavanis, George N. Saridis.
Published/Created: Boston: Kluwer Academic, c1992.
Related Authors: Saridis, George N., 1931-
Description: xix, 252 p.: ill.; 25 cm.
ISBN: 0792392507
Notes: Includes bibliographical references (p. 219]-242) and index.
Subjects: Robotics.
Series: Kluwer international series in engineering and computer science; SECS 182. Kluwer international series

in engineering and computer science. Robotics.
LC Classification: TJ211 .V35 1992
Dewey Class No.: 629.8/92 20

Van Horn, Royal W.
Advanced technology in education: an introduction to videodiscs, robotics, optical memory, peripherals, new software tools, and high-tech staff development / Royal Van Horn.
Published/Created: Pacific Grove, Calif.: Brooks/Cole Pub. Co., c1991.
Description: xvi, 286 p.: ill.; 24 cm.
ISBN: 0534141250 (verso t.p.): 0534141242 (cover)
Notes: Includes bibliographical references and index.
Subjects: High technology and education--United States. Educational innovations--United States.
Series: Brooks/Cole series in educational computing
LC Classification: LC1087.2 .V36 1991
Dewey Class No.: 371.3/078 20

Variable structure control for robotics and aerospace applications / edited by K.-K. David Young.
Published/Created: Amsterdam; New York: Elsevier, 1993.
Related Authors: Young, Kar-Keung David, 1952-
Description: ix, 316 p.: ill.; 25 cm.
ISBN: 0444874461 (vol. 10: acid-free paper) 0444417923 (series: acid-free paper)
Notes: Includes bibliographical references.
Subjects: Automatic control. Robots--Control systems. Space vehicles--Control systems.
Series: Studies in automation and control; v. 10.
LC Classification: TJ213 .V29 1993
Dewey Class No.: 629.8/92 20

Vedder, Richard K.
Robotics and the economy: a staff study / prepared for the use of the

Subcommittee on Monetary and Fiscal Policy of the Joint Economic Committee, Congress of the United States.
Published/Created: Washington: U.S. G.P.O.: For sale by the Supt. of Docs., U.S. G.P.O., 1982.
Related Authors: United States. Congress. Joint Economic Committee. Subcommittee on Monetary and Fiscal Policy.
Description: vii, 34 p.: ill.; 26 cm.
Notes: At head of 97th Congress, 2d session. Joint committee print. "March 26, 1982." S/N 052-070-05707-6 Item 1000-B, 1000-C (microfiche) Includes bibliographical references.
Subjects: Technological employment--United States. Automatic machinery--Economic aspects--United States. Robotics--Economic aspects--United States. United States--Economic conditions--1971-1981. United States--Economic conditions--1981-
LC Classification: HD6331.2.U5 V43 1982
Dewey Class No.: 338.4/7629892 19
Govt. Doc. No.: Y 4.Ec 7:R 57

Vertut, Jean.
Teleoperation and robotics / Jean Vertut and Philippe Coiffet.
Published/Created: London: Kogan Page; Englewood Cliffs, N.J.: Prentice-Hall, c1986.
Related Authors: Coiffet, Philippe.
Description: 2 v.: ill.; 24 cm.
ISBN: 0137821948 (v. 1) 0137822022 (v. 2)
Contents: [1] Evolution and development -- [2] Applications and technology.
Notes: Includes bibliographies and indexes.
Series: Robots. English; v. 3.
LC Classification: TJ211 .R57313 1983 vol. 3
Dewey Class No.: 629.8/92 s 629.8/92 19

Vibrations and dynamics of robotic and multibody structures: presented at the 1993 ASME design technical conferences, 14th Biennial Conference on Mechanical Vibration and Noise, September 19-22, 1993, Albuquerque, New Mexico / sponsored by the Design Engineering Division, ASME; edited by Mo Shahinpoor, Robert Ryan.
Published/Created: New York: American Society of Mechanical Engineers, c1993.
Related Authors: Shahinpoor, Mohssen. Ryan, Robert. American Society of Mechanical Engineers. Design Engineering Division. Conference on Mechanical Vibration and Noise (14th: 1993: Albuquerque, N.M.)
Description: v, 123 p.: ill.; 28 cm.
ISBN: 0791811743
Notes: Includes bibliographical references and index.
Subjects: Robotics--Congresses. Vibration--Congresses. Robots--Dynamics--Congresses.
Series: DE (Series) (American Society of Mechanical Engineers. Design Engineering Division); vol. 57
LC Classification: TJ210.3 .V53 1993

Vision and action: the control of grasping / edited by Melvyn A. Goodale.
Published/Created: Norwood, N.J.: Ablex Pub. Corp., 1990.
Related Authors: Goodale, Melvyn A.
Description: viii, 367 p.: ill.; 24 cm.
ISBN: 0893915548
Notes: Includes bibliographical references and index.
Subjects: Hand. Vision. Robot wrists. Robot vision.
Series: Canadian Institute for Advanced Research series in artificial intelligence and robotics.
LC Classification: QP310.H36 V57 1990
Dewey Class No.: 152.3/85 20

Vision and navigation: the Carnegie Mellon Navlab / edited by Charles E. Thorpe;

with a foreword by Takeo Kanade.
Published/Created: Boston: Kluwer
Academic Publishers, c1990.
Related Authors: Thorpe, Charles E.
Description: xiv, 370 p.: ill.; 25 cm.
ISBN: 0792390687
Notes: Includes bibliographical
references and index.
Subjects: Robot vision. Robots--Control
systems. Mobile robots.
Series: Kluwer international series in
engineering and computer science;
SECS 93. Kluwer international series in
engineering and computer science.
Robotics
LC Classification: TJ211.3 .V58 1990
Dewey Class No.: 629.8/92 20

Visual information processing for television
and telerobotics: proceedings of a
workshop sponsored by the National
Aeronautics and Space Administration
and held in Williamsburg, Virginia,
May 10-12, 1989 / edited by Friedrich
O. Huck, Stephen K. Park.
Published/Created: Washington, D.C.:
National Aeronautics and Space
Administration, Office of Management,
Scientific and Technical Information
Division, 1989.
Related Authors: Huck, Friedrich O.
Park Stephen Keith United States.
National Aeronautics and Space
Administration.
Description: vi, 270 p.: ill.; 28 cm.
Notes: Includes bibliographical
references.
Subjects: Image processing--
Congresses. Television--Congresses.
Robotics--Congresses.
Series: NASA conference publication;
3053
LC Classification: TA1637 .V57 1989

Visual servoing: real-time control of robot
manipulators based on visual sensory
feedback / edited by Koichi Hashimoto.
Published/Created: Singapore; River
Edge, NJ: World Scientific, c1993.
Related Authors: Hashimoto, Koichi.

Description: vii, 363 p.: ill.; 23 cm.
Notes: Includes bibliographical
references.
Subjects: Robots--Control systems.
Computer vision. Servomechanisms.
Series: World Scientific series in
robotics and automated systems; vol. 7
LC Classification: TJ211.35 .V58 1993

Vogt, Gregory.
Space robots / by Gregory L. Vogt.
Published/Created: Mankato, Minn.:
Bridgestone Books, c1999.
Description: 24 p.: col. ill.; 22 cm.
ISBN: 0736801995
Summary: Explains different types of
space robots and their uses.
Notes: Includes bibliographical
references (p. 24) and index.
Subjects: Space robotics--Juvenile
literature. Robotics. Robots. Outer
space--Exploration.
Series: Explore space!
LC Classification: TL1097 .V64 1999
Dewey Class No.: 629.4 21

Vowles, Andrew.
Robotics: how they work and what they
can do / written by Andrew Vowles,
Dan Mackie; editors, Cyril Hayes,
Nadia Pelowich; illustrator, Charles E.
Bastien.
Published/Created: Milwaukee, Wis.,
U.S.A.: Penworthy Pub. Co., c1986.
Related Authors: Mackie, Dan. Hayes,
Cyril. Pelowich, Nadia. Bastien, Charles
E., ill.
Description: 32 p.: ill. (some col.); 29
cm.
ISBN: 0876170092
Summary: Discusses different kinds of
robots, how they work, how they are
being used in industry and research, and
what may lie in the future for robotics.
Subjects: Robotics--Juvenile literature.
Robots. Robotics.
Series: Hayes technology series
LC Classification: TJ211.2 .V69 1986
Dewey Class No.: 629.8/92 19

Vukobratovi´c, Miomir.
  Applied dynamic and CAD of
  manipulation robots / M. Vukobratovi´c,
  Vv. Potkonjak.
  Published/Created: Berlin; New York:
  Springer-Verlag, c1985.
  Related Authors: Potkonjak, V.
  (Veljko), 1931-
  Description: p. cm.
  ISBN: 0387130748 (U.S.)
  Notes: Cataloging based on CIP
  information. Includes bibliographies and
  index.
  Subjects: Robotics. Manipulators
  (Mechanism) Machinery, Dynamics of.
  Computer-aided design.
  Series: Scientific fundamentals of
  robotics; 6 Communications and control
  engineering series
  LC Classification: TJ211 .V836 1985
  Dewey Class No.: 629.8/92 19

Vukobratovi´c, Miomir.
  Control of manipulation robots: theory
  and application / M. Vukobratovi´c, D.
  Stoki´c.
  Published/Created: Berlin; New York:
  Springer-Verlag, 1982.
  Related Authors: Stoki´c, D. (Dragan).
  Description: xiii, 363 p.: ill.; 25 cm.
  ISBN: 038711629X (U.S.)
  Notes: Includes bibliographical
  references and index.
  Subjects: Robotics. Manipulators
  (Mechanism)
  Series: Scientific fundamentals of
  robotics; 2 Communications and control
  engineering series
  LC Classification: TJ211 .V837 1982
  Dewey Class No.: 629.8/92 19

Vukobratovi´c, Miomir.
  Dynamics of manipulation robots:
  theory and application / M.
  Vukobratovi´c, V. Potkonjak.
  Published/Created: Berlin; New York:
  Springer-Verlag, 1982.
  Related Authors: Potkonjak, V.
  (Veljko), 1931-
  Description: xiii, 303 p.: ill.; 25 cm.

ISBN: 0387116281 (U.S.)
  Notes: Includes bibliographical
  references and index.
  Subjects: Robotics. Manipulators
  (Mechanism)
  Series: Scientific fundamentals of
  robotics; 1 Communications and control
  engineering series
  LC Classification: TJ211 .V838 1982
  Dewey Class No.: 629.8/92 19

Vukobratovi´c, Miomir.
  Introduction to robotics / Miomir
  Vukobratovi´c in collaboration with
  Milan Djurovi´c ... [et al.].
  Published/Created: Berlin; New York:
  Springer-Verlag, 1989.
  Description: xiv, 301 p.: ill.; 24 cm.
  ISBN: 0387174524 (U.S.: alk. paper)
  Notes: Translation of: Uvod u robotiku.
  Includes bibliographies and index.
  Subjects: Robotics.
  LC Classification: TJ211 .V8613 1989
  Dewey Class No.: 629.8/92 19

Vukobratovi´c, Miomir.
  Kinematics and trajectory synthesis of
  manipulation robots / M. Vukobratovi´c,
  M. Kir´canski.
  Published/Created: Berlin; New York:
  Springer-Verlag, c1986.
  Related Authors: Kir´canski, M.
  (Manja), 1954-
  Description: x, 267 p.: ill.; 25 cm.
  ISBN: 0387130713 (U.S.)
  Notes: Includes index. Bibliography: p.
  258-264.
  Subjects: Robotics. Machinery,
  Kinematics of.
  Series: Scientific fundamentals of
  robotics; 3 Communications and control
  engineering series
  LC Classification: TJ211 .V839 1986
  Dewey Class No.: 629.8/92 19

Vukobratovi´c, Miomir.
  Legged locomotion robots and
  anthropomorphic mechanisms: a
  monograph / by M. Vukobratovi´c.
  Published/Created: Beograd: Mihailo

Pupin Institute, 1975.
Description: 346, 22, [308] p.: ill.; 25 cm.
Notes: Errata slip inserted. Includes bibliographical references.
Subjects: Robotics. Human locomotion. Artificial legs.
LC Classification: TJ211 .V8413
Dewey Class No.: 629.8/92

Vukobratovi´c, Miomir.
Non-adaptive and adaptive control of manipulation robots / M. Vukobratovi´c, D. Stoki´c, N. Kir´canski.
Published/Created: Berlin; New York: Springer-Verlag, c1985.
Related Authors: Stoki´c, D. (Dragan) Kir´canski, N. (Nenad), 1953-
Description: x, 383 p.: ill.; 25 cm.
ISBN: 038713073X (U.S.)
Notes: Includes bibliographies and index.
Subjects: Robotics. Manipulators (Mechanism)
Series: Scientific fundamentals of robotics; 5 Communications and control engineering series
LC Classification: TJ211 .V845 1985
Dewey Class No.: 629.8/92 19

Vukobratovi´c, Miomir.
Real-time dynamics of manipulation robots / M. Vukobratovi´c, N. Kir´canski.
Published/Created: Berlin; New York: Springer-Verlag, c1985.
Related Authors: Kir´canski, N. (Nenad), 1953-
Description: xii, 239 p.: ill.; 25 cm.
ISBN: 0387130721 (U.S.)
Notes: Includes index. Bibliography: p. [233]-237.
Subjects: Robots--Mathematical models. Digital computer simulation. Real-time data processing.
Series: Scientific fundamentals of robotics; 4 Communications and control engineering series Communications and control engineering series.
LC Classification: TJ211 .V85 1985

Dewey Class No.: 629.8/92 19

Waldman, Harry.
Dictionary of robotics / Harry Waldman.
Published/Created: New York: Macmillan; London: Collier Macmillan, c1985.
Description: viii, 303 p.: ill.; 25 cm.
ISBN: 0029485304
Subjects: Robotics--Dictionaries. Robots--Dictionaries.
LC Classification: TJ210.4 .W35 1985
Dewey Class No.: 629.8/92/0321 19

Walnum, Clayton.
Adventures in artificial life / Clayton Walnum.
Published/Created: Carmel, Ind.: Que, c1993.
Description: xiv, 186 p.: ill.; 24 cm. + 1 computer disk (3 1/2 in.)
ISBN: 1565293568 :
Notes: System requirements for computer disk: IBM-compatible PC; DOS; hard disk. Includes bibliographical references (p. [175]-178) and index.
Subjects: Life (Biology)--Computer simulation. Robotics. Virtual reality.
LC Classification: QH324.2 .W35 1993

Wang, Hang.
Intelligent supervisory control: a qualitative bond graph reasoning approach / Hang Wang & Derek Linkens.
Published/Created: Singapore; River Edge, N.J.: World Scientific, c1996.
Related Authors: Linkens, D. A.
Description: xi, 210 p.: ill.; 23 cm.
ISBN: 9810226586
Notes: Includes bibliographical references (p. 197-206) and index.
Subjects: Intelligent control systems.
Series: World Scientific series in robotics and intelligent systems; vol. 14
LC Classification: TJ217.5 .W36 1996
Dewey Class No.: 629.8/9 20

Warring, R. H. (Ronald Horace), 1920-
Robots & robotology / by R.H. Warring.
Published/Created: Blue Ridge Summit,
PA: TAB Books, 1984, c1983.
Description: 128 p., [8] p. of plates: ill.;
22 cm.
ISBN: 0830606734: 083061673X (pbk.)
:
Notes: Includes index.
Subjects: Robots. Robotics.
LC Classification: TJ211 .W38 1984
Dewey Class No.: 629.8/92 19

Warszawski, Abraham.
Application of robotics to building
construction / by Abraham Warszawski.
Published/Created: Rotterdam, The
Netherlands: International Council for
Building Research Studies and
Documentation, [1987?]
Description: iv, 95 p.: ill.; 30 cm.
Notes: Reprint. Originally published:
Pittsburgh, Pa.: Carnegie Mellon
University, [1984] Bibliography: p. 89-
94.
Subjects: Building--Automation.
Robotics.
Series: CIB report; 90.
LC Classification: TH437 .W35 1987
Dewey Class No.: 624 19

Warszawski, Abraham.
Industrialization and robotics in
building: a managerial approach /
Abraham Warszawski.
Published/Created: New York: Harper
& Row, c1990.
Description: xii, 466 p.: ill.; 24 cm.
ISBN: 0060469447
Notes: Includes bibliographical
references.
Subjects: Industrialized building.
Building--Automation. Robotics.
LC Classification: TH1000 .W37 1990
Dewey Class No.: 690 20

Warszawski, Abraham.
Industrialized and automated building
systems / Abraham Warszawski.
Edition Information: 2nd ed.

Published/Created: London; New York:
E & FN Spon, 1999.
Related Authors: Warszawski,
Abraham. Industrialization and robotics
in building.
Description: xiv, 464 p.: ill.; 26 cm.
ISBN: 0419206205
Notes: Previous ed.: published as
Industrialization and robotics in
building. London: Harper & Row, 1990.
Includes index and bibliographical
references.
Subjects: Industrialized building.
Building--Automation. Robotics.
LC Classification: TH1000 .W36 1999
Dewey Class No.: 690 21

Western European robotics.
Published/Created: San Jose, CA, USA
(4340 Stevens Creek Blvd., Suite 275,
San Jose 95129): Creative Strategies
International, c1982.
Related Authors: Creative Strategies
International.
Description: 155 p.: ill.; 28 cm.
Notes: Errata slip laid in. "March 1982."
Subjects: Robot industry--Europe.
Market surveys--Europe.
Series: Instrumentation industry
analysis service
LC Classification: HD9696.R623 E8558
1982
Dewey Class No.: 380.1/45629892 19

What should be computed to understand and
model brain function?: from robotics,
soft computing, biology and
neuroscience to cognitive philosophy /
editor, Tadashi Kitamura.
Published/Created: xii, 309 p.: ill.; 23
cm.
Related Authors: Kitamura, Tadashi,
1947-
Description: Singapore; River Edge, NJ:
World Scientific, c2001.
ISBN: 9810245181
Notes: Includes bibliographical
references and index.
Subjects: Soft computing. Artificial
intelligence. Neural networks

(Computer science) Brain--
Mathematical models.
Series: FLSI soft computing series; v. 3
LC Classification: QA76.9.S63 W48
2001
Dewey Class No.: 006.3 21

Wickelgren, Ingrid.
Ramblin' robots: building a breed of
mechanical beasts / Ingrid Wickelgren.
Published/Created: New York: Franklin
Watts, c1996.
Description: 143 p.: ill.; 24 cm.
ISBN: 0531113019
Summary: Examines the evolution of
robotics and the efforts of scientists to
develop robots with the abilities of
various animals.
Notes: "A venture book." Includes
bibliographical references (p. 135-139)
and index.
Subjects: Robotics--Juvenile literature.
Robots--Juvenile literature. Robotics.
Robots.
LC Classification: TJ211.2 .W53 1996
Dewey Class No.: 629.8/92 20

Wilson, J. A. Sam.
Control electronics with an introduction
to robotics / J.A. Sam Wilson.
Published/Created: Chicago: Science
Research Associates, c1986.
Related Authors: Wilson, J. A. Sam.
Industrial electronics and control.
Description: vi, 527 p.: ill.; 25 cm.
ISBN: 0574216103 :
Notes: Rev. ed. of: Industrial electronics
and control. c1978. Includes index.
Subjects: Industrial electronics.
Electronic control. Robots, Industrial.
LC Classification: TK7881 .W553 1986
Dewey Class No.: 621.381 19

Winkless, Nels.
If I had a robot--: what to expect from
the personal robot / by Nelson B.
Winkless III.
Published/Created: Beaverton, Or.:
Dilithium Press, c1984.
Description: xiv, 247 p.: ill.; 22 cm.

ISBN: 0880563532 (pbk.) :
Notes: Includes index.
Subjects: Robotics. Robots.
LC Classification: TJ211 .W558 1984
Dewey Class No.: 629.8/92 19

Winkless, Nels.
Robots on your doorstep (a book about
thinking machines) / Nels Winkless and
Iben Browning; with an introd. by John
Peers.
Published/Created: Portland, Or.:
Robotics Press, c1978.
Related Authors: Browning, Iben, joint
author.
Description: x, 178 p.: ill.; 22 cm.
ISBN: 0896610004
Subjects: Robotics.
LC Classification: TJ211 .W56
Dewey Class No.: 629.8/92

Wise, Edwin.
Animatronics: a guide to animated
holiday displays / by Edwin Wise.
Published/Created: Indianapolis, IN:
Prompt Publications, c2000.
Description: viii, 273 p.: ill.; 24 cm.
ISBN: 0790612194 (pbk.)
Subjects: Robotics. Holiday
decorations.
LC Classification: TJ211.15 .W57 2000
Dewey Class No.: 745.594/16 21

Wise, Edwin.
Applied robotics / Edwin Wise.
Published/Created: Indianapolis, IN:
Prompt Publications, c1999.
Description: xiv, 311 p.: ill.; 24 cm. + 1
computer laser disc (4 3/4 in.).
ISBN: 0790611848 (pbk.)
Notes: Includes bibliographical
references and index.
Subjects: Robotics.
LC Classification: TJ211 .W5624 1999
Dewey Class No.: 629.8/92 21

Wolovich, William A., 1937-
Robotics: basic analysis and design /
William A. Wolovich.
Published/Created: New York: Holt,

Rinehart and Winston, c1987.
Description: xv, 393 p.: ill.; 25 cm.
ISBN: 0030061199
Notes: Includes index. Bibliography: p.
353-354.
Subjects: Robotics.
Series: HRW series in electrical and
computer engineering
LC Classification: TJ211 .W64 1987
Dewey Class No.: 629.8/92 19

Wooldridge, Michael J., 1966-
Reasoning about rational agents /
Michael Wooldridge.
Published/Created: Cambridge, Mass.:
MIT Press, 2000.
Description: xi, 227 p.; 24 cm.
ISBN: 0262232138 (hc.: alk. paper)
Notes: Includes bibliographical
references (p. [205]-222) and index.
Subjects: Intelligent agents (Computer
software)
Series: Intelligent robotics and
autonomous agents
LC Classification: QA76.76.I58 W66
2000
Dewey Class No.: 006.3 21

Working safely with industrial robots / Peter
M. Strubhar, editor.
Edition Information: 1st ed.
Published/Created: Dearborn, Mich.:
Robotics International of SME,
Publications Development Dept.,
Marketing Division; Ann Arbor, Mich.:
Robotic Industries Association, c1986.
Related Authors: Strubhar, Peter M.
Description: 266 p.: ill.; 29 cm.
ISBN: 0872632105
Notes: Includes bibliographies and
index.
Subjects: Robots, Industrial--Safety
measures.
Series: Manufacturing update series
LC Classification: TS191.8 .W67 1986
Dewey Class No.: 670.42/7/0289 19

Workshop on CAD/CAM and Robotics:
presented at industrial affiliates meeting
No. 27, SRI International, Menlo Park,

California, 23 May 1984 / edited by
David Nitzan.
Published/Created: Menlo Park, Calif.:
SRI International, [1984]
Related Authors: Nitzan, David. SRI
International.
Description: v, 196 p.: ill.; 27 cm.
Notes: "August 1984." "SRI project
5978."
Subjects: CAD/CAM systems--
Congresses. Robots, Industrial--
Congresses.
LC Classification: TS155.6 .W64 1984
Dewey Class No.: 670/.28/5 19

Workshop on Industry Needs in Robotics
R&D: presented at industrial affiliates
meeting no. 26, SRI International,
Menlo Park, California, 21 September
1983 / edited by David Nitzan.
Published/Created: Menlo Park, Colo.:
SRI International, [1984]
Related Authors: Nitzan, David. SRI
International.
Description: v, 177 p.: ill.; 28 cm.
Notes: "January 1984." "SRI project
5978."
Subjects: Robots, Industrial--Research--
United States--Congresses.
LC Classification: TS191.8 .W68 1983
Dewey Class No.: 670.42/7 19

Workshop on Intelligent Robots:
Achievements and Issues: held at SRI
International, Menlo Park, California,
13-14 November, 1984 / edited by
David Nitzan, Robert C. Bolles.
Published/Created: Menlo Park, Calif.
(333 Ravenswood Ave., Menlo Park
94025-3493): SRI International, [1985]
Related Authors: Nitzan, David. Bolles,
Robert C. SRI International.
Description: viii, 328 p.: ill.; 28 cm.
Notes: "July 1985." "SRI project 7717."
Includes bibliographical references.
Subjects: Robotics--Congresses. Robots,
Industrial--Congresses.
LC Classification: TJ210.3 .W64 1984
Dewey Class No.: 670.42/7 19

Workshop on Robot System Integration: presented at Industrial Affiliates Meeting No. 30, SRI International, Menlo Park, California 94025, 12 March 1986 / edited by David Nitzan. Published/Created: Menlo Park, Calif. (333 Ravenswood Ave., Menlo Park 94025-3493): SRI International, [1986] Related Authors: Nitzan, David. SRI International. Description: v, 139 p.: ill.; 28 cm. Notes: "September 1986." "SRI projects 5978 and 8850." Bibliography: p. 70. Subjects: Robotics--Congresses. Robots, Industrial--Congresses. LC Classification: TJ210.3 .W65 1986 Dewey Class No.: 629.8/92 19

Worldwide robotics survey and directory / Robot Institute of America. Published/Created: Dearborn, Mich. (1 SME Dr., Dearborn 48128): RIA, [1981?] Related Authors: Robot Institute of America. Description: 35 p.: ill.; 22 cm. Subjects: Robotics. Robots, Industrial. LC Classification: TJ211 .W67 1981 Dewey Class No.: 629.8/92 19

Wright, Paul Kenneth. Manufacturing intelligence / Paul Kenneth Wright, David Alan Bourne. Published/Created: Reading, Mass.: Addison-Wesley, c1988. Related Authors: Bourne, David Alan. Description: xv, 352 p.: ill.; 25 cm. ISBN: 0201135760 : Notes: Includes index. Bibliography: p. 307-311. Subjects: Robotics. Artificial intelligence--Industrial applications. Production engineering. LC Classification: TJ211 .W75 1988 Dewey Class No.: 670.42 19

Yoda, T. (Takeo) Rolling bearings for industrial robots / by T. Yoda, N. Obokata, and S. Hioki. Published/Created: New York: Gordon and Breach Science Publishers, c1990. Related Authors: Obokata, N. (Nobou) Hioki, S. (Shoichi) Enu T⁻e Enu T⁻oy⁻o Bearingu Kabushiki Kaisha. Description: ix, 149 p.: ill.; 24 cm. ISBN: 288124744X Notes: Includes bibliographical references (p. 145) and index. Subjects: Robots, Industrial--Design and construction. Bearings (Machinery) Series: Precision machinery and robotics; v. 3 LC Classification: TS191.8 .Y63 1990 Dewey Class No.: 670.42/72 20

Yoshikawa, Tsuneo, 1941- Foundations of robotics: analysis and control / Tsuneo Yoshikawa. Published/Created: Cambridge, Mass.: MIT Press, c1990. Description: x, 285 p.: ill.; 24 cm. ISBN: 0262240289 Notes: Revised translation of: Robotto seigyo kisoron. Includes bibliographical references and index. Subjects: Robotics. LC Classification: TJ211 .Y6713 1990 Dewey Class No.: 629.8/92 20

Young, John Frederick. Robotics [by] John F. Young. Published/Created: New York, Wiley [1973] Description: 303 p. illus. 23 cm. ISBN: 0470979909 Notes: "A Halsted Press book." Includes bibliographical references. Subjects: Robotics. LC Classification: TJ211 .Y68 Dewey Class No.: 629.8/92

Young, John Frederick. Robotics [by] John F. Young. Published/Created: London, Butterworth, 1973. Description: [9], 303 p. illus. 23 cm. ISBN: 0408705221 Notes: Includes bibliographical references and index. Subjects: Robotics.

LC Classification: TJ211 .Y68 1973b
Dewey Class No.: 629.8/92

Young, Kay.
Robotics, 1970-1983 / Kay Young.
Published/Created: Monticello, Ill.:
Vance Bibliographies, [1984]
Description: 93 p.; 28 cm.
ISBN: 0880669640 (pbk.) :
Notes: Cover title.
Subjects: Robotics--Bibliography.
Robots--Bibliography.
Series: Public administration series--
bibliography, 0193-970X; P-1444
LC Classification: Z5853.R58 Y68
1984 TJ211
Dewey Class No.: 016.6298/92 19

Zeldman, Maurice I.
What every engineer should know about
robots / Maurice I. Zeldman.
Published/Created: New York: M.
Dekker, c1984.
Description: xviii, 197 p.: ill.; 24 cm.
ISBN: 0824771230
Notes: Includes index. Bibliography: p.
147-153.
Subjects: Robotics.
Series: What every engineer should
know; v. 11
LC Classification: TJ211 .Z45 1984
Dewey Class No.: 629.8/92 19

Zelikin, M. I. (Mikhail Il'ich)
Theory of chattering control with
applications to astronautics, robotics,
economics, and engineering / M.I.
Zelikin, V.F. Borisov.
Published/Created: Boston: Birkhäuser,
c1994.

Related Authors: Borisov, V. F.
(Vladimir F.), 1961-
Description: xiv, 242 p.: ill.; 25 cm.
ISBN: 0817636188 (alk. paper)
3764336188 (alk. paper)
Notes: Translation from Russian.
Includes bibliographical references (p.
[237]-239) and index.
Subjects: Chattering control (Control
systems) Control theory.
Series: Systems & control
LC Classification: TJ223.C45 B67 1994
Dewey Class No.: 003/.5 20

Zomaya, Albert Y.
Modelling and simulation of robot
manipulators: a parallel processing
approach / Albert Y. Zomaya.
Published/Created: Singaport; New
Jersey: World Scientific, c1992.
Description: xvi, 295 p.: ill.; 23 cm.
ISBN: 9810210434
Notes: Includes bibliographical
references and index.
Subjects: Manipulators (Mechanism)--
Computer simulation. Parallel
processing (Electronic computers)
Series: World Scientific series in
robotics and automated systems; vol. 8
LC Classification: TJ211.47 .Z66 1992
Dewey Class No.: 629.8/92 20

AUTHOR INDEX

A

Abbott, Ben Allen., 140
Abidi, Mongi A., 43
Abodelmonem, Ahmed H., 107
Adams, Laurence J., 179
Adams, Martin David, 7
Adelstein, Bernard D., 11
Adey, R. A., 17
Aebischer, Patrick., 161
Agarwal, Pankaj K., 159
Ager, Richard T., 12
Ahmad, S. I. (Syed Imtiaz), 54
Akay, Metin, 73
Akman, Varol., 12
Alberts, T. E. (Thomas E.), 47
Albus, James Sacra, 12, 13, 106
Aleksander, Igor, 13, 22, 38
Alford, J. Michael (James Michael), 14
Allen, Peter K., 14
Allsup, Randall, 93
American European Consulting Company, 1
American Nuclear Society. Fuel Cycle and Waste
    Management Division, 125
American Nuclear Society. Remote Systems
    Technology Division, 118, 128
American Nuclear Society. Robotics and Remote
    Systems Division, 119
American Society of Mechanical Engineers, 130
An, Chae H., 15
Andeen, Gerry B., 139
Angeles, Jorge, 15, 159
Anstey, David, 89
Arai, Tatsuo, 123
Ardayfio, David D., 18
Arimoto, Suguru, 153
Arkin, Ronald C., 19, 138
Armstrong, Brian Stewart Randall, 19

Armstrong-Hélouvry, Brian, 19
Asada, H. (Haruhiko), 22
Asada, Minoru, 137
Asama, H. (Hajime), 45
Asimov, Isaac, 22
Atkeson, Christopher G., 15
Australia. Parliament., 92
Australian Robot Association. International
    Federation of Robotics, 161
Ayers, Joseph, 108
Ayres, Robert U., 24, 147

B

Babuska, Robert., 57
Bailey, Jacqui, 171
Baillieul, J. (John), 49, 143
Baker, Christopher W., 25
Balafoutis, C. A. (Constantinos A.), 25
Balch, Tucker, 13, 1397
Balchen, Jens G., 65
Baldwin, Margaret, 25
Barbera, Anthony J., 106
Barrett, Norman S., 25
Basañez, L., 139
Bastien, Charles E., 193
Bastin, G. (Georges), 185
Battrick, B., 72
Becquet, Marc C., 180
Beer, Randall D., 27
Beheshti, M. R. (Reza R.), 8
Bejczy, A., 111
Bekey, George A., 24, 107, 138
Ben-Daya, M. (Mohamed), 53
Bendehmane, Diane B., 147
Beni, Gerardo, 134
Berger, C. S. (Clive S.), 186
Berger, Fredericka, 25
Berger, Phil, 26

Berk, A. A., 26

Berkstresser, Gordon A., 24

Bernard Hodes Advertising (New York, N.Y.),
    151

Bey, I. (Ingward), 108

Beyer, Mark (Mark T.), 26

Bhanu, Bir, 26, 30

Bianchi, G. (Giovanni), 114, 185

Billard, Mary, 26

Billingsley, J. (John), 95

Blankenship, John, 27

Bode, M. F., 141

Boissonnat, J.-D. (Jean-Daniel), 58

Bolles, Robert C., 198

Bolz, Roger William, 190

Bone, Jan, 27, 28

Bonivento, Claudio, 12

Book, Wayne J., 152

Borisov, V. F. (Vladimir F.), 200

Boss, M. Kenneth, 139

Bourbakis, Nikolaos G., 18

Bourgine, Paul, 187

Bourne, David Alan, 199

Bowker A&I Publishing, 144

Bowling, Charles M. (Charles McKaughan), 28

Boyer, Kim L., 16

Boyet, Howard, 28

Brady, Michael, 145, 153, 155

Breguet, Jean-Marc, 99

Bringsjord, Selmer, 28

Briot, M., 123

Brooks, Rodney Allen, 28

Brown, M. (Martin), 63

Browne, Antony, 106

Browning, Iben, 197

Brownstein, Barry, 157

Bryne, Raymond H., 144

Buchanan, David R., 24

Buchsbaum, Frank, 29

Bundy, Alan., 29

Bunke, Horst, 16, 21, 168

Burdick, Joel Wakeman, 29

Bürger, Erich, 29, 30

Burger, Wilhelm, 30

Burke, W. R., 173

Burks, Arthur W. (Arthur Walter), 30

Burnett, Betty, 30

Burnett, Piers, 13

Burns, William C., 30

Business/Technology Books (Firm), 36, 158

C

Caldwell, D. G. (Darwin G.), 7

Cameron, Stephen, 7

Campbell, Wendy, 112

Canadian Institute of Metalworking. Computer
    and Automated Systems Association of SME,
    114

Canadian Nuclear Society. Canadian Society for
    Mechanical Engineering. American Nuclear
    Society, 127

Canny, John., 109

Canudas de Wit, Carlos A., 7, 185

Carnegie-Mellon University. Computer Science
    Dept. Carnegie-Mellon University. Robotics
    Institute, 21

Carnevale, M., 103

Casals, Alícia, 169

Casasent, David Paul, 76, 77, 147, 149

Caudill, Maureen, 33

Cetron, Marvin J., 33

Ch`en, Han-fu, 119

Chacko, George Kuttickal, 33

Chamberlain, Denis A., 24

Chen, Su-shing, 175

Cheng, Dai-Zhan, 119

Chester, Michael, 34

Chiacchio, Pasquale, 35

Chiaverini, Stefano, 35

Chin, Felix, 34

Chinni, Michael J., 128

Chirouze, Michel., 34

Chmielewski, Thomas A., 84

Chorafas, Dimitris N., 34

Choudhary, Alok N. (Alok Nidhi), 34

Christensen, H. I. (Henrik I.), 168

Chudý, Vladimír, 80

Clarke, Vanessa, 64

Cliff, Dave, 55

Coiffet, Philippe, 34, 192

Cole, Corey, 33

Colgate, J. Edward (James Edward), 11

Colombetti, Marco, 46

Computer and Automated Systems Association
    of SME, 23, 52

Connell, Jonathan H., 39

Conrad, James M., 39

Consi, Thomas R., 27

Copyright Collection (Library of Congress), 74

Corke, Peter I., 40, 49

Cory, James F., 133

Cousineau, Leslie, 41

Craig, John J., 41

TITLE INDEX

#

A

C

D

H

I

Q

R

SUBJECT INDEX

E

F